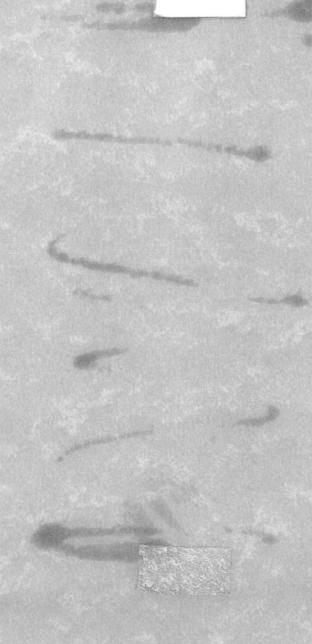

KRUGER
Portrait of a
NATIONAL PARK

KRUGER
Portrait of a
NATIONAL PARK

David Paynter with Wilf Nussey

M

ISBN 0 86954 208 7

First edition, first impression 1986

Published by
Macmillan South Africa (Publishers) Pty Ltd
P O Box 31487, Braamfontein, 2017
Johannesburg

Associated companies throughout the world

Cover and design consultant: Mel Miller
Layout by David Paynter
Set in Andover 10 on 11½ pt
by Unifoto (Pty) Ltd
Printed and bound by
Printpak Books, Cape Town

PREFACE

This book was conceived on a warm September evening in 1980 at the lovely Shingwedzi camp in the northern part of the great Kruger National Park. The Wildlife Society of Southern Africa was having a press conference there and one of those present was David Paynter, a professional photographer with many years of experience covering news and features in Africa, and also then chairman of the Society's publications committee.

One of the others there was Piet van Wyk, head of the Department of Research and Information at the National Parks Board in Pretoria. Van Wyk is a biologist who specialises in botany and has written the definitive two-volume work on the trees of the Park. While chatting to Paynter he mentioned that many books had been written on particular aspects of the Park but, at the time, that there was no comprehensive book about the Park as an entity.

This thought was to set Paynter on a photographic pilgrimage that took years. First he obtained the National Parks Board's approval and cooperation and then began to visit the Park regularly, devoting all of his annual leave allocation to the project, until he had accumulated a huge collection of photographs of animals, birds, insects, views and a multitude of other facets of the Park's life and times.

David Paynter

In a total of more than eight months there he filled 17 notebooks with information, mostly details of the network of roads, the waterholes, camps, what they offer and where which animals can be found.

Travelling usually in an elderly Volkswagen camper, sometimes in other vehicles, Paynter cruised every road, side road and detour in the Park at least three times and some roads before they were opened to the public. There is not a camp at which he has not stayed, not a watering point he has not visited, not a view he has not seen and hardly an animal he has not photographed.

He has interviewed or photographed scores of the Park's rangers and research staff, whose help, guidance and information were indispensable to this book, and has become almost as familiar to them as one of their own.

He has shared in their adventures, travelling with them to some of the remotest parts of the Park. In July 1985 he went to Malawi with a game-catching team to record the capture and translocation of Lichtenstein's hartebeest from that country to the Park. He has spent many an evening at a campfire with them, soaking up their knowledge and wisdom and their tales of life in the wild, the humour and drama – and sometimes the tragedy.

In his quest for information and illustrations, he spent many hours in various libraries and museums and made contact with people all over Southern Africa, including Zimbabwe – and even in Scotland.

In 1983 Paynter sought out Wilf Nussey, a newspaper editor and former specialist for 20 years in covering African affairs, to help him shape a book around his core of words and pictures. A lifetime conservationist who spent much of his youth in the then animal-rich Lowveld not far from the Park, Nussey digested the small mountain of information Paynter had collected, added further research into the Park's history, prehistory and operations, garnered more about the Park's life, and wove it all together as a book. Paynter himself wrote the detailed descriptions of the camps and where to go and what to look for in the Park, plus

Wilf Nussey

the advice to photographers.

The result is this first fully rounded book giving an insight into all aspects of what is probably the world's richest wildlife reserve and certainly one of the best managed.

One of the book's most important features is that almost all the photographs were taken at places accessible to any visitor to the Park. The only exceptions are some taken on the banks of the Luvuvhu river at Pafuri, in a rhino enclosure at Pretoriuskop, of a suni adopted by an honorary ranger at Pafuri, and of a black eagle and a martial eagle that were in temporary captivity at Skukuza, the former because it was injured and the latter because it was needed for research purposes.

The authors do not pretend that this book is comprehensive; that would require a work of encyclopaedic proportions. It is intended as a broad profile of the Park for enthusiasts, both young and old, to help them gain greater satisfaction from the Park, to inspire more people to visit it, to help promote knowledge of wildlife and understanding of conservation.

CONTENTS

INTRODUCTION

Part 1: THE SPIRIT OF CONSERVATION

CREATURES AND PLANTS OF THE PARK 5
Mammal characters 5
Birds, big and small 7
The cold-blooded denizens 13
The amazing insect world 17
The wonder of trees 24
Flowers of the veld 30
Gills and fins 34

HISTORY: FROM PRIMEVAL TO PRESENT ... 37
Early man, the hunter 37
Explorers, pioneers and fortune-seekers 40
A man and his Eden 47
The Park – early years 57
After World War II 66
Guarding the future 67

THE FUSION OF SCIENCE AND NATURE 69
Maintaining an ecosystem 69
Taking stock from the air 70
Killing to save 71
Bringing them back alive 75
The menace of disease 79
Water – life-blood of the Park 81
The rains 83
Fire as nature's aide 84
Rangers – the guardians 85
The war against poaching 88
Flying for science 94

SAGAS OF THE PARK 101
Tragic encounters 101
Rescue from the crocodiles 101
Death in the river 106
Wolhuter and the lion 107
Mdluli – a Park legend 110
Animal oddities 115
The "no option" option 118
Great tuskers of the Park 119
The white lions 128

Part 2: THE PARK AND THE TOURIST

PRACTICAL INFORMATION FOR VISITORS 139
Topography and climate 139
Accommodation 140
Restaurants 140
Shops 140
Other amenities 141
Your health 141
How to get there 142
The motorist 142
Travelling hours 143
Caravanners and campers 143
Roads and by-ways 144
Rules of the Park 145
Tourism in the future 147
Understanding binoculars 148

THE FUN OF PHOTOGRAPHY 151
Hints for better wildlife photography 151
Favourite picture places 162

GUIDE TO EXPLORING THE PARK 167
Pretoriuskop/Berg-en-dal area 169
Skukuza/Tshokwane area 181
Lower Sabie/Crocodile Bridge area 195
Satara/Orpen Gate area 209
Olifants/Letaba area 225
Shingwedzi area 243
Punda Maria/Pafuri area 259
Wilderness trails 272

KNOW YOUR TREES BY NUMBERS 277

ACKNOWLEDGEMENTS 280

BIBLIOGRAPHY 281

INDEX 283

INTRODUCTION

In the far north-eastern corner of the Republic of South Africa lies one of the world's great natural treasures – a huge segment of pristine Africa preserved virtually intact where an enormous variety of wild creatures live the same life that their kind have known for untold thousands of years.

This is the giant Kruger National Park – bigger than Israel, about as big as Wales – which the foresight and dedication of a few wise men saved from the development, politics and greed that have ravaged much of the rest of Africa.

One of the world's 14 major nature reserves, it ranks among the best as a conservation model and natural laboratory and contains an unequalled diversity of wild animals, bird and plant life. It is a refuge where modern man can escape the pressures of today's life-styles and commune with unspoilt nature.

Over 380 km long and on average 60 km wide, the Kruger National Park covers 19 485 km² (7 523 sq. miles) in a long slice of land lying north-south. In the north it shares a short border with Zimbabwe, its long eastern border is against Mozambique and to the south and west it is flanked by South Africa's Transvaal province.

Conceived by the genius of a great leader of the last century, born in controversy and tribulation, the Park has survived seemingly impossible odds to reach its present status and renown.

In this segment of subtropical to tropical Africa mankind is a carefully controlled intruder. Its meticulously planned network of roads and strategically sited rest camps afford a window to an Africa that no longer exists elsewhere; an Africa wiped out by population explosions, guns, hunger and human settlement.

The view is of wooded plains, thickly forested river banks and hills whose more than 2 000 different forms of vegetation sustain an unmatched richness of wildlife: 146 species of mammals, over 490 different kinds of birds, 114 species of reptiles, 49 kinds of freshwater fish, 33 species of amphibians and innumerable insects and other forms of life – many of them found nowhere else.

Today the Kruger National Park is en-trenched, intact, although there still lurks some danger from those who would rape it for minerals. It is an asset not only to South Africa but to people everywhere. Its already immeasurable aesthetic worth is complemented by a growing international awareness of its value as a giant open-air laboratory – with a number of not altogether expected spin-offs.

Guarding and managing this priceless heritage is a permanent staff of nearly 1 500 people who represent a formidable array of scientific, managerial and technical skills. Most visible are those who run the facilities for visitors to the Park – the 17 camps (with another six scheduled for long-term development) with their accommodation, restaurants and shops, and the wilderness trails where people can walk through the wild with expert guides. Behind them, largely unseen, is a host of administrators, game rangers, road crews, pilots, scientific researchers and others who conduct a continuous and highly specialised operation to maintain the Park's structure and life.

The creed of the Kruger National Park is that its internal world should survive untouched; that its normal life cycles should continue untrammelled by outside agency or influence as far as practically possible. People come second to wildlife, but only so that they can experience and appreciate its infinite richness and benefit from the fund of knowledge it yields.

wildlife but with facilities so raw and undeveloped for lack of income that they are beyond the reach of all but the very wealthy who can afford to travel in their own cocoons of comfort.

The men who guide the destiny of this Park – the directors and planners of South Africa's National Parks Board of Trustees and the senior officials of the Park itself – will soon have to make a crucial decision for its future, forced upon them by the Park's fast-rising popularity and thus heavier public demands upon it. The decision will be whether to set a final limit on the public's use of the Park, or to expand its facilities to cater for more people, at the risk of its becoming so crowded and developed that it loses its intrinsic wilderness character and becomes just another holiday resort.

Therein lies the vulnerability of the Kruger National Park. From its inception many years ago as the Sabie Game Reserve, its successive masters have laid stress on the need to strictly control the number of people admitted to safeguard its character as a uniquely unspoilt African environment. And yet...

The Park's midwife and mentor for many years, who more than any other individual shaped its personality, Colonel James Stevenson-Hamilton, had this to say when he retired in April 1946, after 44 years as warden there:

"I had at least brought up Cinderella and launched her on her career. I loved her best when she was a pathetic and

"I loved her best when she was a pathetic and dust-covered little wench . . ."

There are many invaluable national parks elsewhere in the world, including Africa, of an importance parallel to that of the Kruger National Park. But as far as the authors can determine, none has reached so fine and precarious a balance between mankind and the wilderness as has this one. It has become a dogma in this day of profit and loss and of cost to the taxpayer that a national park cannot truly survive unless it justifies its existence in benefits to the populace.

There are parks, particularly in Africa and the Far East, which are superb in dust-covered little wench, derided and abused; always I felt that, given her chance, and her attractions recognised, unlimited possibilities lay before her. Now that she had become a Great Lady it was fitting she should be provided with custodians perhaps better suited to provide her new requirements.

"Might her success, and the gifts increasingly showered upon her, not at last permanently affect her character, and transform her into a dame so bedecked by human art that her natural loveliness would be hidden, and her

1

simple nature spoiled? Might those holding her future in their hands, realise the true nature of their Trust, and not, by estimating her worth at artificial values only, cause her to languish, and ultimately perhaps to perish?"

Prophetic words. Halfway through his own career as the ruler of the Park, Stevenson-Hamilton himself saw the inevitability – the necessity and the value – of admitting tourists to it. Otherwise, already in those early days, there was not much point in having such a conservation area. Why keep animals in an intact environment except to educate people about what used to be? So he promoted tourism to confound his opponents who wanted to kill the Kruger National Park, and did so with great success. He made the public his allies.

Since then his successors have echoed his sentiments. The words have been different but the message has always been the same: the Park must survive for the benefit of the people, but people must not be allowed into it in such numbers that they destroy it.

In Stevenson-Hamilton's days that was fairly easy. One had to be prepared to "rough it" in the toughest sense of the phrase to visit the Park. So wild was the whole Lowveld then that as recently as after World War I large sections of the north were tamed only by giving tracts of virgin land to returned soldiers as war gratuities. The Park was not a place to which the public could flock in large numbers, or would have wanted to had transport to it been less primitive.

But after World War II, South Africa boomed. The Witwatersrand goldfields swelled into the largest industrial complex in all Africa. Urban growth exploded and as the crowding and pressures of city life grew, as roads developed from tracks to modern freeways and the use of aircraft became commonplace, the number of people seeking periodic escape rose from thousands to millions. Simultaneously the spectacular expansion of international communications which saw ocean travel succumb to the airlines brought a fast-rising flood of visitors to this country – and to almost all a visit to the Kruger National Park became *de rigueur*.

There are those, including independent conservation experts, who argue that the Park is already too crowded, too civilised, and that no more camps should

be built and ways found to reduce the human pressure.

They point at some concessions that have already been made. People or companies with enough money may sponsor accommodations at camps within the Park as luxurious as they see fit, but subject in concept and design to approval by the National Parks Board and Park administrators. They have only one month per year free use of such cottages or private camps and own neither the buildings nor the sites. Several private camps have been erected under this scheme.

Purists point a finger at the new Berg-en-dal camp in the far south because its architecture is quite different from other camps, although it harmonises well with its surroundings, and to the fact that it has conference facilities for business seminars, complete with modern audio-visual services – which they charge is foreign to the Park. The conference hall, however, is also used for seminars on wildlife and its preservation.

"We will most certainly not under any circumstances allow our greatest natural asset to be swamped by masses of people . . ."

When a new camp like Berg-en-dal is built, the authorities state, additional roads are cut to avoid overcrowding and wilderness trails are opened to offset the camp sophistication.

In fact the Park authorities are keenly aware of the vital need to preserve the "wilderness mystique" and they have imposed definite limits on the number of visitors.

"We will most certainly not under *any* circumstances allow our greatest natural asset to be swamped by masses of people," states chief warden Dr U. de V. Pienaar. "Only a very small part of the Park has been zoned for development and at present only 3,3 per cent of its total area is affected by all aspects of development, such as camps, roads, firebreaks, dams, etc., whereas 32,3 per cent has been zoned as wilderness area

in which any form of development will be prohibited in perpetuity."

As to the criticism that the Park is becoming "too civilised, too full of amenities" the Park authorities make the salient point that it is important to make conservationists of people for whom facilities such as air conditioning, restaurants and swimming pools matter – the denizens of Manhattan and Hillbrow – as it is to provide wilderness camps for the already committed.

Park officials have been warning for years that the Park has a finite capacity for tourist development and that the day will come when saturation point would be reached. Now that point is very close with the annual total of visitors peaking at over half a million.

Just as they determine the Park's animal capacity by careful management founded upon research, so the Park authorities are now determining its human capacity. They have for years been limiting the influx of visitors during peak periods. Those admitted enjoy probably the cheapest amenities of any national park in the world – considerably cheaper than most of South Africa's holiday hotels.

One body of opinion holds that the Park should increase its fees for the more luxurious accommodation in particular, thus helping to curb the flood with tariff barriers, and if this cuts Park revenue, the government should then step in and subsidise it. But, as Dr Pienaar asserts, this is "a *national* park which must be accessible to *all* members of the public. It is our policy to keep pace with the inflation rate but certainly not to raise fees to a level that would make the Park accessible to only the affluent and a place out of bounds to the less privileged."

Up to this point the Park's evolution represents a massive and unique human effort against seemingly insuperable odds.

As Dr Pienaar puts it: "When we look back today and contemplate the achievements of the conservation movement in South Africa in all respects, the Kruger National Park saga presents a tremendous success story – the story of a dedicated band of conservationists and their faithful public supporters who have ensured that some of the natural splendour of yesterday will remain a part of today and, hopefully, also of tomorrow."

The natural life in the Park is an incredibly complex interweave of thousands of plant and creature species in such abundant array as to provide satisfaction to virtually any kind of specialist, be he entomologist, ornithologist, zoologist, botanist or simply a keen photographer. But the star attractions are always those big mammals – among the Park's 146 kinds – which represent power and dominance.

Mammal characters

The king among these is the elephant, the seemingly gentle giant which in an instant can turn into a destructive behemoth of bone and muscle. Nearly 7 500 are scattered all over the Park, mostly in the north, and they include several which are internationally renowned for the impressive size of their tusks (see p. 119).

Far more popular, though less impressive than the elephant, is the lion – the great lure for the tourist in almost any African game reserve. Some 1 500 are resident throughout the Park, and can commonly be found lazing by the roadside or gathered around a kill.

Less common than the lion but equally spectacular are the other two big cats: the leopard, quite easily found by those who know where to look – in dappled shadows or in trees, and the cheetah, the most superbly graceful of all the big cats.

The white rhino, which is becoming quite common in the south, and the truculent black rhino draw crowds when they appear feeding near the roads. Buffalo are everywhere in the Park, in herds of up to 1 400 though usually not more than 500 to 800, churning up mists of grey dust as they move across the plains beneath a forest of big, curved horns.

Many other predators appear unexpectedly to round off the view of Park life: a pack of huge-eared wild dogs loping through the grass; hyenas slouching near lion kills or laughing maniacally at night; the African honey badger or ratel trotting purposefully through the bush or through a camp like a miniature tank; the irrepressibly inquisitive dwarf mongoose coming ever closer to stare with unblinking little black eyes, his glossy brown coat shining in the sun.

The Park's equivalent of the crowds thronging city streets is the ubiquitous impala, which with its numbers in the region of 150 000 is by far the commonest resident. So common are they, that after a day or two many visitors hardly seem to notice them. This is unfortunate, for the behaviour of a herd of impala – the mock fights, courtships, grooming, and amazingly graceful leaps – can entertain for hours.

Apparently not much less common, though fewer in number, are the skittish zebra and the clown of the veld, the blue wildebeest, which are often seen together. Nearly as numerous, but seen less frequently owing to its shyness and its habit of moving in smaller herds and favouring thicker bush, is that most gracefully imperious of all antelope, the kudu, with its magnificently dappled striped coat and splendid spiral horns.

Almost as splendid as the kudu bull (some say more so) is its cousin the nyala – smaller, darker and more delicate in its movements than the kudu and with white-tipped horns. Nyala are found mainly in the north, with some wandering the Park's perennial river bush even as far south as the Sabie river.

Perhaps the most dramatic of the antelope, and reputedly the doughtiest fighter, is the sable antelope. Predominantly black with brown and white facings, it carries a pair of great, back-curving horns which it can wield like scimitars with the thrust of its powerful neck. Sable are most common in the western half of the Park, appearing sporadically in the north but seen as far south as Pretoriuskop. Its cousin the roan antelope, mainly grey with donkey-like ears projecting from a black-and-white mask, is much less common and also appears chiefly in the north.

There are several other frequently seen antelope species which are too numerous to list here but can be identified easily from the many specialist publications available in the Park. Some of these antelope are rarities in Africa: the terrier-sized suni, an exceptionally shy and delicate creature with a brown coat and stubby, ringed horns which is found only in the far north; the graceful mountain reedbuck; the klipspringer, often seen poised on high rocks like a ballerina *en pointe*; the grey duiker, seen everywhere; and the tiny steenbok, which mates for life and is seldom seen in more than pairs. A shy creature but a delight when found is the bushbuck, so elegantly beautiful with its velvety reddish coat speckled with silvery highlights along the flanks, and with white along the chest and underbelly. The males sport superbly shaped horns.

Last but not least of the antelope is the largest: the huge eland, which despite its 800 kg bulk can clear a 2,5 m (8 ft) fence with ease. There are just on 800 eland in the Park, all in the north except for a small group at Pretoriuskop and a few stragglers in the central area.

One of the smallest of the Park residents is the dwarf mongoose (below) and the biggest is the elephant, here in militant mood (facing page). But in charm they are equals.

Chacma baboons are seen all over the Park. They captivate visitors with their antics but carry huge fangs and are proportionately about twice as powerful as a man. Another common primate is the African vervet monkey, while a rarity, the samango, is found only in the forests along the Luvuvhu river in the north.

Giraffe are among the most popular and photogenic characters in the Park. They are widespread and quite common, and can entertain for hours as they browse, stripping branches clean of leaves with their semi-prehensile tongues, oblivious of thorns.

Many of the larger river pools yield the gregarious hippopotamus, the tubby bushveld submariner that spends most of the daylight hours almost completely submerged, snoozing, mating, suckling calves and often blowing loudly or voicing a great honking roar that can be heard for miles around. It has huge tusk-like canine and incisor teeth sometimes as long as 30 cm from the gum line (70 cm in all), which are used purely as weapons. It leaves the water to graze at night, cropping the grass with its rather horny lips. It has a smooth, hairless and very sensitive skin which excretes a thick, red fluid at times which gives an effect of "sweating blood".

Birds, big and small

For a great many of the people who visit the Kruger National Park, the lure is not the big cats, the elephant or the antelope; their speciality is bird life.

The Park has an immensely rich variety of just over 490 species – more than half the total found in all of Southern Africa. Of these some 280 are permanent residents and the rest seasonal migrants.

The variety ranges in size from the flightless ostrich, the biggest bird in the world (incidentally, those resident in the Park represent some of the few remaining truly South African members of the species surviving today), down to tiny waxbills and weavers, cisticolas, long-beaked sunbirds flashing in the light like jewels, little white-eyes and pygmy kingfishers. They occupy every kind of environment, living on seeds, insects, nectar or flesh and sometimes on each other. They include oddities like a big, brown owl that catches fish by night and a bird of prey that can contort its legs backwards. Their songs, shrieks, screeches, hoots and rattles spice the air with living sound.

To highlight just a few; one often sees a medium-sized grey bird land clumsily on a tree top, teeter back and forth a bit

The little bee-eater is a widespread resident easy to photograph because it often returns to the same perch to eat its prey.

with its broad tail fanned out and a crest erect on its head, and utter a raucous cry which sounds like *kweh-h-h* or, to some, like a distortion of *go 'way*. It is the grey lourie or turaco, commonly known as the go-away bird – not only for its call. Being an inquisitive creature, it often used to follow hunters in the bush, perching on the top of a nearby tree and crying *kweh-h-h . . . go 'way*, a sound which can be heard from a considerable distance. The legend arose that the grey loerie was the self-appointed sentry for all animals and deliberately warned of a hunter's presence – which led to many a loerie being shot off its perch.

In sharp contrast to the drabness of the grey loerie is its spectacular cousin, the purplecrested (actually blue with a purple sheen) variety, fairly common throughout the Park. Its call is unremarkable but what distinguishes it is the flash of brilliant scarlet on the broad trailing edges of its wings as it flies through the trees in dips and swoops. The scarlet is mostly hidden when the bird is perched in trees, where it likes to leap and skip high up in the branches, and then its visible coloration is an electric blue and green body shading to a rust in front, a bright red eye, and a bright blue crest which it constantly raises and drops.

The commoners of the Park birds, the everyday plebeians, are the red and the yellowbilled hornbills, the francolins and the glossy starlings. All have one prominent characteristic in common: they are irrepressible beggars and, given half a chance, thieves too. All are likely to

The impala (left) is the most common resident of the Park. Tallest is the giraffe (above), seen here resting in the noon-day heat near Tshokwane picnic site.

come running to a car as soon as it stops at almost any picnic site or shady riverside road where visitors pause to rest. The plump francolins, also termed African pheasant, can become so tame they will cheekily perch on a car window to beg food. The hornbills are like flying clowns with their front-heavy beaks and ungainly landings, and will rise to snatch out of the air a morsel tossed at them. The starlings, iridescent blue in the sunshine, flit back and forth like scraps of silk, landing right at a visitor's feet to expectantly cock an impertinent yellow eye.

The Park is a paradise for water birds and abounds with herons, storks, snipes, coots, crakes, wild ducks and geese, ibises, hamerkops and others. Some are very specialised, like the rare black egret which spreads its wings forward to form an umbrella over the water, either to attract fish to within beak-reach or to enable it to see clearly without the sun's reflection. Another is the African jacana or lilytrotter, distinguished by its blue forehead, yellow and red body and black and white neck, which lopes busily across lily patches with its very long, splayed toes, pausing to turn the leaves over to feed on insects beneath.

Sometimes very common in the Park, depending on how much food is available, is a tall, heavy-beaked, charcoal grey and white stork with the stately gait and doleful mien of an undertaker officiating at a funeral. Which it is, in a sense. This is the marabou, a carrion eater-cum-killer which competes with the vultures for the leftovers of a lion's meal, sometimes eats fish trapped in pools and has the unendearing trick of visiting the colonies of the redbilled quelea to shake the young out of their nests like fruit and eat them on the ground. Like the vultures and the hyenas, the marabou serves a valuable role in the ecocycle of helping to keep nature clean. A more dramatic-looking cousin is

The predatory gymnogene has double-jointed knees enabling it to reach into holes and crevices for small reptiles and young birds.

The beautiful green pigeon is difficult to see because it freezes into immobility when threatened.

the saddlebilled stork, with its striking red-black-red barred beak topped by a canary yellow saddle.

Another large and beautiful bird with not so endearing habits sports the unusual name of gymnogene and is quite common. It is a glossy grey with a very noticeable striped black-on-white chest and underparts, almost like a Fair Isle jersey, and a bright yellow face which has the ability to blush. In flight it is identifiable by the black "fingertips" and trailing edges on its wide wings and the white bar across the underside of its black tail. It eats all kinds of creatures, such as frogs and insects, and some wild fruit, but preys chiefly on the young of other birds. For this nature has so adapted its legs that its knees are double-jointed and can bend backwards to enable it to reach down into nesting holes in the trunks of trees and into corners no other bird can get at.

Totally innocuous by contrast is the green pigeon of blue-tinged plumage which behaves in trees more like a parrot, which accounts for its Afrikaans name *papegaaiduif* (parrot dove). Flocks of them clamber around the branches of wild fruit trees, often hanging upside down to feed, and when suspicious will suddenly sit dead still and become almost invisible. When frightened they explode out of the trees in a rush of beating wings.

In the eyes of many the most beautiful bird by far in the Park is the fairly rare and oddly named Narina trogon, found mostly in the north. There is no mistaking it: a combination of peacock blue and crimson with a liquid-smooth flight. Another contender for beauty honours is the no less spectacular and very common lilacbreasted roller, which performs vigorous aerobatics in a symphony of lilac and electric blue. Yet another – also fairly common in the summer – is the carmine bee-eater, which arrives from its breeding grounds in central Africa to join its cousin, the equally attractive European bee-eater with its contrasting vivid blue and green plumage, which wings in from southern Europe, north Africa and the southern USSR.

All over the Park there are birds with extraordinarily long tails trailing out behind them as they fly. They belong to various species, of which the paradise whydah male is the most striking, and then only in the summer breeding season. It drags along behind it a great regal train of lush feathers which so weigh it down that its flight is a series of rollercoaster loops and dips. Its body is not much larger than that of a weaver.

Another commoner quick to make the acquaintance of visitors is a lumbering, turkey-sized and fascinatingly ugly character – the ground hornbill. It has wings but prefers to walk, unless fright-

ened. It makes a deep booming sound, rather like a lion roaring in the distance and eats virtually anything it can hold long enough to toss into the air and swallow. In some parts of the Park, especially around Skukuza, the ground hornbill has recognised passing motorists as an easy source of food and begs quite unashamedly, coming close to cock a big black eye – surrounded by garish red and fringed by luscious lashes – at the visitor until given a titbit (which is, in fact, illegal).

Perhaps the most fascinating group of birds in the Kruger National Park is the one given least attention by most visitors. No doubt this is because they are uncommon in the camp areas, they nest more remotely than other birds and they fly high. When they are seen it is very often when flying against the eye-numbing brilliance of the sky, or perched in the tops of trees. These are the raptors, the airborne equivalents of the lion, leopard and cheetah – the hunters: the eagles, hawks, falcons and kites.

Fifteen of the 17 eagles in South Africa are found in the Park, and nine of these are permanent residents that do not migrate seasonally.

The martial eagle is the king of the African raptors – a giant of about 6 kg (13 lb) in the female and 3 to 5 kg (7 to 11 lb) in the male with a wingspan of nearly 2 m (some 7 ft). They mate for life, produce a chick every two years and rule in territories of nearly 100 km² for each pair. The martial eagle is a bird of astonishingly military bearing and from a perch high in a large, dead tree will stare back at a visitor with Prussian arrogance from beneath great, beetling brows. They hunt quite large animals such as young antelope, mongooses, leguaans and guinea fowl, often by lying in ambush, carefully concealing themselves in thick tree foliage above open ground.

The crowned eagle is a magnificent creature found in many parts of Africa, as far afield as Guinea in West Africa and Ethiopia, as well as in South Africa. It is a bird of the forests and is not common in the Park, where it is largely restricted to the Luvuvhu and Limpopo rivers in the north. It is clearly identified in flight by its short, broad wings which

Fluttering its vamp's eyelashes is one of the Park's big birds, the ground hornbill. They are seen often at the roadside just north of Skukuza.

are strongly barred underneath in black and white except for the deep leading edges, which are brick-red. Perched, adults can be spotted by their dark bodies, mottled chests, yellow feet with black talons and prominent crests. The crowned eagle is a powerful hunter which lives on monkeys, mongooses, dassies (rock hyraxes) and even small antelope, for which it waits in ambush, sitting high in trees and watching open glades. Because it is a forest dweller, this raptor has the ability to fly almost vertically upwards. It uses the same nest year after year, rebuilding and adding to it until it becomes huge.

The black eagle, unmistakable for its colour and a big white V in the middle of its back, is smaller but no less impressive than the martial eagle, and more so when it flies with huge dives and climbs during courtship. They nest along the cliffs and koppies close to the Luvuvhu river – the habitat of their staple diet, the dassie – and have the charming characteristic of a very close, lifelong pair-bond which includes much time spent together just soaring over their territory, seemingly purely for enjoyment.

The steppe eagle is a large, brown bird sometimes confused with the tawny eagle and Wahlberg's eagle and which migrates all the way between the Park and the Siberian and Kazakhstan regions of the USSR – an enormous distance for which they have to fatten themselves on frogs, rats, mice, small reptiles, and also termites, which although high in food value have to be consumed by the thousand.

The Wahlberg's eagle is an impertinent fellow which will not hesitate to attack even a martial eagle invading its terrain. The most common eagle in the Park during the summer months, this rather smaller migrant varies a good deal in coloration but is basically brown. It soars high and is notable for its tremendous swoops in which it will fall out of the sky like a ball and then open its wings and brake moments before it strikes its prey, usually a rat or lizard or similar small animal. Nests are easily seen along the Park roads: there is one in a leadwood tree next to the H4-1 tarred road about 10 km from Skukuza on the way to Lower Sabie.

The fish eagle is undoubtedly the best-known African raptor – a beautiful

bird often seen at rivers and dams in the Park, striking in its crisp black-and-white plumage. Its fame stems from many television films showing it skimming the water's surface to pluck out a fish with a swift forward strike of its talons. Its cry – a clear, piercing call which can be heard over long distances – has almost become the hallmark sound of Africa. Although fish is its staple diet, the fish eagle will readily hunt other prey such as ducks, cormorants and herons and will stealthily stalk and steal from other fish-catching birds like pelicans and herons.

The bateleur eagle is said to be one of the most striking birds of prey in the world. Its head and chest are black and its back a rust brown, though in some birds it is a lighter colour. The legs and the skin around the eyes are usually red, although sometimes orange or even yellow. It is renowned for its ability to soar at high speed, about 80 km/h, 50 to 100 m above the veld, rocking from side to side like a tightrope walker – hence its name. It is quite common in the Park and lives by eating carrion and hunting.

The blackbreasted snake eagle and the brown snake eagle, as their names imply, live off snakes. Both have talons so adapted that they can grip an object as slim as a pencil. They float down upon their prey, pick it up alive and usually kill it in flight and swallow it whole.

Owls are also raptors and one of the most intriguing in the Park has the same tastes as the fish eagle. This is the large (about 60 cm or more than 2 ft tall) Pel's fishing owl, a reddish brown bird without the typical owl "ear" tufts and with big, dark pools for eyes, which is found mainly in the forests of the north, along the Limpopo and Luvuvhu rivers but also along the Olifants river. It is usually difficult to see, but can be found in the late afternoons, singly and in pairs, as it sets out to seize fish in floodpans and shallow river pools. Mostly it fishes by night with specially adapted long claws, naked feet and rough soles and it is reported that when several of them are gathered at the same pool they make a raucous chorus of hoots and screeches.

The Park has a variety of other owls, small and large, but they have to be sought with determination as they are mainly nocturnal.

Of course, the Park also has vultures,

and these are a subject worthy of study by the visitor in their own right. There are five species in the Park: the white-backed vulture, the Cape vulture, the hooded vulture, the whiteheaded vulture and the lappetfaced vulture, and they occur in large numbers everywhere. After the hyenas, these are the second line of nature's refuse removers. They get themselves into an appallingly messy state when feeding on a carcass but somehow reappear later quite clean; they have the decorous custom of bathing in rivers and waterholes, apparently to keep cool as much as clean. No fools, obviously, the vultures have learned to follow the Park's helicopter during times

The Cape vulture is declining alarmingly in numbers throughout South Africa.

when surplus animals are being culled (see p. 71).

They have also provided Park staff with some amusing anecdotes. One such story involves a vulture and a hyena. Generally, vultures will wait for hyenas to have their fill before they move in on a kill. On this occasion, a whitebacked vulture and a hyena had a slight misunderstanding. The hyena seized the vulture in its enormously powerful jaws, carried it about 50 m from the carcass and dumped it on the ground. The vulture did nothing to free itself while being ignominiously carted off but when dropped, shook out its feathers indignantly and took long strides back to the carcass – but now waited patiently for a chance to feed.

On another occasion rangers spotted

10

a buffalo carcass covered in pecking, ripping, tearing vultures, a common enough scene. Suddenly all the vultures bounced into the air and flew off in fright. The rangers investigated and found that one incautious vulture had become trapped inside the chest cavity of the dead animal, imprisoned by the ribs. It panicked and the noise of its flapping drove off its peers. The men lifted up the carcass to free the bird, which then lost all interest in feeding and spent the next few hours sulking in a nearby tree while it dried off.

The hooded vulture, easily identified by its slender hook-tipped beak, is found in many parts of Africa and is one of the smallest vultures, feeding largely on the kill leftovers of jackals, eagles and small cats and on the dung of lions and other carnivores.

The lappetfaced vulture, white-legged and bare on the neck and head, is Africa's largest vulture with a wingspan of nearly 3 m (10 ft) and has a particularly powerful beak which makes it the only one able to tear thick animal hides. Other vultures will cautiously make way for this one.

The whiteheaded vulture is the prettiest vulture with a distinctive red beak and identifiable also by the white secondary feathers seen in flight. Like the lappetfaced and hooded vultures and the gymnogene, it blushes, changing the colour of the neck and face. This one is a hunter as well as a scavenger.

The whitebacked vulture is the most common and there is never a kill without these on it.

They are often confused with the rarer – in fact endangered – Cape vulture, which breeds only on cliffs and comes into the Park from the rocky faces of the nearby Escarpment. The Cape Vulture has an extremely good wing loading (that is a high wing area in proportion to body weight) which enables it to fly great distances.

These latter two vultures belong to the group called griffons, one of whose characteristics is that they usually soar very high above the ground and have to wait until mid-morning on most days to leave their roosts, when there are enough air currents for them to ride.

A beautiful bird found along all the perennial rivers in the Park is the whitefronted bee-eater, which nests in holes in mudbanks.

The cold-blooded denizens

The mere word "reptile" makes many people shudder. Since Genesis they have been associated with things evil, poisonous, sinister, cold and clammy. A serpent misled Adam and Eve. An asp killed Cleopatra. Tarzan fought pythons.

It is a totally unfair reputation: reptiles fill a major slot in nature's scheme of life and include some of its most efficient designs for survival. A crocodile dragging an antelope into the water may be a grim sight but the crocodile is fulfilling a purely natural need – to feed itself, as is the case with lion and leopard. In any case, crocodiles subsist chiefly on fish. Similarly, the mamba's poison is not for use on man but to kill its prey quickly.

The "dragon" of the Lowveld, the leguaan or monitor lizard (above) is a favourite meal of the most handsome of African eagles, the martial (facing page).

The Nile crocodile is certainly Africa's most renowned reptile and the Park's waters are home to a great many. They are easily seen at rivers' edges and on sandbanks, sometimes lying dead still basking in the sun with their great mouths gaping and yellow while bold sandpipers peck insects from their bodies and even pick clean their teeth. The crocodile is an exceptionally efficient hunter and respects only its constant neighbour in the water, the hippo, and the elephant, a frequent bather. Crocodiles that have been unwise enough to latch onto elephants' trunks have been hauled bodily out of the water and stamped to death on the banks. Sometimes a small crocodile with more greed than sense seizes the snout of a large

drinking giraffe – to be lifted high in the sky and shaken like a dishrag until it lets go.

The crocodile is said to be the killer of more people in Africa than any other animal, although others give the prize to the hippo or the buffalo. While the crocodile's diet is mainly fish and turtles, it displays its stalking skill and its great strength in catching animals that have come down to drink. Crocodiles have been seen dragging full-grown buffalo bulls and even male lions to a watery death – not surprising when they grow to more than 5 m (16 ft) in length and can weigh nearly a ton. The biggest crocodile ever measured in Southern Africa was caught alive in Venda near the Park's northern end and weighed 905,7 kg (1 996 lb) with a length of 5,5 m (18 ft). But there are others at liberty in the Park believed to be as big, especially in the Luvuvhu, Olifants and Sabie rivers.

The crocodile plays a key role in maintaining the balance between animal populations. For instance, scientists have noted that where crocodile numbers have been reduced there have been population explosions of barbel, or catfish, which in turn have disturbed the balance between barbel and other fish.

Similar to the crocodile, for the un-initiated, are the leguaans, of which there are two kinds. One is the white-throated monitor or rock leguaan, which reaches about 1,5 m (5 ft) in length, wanders about the veld and the koppies and if threatened, raises itself up on its stumpy legs like some anachronistic dragon and hisses loudly, with its forked tongue flicking in and out. It is a grey-brown reptile with a spotty, banded body and its only defence, apart from a strong bite, is to lash out with its tail, which is longer than its body. This leguaan easily climbs trees and lives chiefly on small mammals, birds and eggs.

The Nile monitor or water leguaan is a larger fellow with a body length reaching more than 2 m, of which two thirds is tail. It is a common river dweller and swims swiftly and easily by propelling itself with its powerful tail while its limbs are clasped to its sides. The tail is also its weapon and can cause considerable damage if the leguaan is cornered. It is grey-brown to dark brown with yellowish spots and bands and lives on

shellfish and fish, but readily eats snails, birds and the like. It has been seen to use its own body to shape a dam in the shallows, trapping small fish which it eats at leisure.

Perhaps the most easily observed reptile in the Park is a striking character which looks like a relic from the age of the dinosaurs and is found in every tourist camp. It is the tree agama or, in Afrikaans, *boomkoggelmander*, with a bright blue head and neck and a tail of equally bright yellow. They will bite hard if handled and are feared by many indigenous people because of their garish appearance. In the camps they have become fairly accustomed to humans, although outside the Park they are shy, and are commonly seen skittering among rocks, up trees or lying along branches with their heads bobbing up and down. They are usually about 35 cm (14 in) long, and live mainly on insects.

Geckos are entrancing little nocturnal lizards with splayed, five-toed feet which seemingly can take them anywhere, even upside down across ceilings. More than a dozen kinds live in various habitats in the Park but the one most often seen is the tropical house gecko because it inhabits most camps. It is a 5 to 7 cm (2 to 3 in) greyish to brownish fellow with dark bands or spots, and has the cunning trick of dropping its tail off when frightened, presumably to distract a pursuer. Geckos eat the insects that pester people, such as flies, cockroaches, mosquitoes and small beetles.

The leopard or mountain tortoise, common all over southern and eastern Africa, appears in considerable numbers in the Park on the roads after rain, when they trundle on to the tarred surfaces to drink from puddles – making them something of a hazard to motorists trying to avoid running them over. They grow to over 25 kg (55 lb) and live a very long time.

The serrated hinged terrapin is a freshwater turtle seen just about wherever there is water in the Park. They are brown to black and about 30 cm (1 ft) long at maximum, and lie half in and half out of the water basking in the sun, or swim about with just their nostrils above the surface. They like resting on stones, logs or anything else that happens by – as a wandering hippo discovered when it overnighted in a pool near Letaba (see p. 234).

Without doubt, the most formidable and justly respected snake in South Africa is the black mamba, whose name is slightly misleading as its colour varies from pale greyish-green in youth to gun-metal at maturity. It is a most impressive reptile though not often seen because it is shy, swift and quick to avoid contact – unless threatened, when it is pure danger. Though very slim, it averages some 2 m (6 ft 6 in) in length but specimens of up to 4 m (13 ft) have been recorded. It lives in open veld in a hole or under an old tree and hunts far and wide for dassies and other small mammals. If disturbed, the black mamba characteristically heads straight for its lair and anybody in the way is in trouble. It moves at a considerable speed (some 16 km/h – about 10 mph), can raise most of its length from the ground, and can strike extremely fast in any direction except backwards. Its paralysing poison works very fast and the mamba does not strike once, but again and again. Fortunately, tragic encounters between humans and mambas have seldom occurred in the Park.

Other snakes include the beautiful, chocolate brown, black-collared Mozambique spitting cobra, the Egyptian cobra, garter snakes and various adders, including the big, thick puffadder – common all over Africa and reputedly the biggest killer of people because it is so sluggish it is often stepped upon. It strikes with blinding speed and has huge, curved fangs which stab like daggers.

Frogs are legion in the Park, numbering some 33 species, and when they emerge in ponds, puddles, rivers and dams after the rains, their nightly chorus of croaks, tweets, pings and whistles can drown out all other sounds of the bush. They too play an important ecological role: without their voracious appetites the environment would be overrun by insects. They come in many shapes and forms – and sounds. The biggest is the bullfrog, which grows up to 200 mm in length (nearly 8 in) and can live for more than 20 years. One of the most curious is the grey tree frog, which in the rainy season builds nests of foam carefully situated on low branches hanging over pools of water. Eggs are laid within the foam, resembling clusters of snow white candy floss, and the tadpoles drop into the water after hatching. This operation is always heralded by their exuberant squeaking night and day.

The smallest is the dwarf puddle frog, a lively little midge-eater which buzzes incessantly in the rainy season. The most colourful is the ornate frog, aptly named for its brown, orange, green, yellow and black abstract markings.

Two completely new species of frog, the portly sandveld pyxie, or officially *Tomopterna krugerensis,* and the golden reed frog, *Afrixalus aurens*, were first discovered recently in various pans and pools within the Park.

The unmistakable tree agama can be photographed easily at Skukuza and Satara camps although they are often difficult to see because of their camouflage **(below)**.

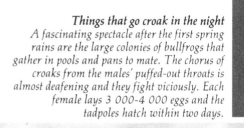

Things that go croak in the night
A fascinating spectacle after the first spring rains are the large colonies of bullfrogs that gather in pools and pans to mate. The chorus of croaks from the males' puffed-out throats is almost deafening and they fight viciously. Each female lays 3 000-4 000 eggs and the tadpoles hatch within two days.

The amazing insect world

The insects of the Park form an enormous universe of their own, vastly outnumbering the variety and numbers of all other living species – a universe of immense range – from creatures near microscopic in size to others that, for most of their lives, never see the light of day because they are internal parasites in other animals, to those that live underwater, to those that fly and those that crawl, to the poisonous and the simply big and aggressive. Some are dangerous to man, because of the diseases they transmit; others are edible by man, like the mopane worm and the flying ant. Insects make up nearly three quarters of all living species on earth if plants are excluded, and more than half if they are included.

So gigantic is this spectrum of insect life in the Park that it is still very much in the process of being explored by scientists. Only a few, like the *Anopheles* mosquito, house fly, tsetse fly and horse fly, are pests or disease-carriers. They all have specific roles in nature's order and to carry these out they come in a nearly infinite range of shapes, sizes, colours, habits and abilities. They help keep the world clean, like the dung beetle. They help vegetation to procreate, by spreading pollen. They control one another's populations, by eating one another. They sustain bird, animal and other insect life, by being eaten. Without insects nature would come grinding to a halt.

It is not possible here to do more than describe some of the more interesting examples from this world-within-the-world.

In the African wild nothing is wasted, not even dung. Hyenas, vultures, jackals and maggots clean up the debris of animals killed by predators. Part of the vast team of scavengers which daily scours away the detritus of life and death is the family of dung beetles – seen easily enough by anyone who takes the trouble to study so seemingly unlovely a subject as a lump of dung.

This is the home of the scarab beetle, the industrious little creature so revered by the ancient Egyptians. There are many kinds of African dung beetle – more than 2 000 – and some are so specialised that they will use only the dung of certain animals, like the buffalo, rhino or elephant. They will gather at a

This scarab beetle dining on fresh rhino dung was photographed near the Wolhuter trail camp in the south of the Park.

fresh lump of manure within minutes of its being dropped, and within hours it will have been broken up, scattered and effectively returned to the earth.

The noted South African wildlife writer James Clarke once wrote: "The scarab beetle is a little like Prince Charming in *Cinderella*. But only a little. He meets his bride-to-be at the ball. What tends to take the magic out of the situation is that the ball is actually a ball of dung. Entomologists call it the 'nuptial ball'. Such a pretty name for a ball of 'you-know-what'."

Watching a pile of, say, fresh rhino droppings at a roadside (not on the tarred surface), one will notice a typical scarab a few centimetres long gathering the grassy dung and then working and rolling it with its hind feet into a steadily enlarging sphere which can reach the size of a tennis ball. When it is satisfied with the size of the ball the scarab will trundle it off, still moving backwards at a frenetic pace, until it finds a reasonably soft patch of earth. The beetle then digs around and under the ball, slowly sinking it into the ground. When it has vanished from sight the ball becomes a place in which either the scarab's egg will hatch, when the new larva will find itself surrounded by food, or in which the beetles will mate, fairly safe from the mongooses and other creatures which like to eat them.

In its own inimitable way the dung beetle fulfils a huge purpose in nature. But for its efforts the Park might be blanketed in a layer of dung or made intolerable by the hordes of flies which would breed in it.

Most prominent among the many kinds of ant in the Park are the Matabele ants, which move in columns sometimes many metres long. If the column is disturbed they immediately fan frantically out looking for the disturber, and can bite painfully if they find him (see p. 22).

But there are many other invertebrate creatures aside from insects; not least among these are the spiders and scorpions.

Were scorpions the size of lions the Park would indeed be a horrifying place. These members of the invertebrate world are formidable predators; the terror of the arthropods. But fortunately they are retiring, mainly nocturnal creatures which avoid people and most are small, although the larger kinds can reach nearly 20 cm.

About 160 of the 600 species of scorpion known throughout the world are found in Southern Africa and many inhabit the Park. All belong to two families that are easily distinguishable. *Scorpionidae* have very big, lobsterlike claws, long, thin tails and are not very poisonous. *Buthidae* have small claws, very thick, strong tails and the poison of some of these species is potent enough to cause excruciating pain and, in small children and sick adults, even death.

Curious creatures which give birth to live young and carry them around on their backs, the scorpions do use their stings (sharp, curved like a cat's claw and mounted at the tip of the tail) in self-defence but their main function is to paralyse their victims, which include the whole range of the invertebrate world, including other scorpions. The quicker, thin-tailed scorpion relies on the power of its great claws to crush its prey, whereas the more sluggish thick-tailed kind depends on its highly toxic poison for the *coup de grâce*.

Best known of the latter is *Parabuthus* (shown here), a medium to large member of the buthid family which nests in tunnels up to a metre deep. A member of the Park staff, author and entomologist L.E.O. Braack, gives a vivid description of this group: "These are the rogues of the scorpion world, small-clawed but with massive tails which they lift menacingly towards any potential enemy. Grey-black, they strut about in silent arrogance in the instinctive knowledge that few animals will dare to tamper with them."

One animal that does dare is the irreverent baboon.

Scorpions form a part of their widely varied diet, like canapés between meals, and sometimes a troop of baboons can be seen moving slowly across the veld lifting rocks and snatching up the scorpions by the tail between finger and thumb and hoisting them aloft, thereby frustrating their efforts to sting. They then break off the stings and accompanying poison sacs and gulp down the rest, crunching them with evident satisfaction.

One of the oddest scorpions in the Park is also the biggest: an extremely flat species which rejoices in the name *Hadogenes* and lives in fissures in rocky outcrops – hence its flatness. It is a sluggish mover with a very long, thin tail in the males.

While there are many scorpions in the Park, they tend to keep away from human habitation and cases of people being stung are very rare, except among those foolish enough to go probing under stones or behind the dry bark of trees.

Unlike the scorpion with its tough, chitinous armour, the centipede has a soft, flexible body adapted to its role as a fast-moving nocturnal hunter. It is a rather unnerving character to encounter but most species are fairly harmless; the poison injected through the hollow front pair of legs is no worse than a bee sting. Centipedes can grow to fairly impressive sizes, some as long as a man's hand.

Not to be confused with centipedes are the totally harmless millipedes, identifiable by the multitude of legs which move beneath their long, cheroot-like bodies like side curtains being blown by the wind. Trundling across the road or patch of dust like a miniature freight train, the millipede has provided entertainment for countless generations of children as it goes about its way seeking its diet of rotting vegetation. If disturbed it will usually roll up into a catherine wheel, but sometimes, if it lands on its back, it will contort violently in simulation of a snake to deter the intruder. This very common hard-shelled Park resident also grows to an impressive size for the invertebrate world, some reaching between 20 and 30 cm (8 to 12 in). Lore has it among certain African tribes that to kill millipedes is to bring on a deluge.

For those with an eye for beauty in detail the Park can be a treasure house in the early morning before the heat of the new sun has burned off the moisture of the night. Then, in the bushes and the trees, strung between the tracery of branches, can be seen symmetrical diadems glittering in the slanting sunlight: spider webs decked with dewdrops more clear and perfectly shaped than any diamonds.

The Park is alive with spiders of all shapes and sizes, some of great beauty and brilliant colour and none particularly dangerous, however formidable they appear. Many spin fine skeins of silk to catch flying insects; others, like the long-legged, colour-banded orb-spinners of the genus *Nephila*, string webs of extraordinarily tough, thick, golden strands.

Some webs are haphazard, others so meticulously patterned they appear to have been designed by a computer. One of the few communal species of spider in Southern Africa is responsible for large, untidy clumps of webs as much as 30 cm long, which envelop the branches of small bushes – nests within which several hundred spiders share their prey, leaving the debris of empty shells and wing cases strewn about their tiny tunnels inside.

Other spiders make no webs but hunt, such as the jumping spiders, some of which are less than a centimetre in length, with huge headlight eyes to enable them to stalk unsuspecting insects and suddenly leap upon them. Among these are the big baboon spiders – ready-made characters for a horror movie with their hirsute bodies, thick, strong legs and great jaws below shining eyes – which live in silk-lined burrows where they raise their young. These night raiders can reach 5 cm (2 in) in length.

Harmless spiders are often found in huts in the Park's camps where they serve a purpose: culling the insects which would otherwise become a nuisance to people. It has long been the practice on many South African farms to let spiders spin their webs and make their nests in homes to combat mosquitoes and similar pests.

If scorpions are the lions of the invertebrates and spiders the leopards, then the cheetahs are the fascinating creatures called the sun spiders, solifuges or any of a variety of nicknames. They look like spiders but are a completely different group of arachnids, highly specialised for their particular way of life. Orange

Contrast in colour: a predominantly green scutellrid bug feeds on the bright blossom of Gridia seriocephalus *poking through a thicket at Pretoriuskop camp. The leaf in the picture is of a different plant.*

to red and very hairy with long legs and fat abdomens, they are not poisonous but equipped with huge, powerful jaws sticking out of the front of the head and containing sharp teeth which can draw blood.

Sun spiders are chiefly nocturnal although they can be seen by day, and their wont is to rush about at high speed in any direction until they literally bump into something edible. Twin eyes like spotlights in the middle of the head give them reasonably good vision, but their main catching tool is a pair of sensory feelers which are held out before them. Once touched by these, the fate of the victim is usually sealed; once those great jaws clamp shut, there is no escape.

Because of its erratic, seemingly aimless running about, the sun spider is known in Somalia as "the old man who has lost his camels". In German it is called the "waltzing spider", and in Afrikaans the "hunting spider". In Namibia it has the curious Afrikaans name of *baardskeerderspinnekop* which means "beard-shaving spider", presumably because its mandibles are vaguely akin to a barber's shears.

These strange little beasts will dash into the light of a campfire to seize an insect lured there or even a titbit dropped from the barbecue, such as a piece of gristle or fat.

The sun spider has another distinguishing feature: it breathes visibly. After running about vigorously it can be seen, when it pauses, to inhale and exhale like a panting dog. Its breathing is through simple tubes like those of insects but by this device it gathers enough oxygen to maintain its frantic pace.

Returning again to the insect world, it is appropriate to briefly discuss those spectacularly colourful squadrons of aerial dancers, the butterflies. More than 219 species, a third of all those found in South Africa, have been recorded in the Park and many of the most colourful are also among the most common, like the big African Monarch which can be seen at any time of the year, the brilliant Yellow Pansy, the darker Charaxes and the bright and diminutive Pale Ranger. Their names are often as colourful as their wings: Angola White Lady, Osiris Blue, Constantine's Swallowtail, Bushveld Emperor, Woolly Legs, Palmtree Nightfighter, Brown Commodore ... to name but a few.

*An orb-spinning argiopid spider **(inset)** and a pisaurid or nursery web spider, both photographed in camps.*

THE MILITANT ANT

One of Africa's least known, least understood groups of creatures is that of the ants – true ants, as distinct from the termites, with which they are often confused – because it is so difficult to study them in the wild. There are thousands of different kinds, belonging to the scientific order *Hymenoptera*, which includes bees and wasps and numbers more than 100 000 species world-wide.

Many are highly specialised. One kind lives only in the long, white thorns of a particular *Acacia* tree. Some weave, others harvest leaves, several sorts "herd" and nurture aphids like cattle for the sugary fluid they exude when caressed. Some sting, others squirt formic acid in self-defence, yet others spit an unpleasant fluid which quickly hardens into a resin-like substance that immobilises enemies and prey. They form a vast family whose characteristics have changed little through the aeons. Some naturalists claim that if nuclear storms ravaged the earth, the ant would be dominant among the survivors.

The family includes many kinds of predators – ants which live by hunting, as do the lion, eagle and crocodile. But proportionately, gram for gram, ants are by far the most powerful of the hunters, able to lift with ease many times their own weight and move at a speed which relatively would be in the region of 100 km/h for human beings.

Intriguingly, many species of ant give the impression of having some sort of common feeling ... "mutual mind", some observers term it. But to label it thus or in any way suggest a common intellect would be attributing too complex an intelligence to the ant.

Harass a single ant in the vicinity of a busy nest and within seconds all others within a sizeable radius will suddenly become agitated, alert and aggressive, and start running towards the harassed one. This communication is not achieved by telepathy, however, but by pheromones – chemical substances which when secreted by an ant immediately vaporise and swiftly spread, alerting other ants. These chemical signals play an important role for many species of ant in reproduction, giving the alarm and laying trails.

The most ferocious African members of the ant family belong to two groups, the subfamily *Dorylinae* commonly termed army ants or driver ants, and the subfamily *Ponerinae*, which includes the Matabele ant. These subfamilies have one prominent characteristic in common: they go forth on raids in long columns, and woe betide anyone foolish enough to interfere with or step on such a column.

Apart from a few other superficial similarities, the two are quite different. The driver or army ants inhabit the tropics of Africa and South America – in East Africa they are known as safari ants for their custom of trekking long distances in columns. These are fairly broad spectrum feeders and will eat most living things. The story is told that if a farmer finds a column of ants heading for the farmhouse, he must let them enter and when they are in, thoroughly disturb the column with a broom . . . and get out quickly. If he gives the ants a few hours and comes home when the column has moved on, he will find the house scoured clean of roaches, bugs and all other pests including other kinds of ants.

In South America driver ants are said to move in columns of up to half a million individuals which will kill and eat anything living, including animals as big as cattle, which is too slow or otherwise unable to avoid them. Such columns extend for scores of metres.

The Matabele ant is found in most of Africa – under various names – and while it is no less ferocious than the driver or army ants, it fortunately specialises in only one kind of prey: termites. The Southern African name "Matabele ant" is derived from the group of people who broke away from the mighty Zulu nation led by the warrior Shaka Zulu, under one of his generals, Mzilikazi. They moved west and north through what is today the Transvaal and Botswana, carving a bloody path through other tribes until they finally settled in present-day Matabeleland in south-western Zimbabwe.

The Matabele ant of the Lowveld and the Kruger National Park has the scientific name of *Megaponera foetens*. It is quite commonly seen as the columns cross Park roads – dark ribbons about a hand's breadth wide and a metre or two long (although sometimes longer) probably on the way to raid a termite nest or possibly moving house.

Matabele ant scouts seek out termite nests and when a scout finds one, it returns to its own nest while laying a scent trail on the ground. Hundreds of worker ants will then follow this trail and at the termite nest they will catch and sting as many termites as they can with the stings on the tips of their abdomens. The poison they inject paralyses, but does not kill the prey.

As the termites are stung, they are piled up outside their mound. When the raid is over, each Matabele ant picks up several termites and carries them back home, where they are eaten by the ants and their larvae.

Anyone who has been stung by a Matabele ant will confirm that its venom is potent and painful, though not fatal to humans unless stung many, many times.

The subfamily *Ponerinae* are regarded as among the most primitive of ants and several species are more so than the Matabele ant, like those which hunt singly because they cannot communicate by scent. Several species are so highly specialised that they hunt only particular kinds of insects.

A fine description of the Matabele ants comes from regional ranger Flip Nel of Shingwedzi in the north of the Park:

"In a world roamed by elephant, buffalo, lion, rhino and other animals in the heavy category, one tends to overlook the small things. The so-called Matabele ants, *Megaponera foetens*, do not allow you to step over them casually, because, in the first place, it hurts when they bite and, secondly, because they are truly interesting to watch.

"A little while ago I once again had the opportunity to watch these ants, which are a dull black in colour and as much as 12 mm in length. They are particularly active on cool, overcast days, during late summer afternoons after it has rained, or early in the morning when the veld is still wet with dew.

"It is hard to tell how they know where to go, but they invariably head straight and purposefully for a white ant (termite) nest which is visible only as a layer of earth glued around a piece of dead wood or on a bare spot in the grass.

"This particular group I was observing consisted of about 800 Matabele ants walking in ranks of five to 10 abreast – very orderly, like a well-trained army battalion. The group formed an almost solid line which, in a way, resembled a large, black snake winding through the grass.

"Usually thousands of Matabele ants are involved in a raid like this and the average length of such a mobile army measures almost 4 m. Recently Leo Braack, who is at present doing research for a doctorate in entomology in the Kruger Park, and myself, measured a Matabele ant army which stretched some 23 m in length.

"When the prey is the large harvester termites, the raid is not so easily carried out. Presumably the termites put up some resistance, because only the biggest of the Matabele ants enter the termite holes and it takes some time before they reappear with a victim. Each Matabele ant takes hold of a termite and they return to the nest which, in one case, was 10 m away.

"Sometimes the Matabele ants move lock, stock and barrel to another spot, probably as a result of a food shortage in the vicinity of the old nest. When they move, the pupae and larvae are taken along. One particular group moved over a distance of about 12 m. They carried the larvae and purpae for about a metre at a time, put them down and went back for more. The group's organisation was impressive: they seemed to know when everything had been moved, for they do not turn back unnecessarily. As soon as all the pupae and larvae were gathered at a

"If you venture too close and one of the guards scouting the area in the vicinity of the marauding column finds you and inflicts a bite with its sharp jaws, you simply have to react fast!

"As soon as the ants reach their goal, the soft rustling sound they make while they are on the move becomes louder, and they spread out to collect as many white ants as possible. The ants actually turn the earth over in their search for their prey which they gather together in one spot.

"All this happens very rapidly and soon each Matabele ant takes four to five white ants in its jaws and they start retreating – along exactly the same way they came. They present quite a spectacular procession on their way back with the loot. If the raid was successful, they soon return to make a second and even a third attack.

particular point, the ants in front immediately started moving forward again. A few ants remained behind to make sure that everything had been cleared up. They move fairly rapidly, considering the distances they cover.

"I noticed another interesting phenomenon: some reddish flies, somewhat larger than house flies, showed a keen interest in the moving columns of Matabele ants. It soon became clear why – they were trying to reach the ant larvae. They did not attack blindly, however, and were obviously wary of the ants. So they lay in ambush where the ants had to move over difficult ground and their numbers were thinned out. Then the flies made quick attempts to take hold of a larva, or suck it out, or lay eggs in it. Sometimes the flies succeed in reaching the larvae, but they usually have to move away quickly when the Matabele ranks close again."

The wonder of trees

So vast and varied is the plant life of the Kruger National Park that studying it has become the life work of several scientists; the trees alone are far too large a subject to be described here in detail.

Broadly, however, the Park can be divided into four main areas of vegetation:

- Between the western boundary and the centre of the Park south of the Olifants river, the veld is dominated by the red bushwillow and *Acacia* thorn trees, with many marulas among them.
- The region south of the Olifants river from the centre to the eastern boundary is important grazing land with red grass and buffalo grass and with knob-thorn, leadwood and marula trees.
- In the Park's western half north of the Olifants the dominant trees are the red bushwillow and the mopane.
- The ubiquitous shrub mopane blankets almost the whole of the Park's north-eastern region.

Within this framework the vista of trees changes constantly, especially in valleys and along the rivers. In autumn the endless mopane of the north-east carpets the land with glowing red which on the approach to a river or dam suddenly gives way to the rich, almost fluorescent greens of riverine growth rank with creeper and vine.

On the roads around Olifants camp the off-white, fine-seamed bark of the leadwood trees glows palely against the backdrop of the mopane's yellow and red and the verdant green of other growth. The fat-bodied trunks of the huge baobabs tower above the vegetation like Don Quixote's imagined windmill monsters, stubby arms reaching to the sky. Most dramatic and beautiful, to many, are the giant sycamore figs along river banks, their high domes of thick foliage shading massive branches and trunks bolstered by flying buttress roots. The scene is ever-changing, ever-new – through the seasons and through the changing shades and angles of light from dawn to dusk.

Many trees are known not only for their beauty and size but for purely practical reasons. The wood of some is especially good for furniture, long-lasting

VEGETATION MAP

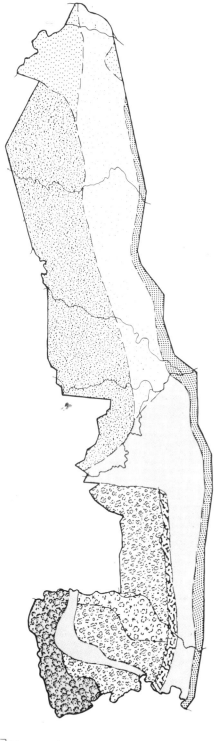

- Mopane shrub veld
- Mixed Red bushwillow-mopane veld
- Sandveld communities
- Lebombo Mountain communities
- Knobthorn-marula veld on basalt and dolorite
- Mixed Combretum veld
- Mixed Combretum-Acacia veld
- Silver Clusterleaf-sicklebush and Malelane Mountain communities
- Delagoa thornveld

log fires, medicines or for making wagon wheels. The fruits of many are edible, though not always palatable.

The marula is probably the Lowveld's best-known tree because of its attractive shape and rich green when clothed in summer foliage, and particularly because of its small round fruit which encloses a very hard, misshapen nut. The fruit has long been used by indigenous people to make beer, which becomes more potent the longer it is kept. Farmers' wives use the fruit to make a tart or a sweetish jam or jelly to spread on bread or which gives zest to venison in the same way that apple does to pork, or mint to mutton.

Wild animals – from birds to elephants – love the fruit, which falls to the ground when ripe and draws hordes of creatures. It also ferments when over-ripe and Lowveld lore is rich with stories of elephants stuffing themselves with the fallen fruit and then rolling drunkenly away to bounce from tree to tree. Baboons are said to suffer the same fate but in neither instance is there any firm record of hangovers.

The marula's typical shape is a straight single stem supporting a large, round canopy of dense, small leaves, and is common throughout the Park. It is quite a large tree, averaging about 15 m (50 ft) in height in full growth.

African tribal legend has it that when God became angry with the baobab He kicked it out of Heaven and it landed head down on earth with its trunk and roots sticking up in the air. It is an apt description for this strange, bare giant often more than 20 m (70 ft) tall and 10 m (34 ft) thick, whose twisted, tapering branches end in ragged clumps of foliage. In some parts of Southern Africa, such as the Tete region of central Mozambique, there are entire forests of nothing but baobabs, some immense. The largest recorded in South Africa is about 26 m high and 19 m in circumference. So big are they that a wayside pub was built into the hollow trunk of one in the Lowveld goldfields late in the last century. Another, in Namibia, contains a flush toilet.

The baobab is unmistakable, even without knowing that its pillarlike trunk is greyish and fluted. Here it is also known for practical reasons: the pulp of the big, oval fruit makes a cooling drink for the hot bush traveller; the pulp

*Above: The best-known landmark in the north of the Park is Baobab Hill, situated 8,8 km south of the Luvuvhu river on the main road to Pafuri. Early travellers to and from the east coast used this tree as a navigational aid. **Overleaf:** A view of the magnificent riverine forests along the Luvuvhu river with the luminous green of fever trees prominent. Across the rivers is the historic Crooks' Corner, then there is the Limpopo river and, in the distance beyond that, lies the most south-easterly point of Zimbabwe.*

contains various useful chemicals and in the old days was used to make cream of tartar; bulbs on the roots can be ground to make a porridge; and the bark can be cut and peeled off to make a shelter. Elephants and apes like the fruit and elephants cause extensive damage exercising their tusks on the bark.

Baobabs can live to a great age – up to 4 000 years, according to the latest scientific investigations – and have the survival capacity of continuing to grow even if they fall over. They are found mainly in the northern half of the Park, with a few growing as far south as the Nwaswitsontso river in the central area of the Park.

Many a hunter, transport rider or game ranger trekking through the Lowveld in the furnace-like heat of midsummer has offered up a heartfelt prayer on coming upon the wide, welcoming giant called the "wild fig" or "river fig" – properly the sycamore fig. This huge tree which carries its thick wrap of leaves throughout most the year, is one of the loveliest in Africa. So wide does its canopy spread, that it is able to offer cool shade for most of the day. It likes water and is found all over the Park wherever rivers flow – the massive central trunk dividing into a forest of branches whose high forks are favoured by leopards for storing their catches. Often the sycamore fig is festooned, as at Skukuza camp, with thousands of dangling weaver nests.

The fig it bears is a hairy, small, round "fruit" which grows straight from the branches in clusters. Some animals eat it but it is of little use to people. The indigenous people and the early settlers made an infusion of the bark and leaves to treat a variety of ailments from chest complaints to diarrhoea.

All over the Park, but concentrated particularly in the Timbavati, Luvuvhu and Tsende areas, is the tree which more than any other – more even than the baobab – is the symbol of the great African savannah: the umbrella thorn. Like the sycamore fig, this tree was much favoured by the early travellers for its shadiness – and, unfortunately, also by elephants seeking its leaves and its quaintly curled pods for food.

The umbrella thorn seems to be the tree most often used by artists depicting Africa. From its short trunk its branches spread out wide to carry a flattish, umbrella-shaped crown of shading foliage. During the summer months the tree is covered with fine, whitish clumps of flowers like powder puffs that give some people hay fever.

Elephants think nothing of reaching up to tear down branches to get at the leaves and have caused much destruction among the Park's umbrella thorns. This is why in some places, such as the Timbavati river road, better examples of this tree have carpets of jagged boulders laid down by the Parks Board for several metres around their stems. Elephants dislike walking on the boulders and so avoid the trees.

Apart from its name, the most striking feature of the fever tree is its long, straight, yellow-green trunk – especially in the early mornings and late after-

The blossom of the sickle bush.

noons when it glows in the slanting light of the sun. Its name is derived not from the fact that it causes fever, but from the fact that the old travellers learned never to camp in its neighbourhood. This is because the fever tree grows in low, damp regions – also the chosen place of the mosquito, the carrier of malaria. With malaria controllable now by prophylactic drugs, that danger has vanished, but in the evenings the fever tree is still synonymous with the whine of mosquitoes. The tree reaches about 15 m and is topped by a high, rather sparse network of branches and leaves. Also an *Acacia*, it bears powder puffs of bright yellow flower clusters in the late spring and early summer.

The leadwood tree is thus named because its wood is so heavy it sinks in water and so hard it quickly blunts tools. Up to 20 m high, the leadwood is found all over the Park, mostly in low-lying areas, and has a tall trunk which is usually straight and clean and topped by a slightly spreading crown of leafy branches. Older trees have very pale branches and the bark is seamed by cracks into a mass of small, irregular shapes like the wrinkles on the back of an old man's hand.

The leadwood is unusual for this climatic region because it grows very slowly and probably takes hundreds of years to reach full maturity. The long-dead trunk of one found elsewhere in Southern Africa was carbon-14 dated and found

to have been 1 700 years old when it died.

Another striking tree which reaches about 20 m and is common all over the Park enjoys the unusual name of jackal-berry, perhaps because the seeds have been found in jackal dung. The berries taste quite pleasant and are eaten by both man and animals. The tree is straight and thick stemmed, its top an explosion of growth in all directions. It is found in a number of rest camps where it is usually identified by name or by a number. The small, pretty flowers are borne on separate male and female trees and the jackal-berry has the useful characteristic of being disliked by termites.

Inside Letaba camp, and also at Shingwedzi, are groves of tall, pole-stemmed lala palms with shaggy tops like blonde

Rooihaarbossie *and lunch guest.*

hair ruffled by the wind. These palms reach up to 15 m and are found chiefly in the eastern half of the Park – attractive reminders that this is near the tropics. The fruit of these palms is inedible but they have a range of other uses. Africans tap the sap from their stems to make a fairly potent brew. The fibre of the big, fan-like leaves is used to make hats, rope, mats and thatching.

The wild date palm is not to be confused with the lala. It is much shorter, reaching only about 6 m, and occurs throughout the Park in bushy clumps close to water. Its appearance is much more raggedy than that of the lala and it has more uses: its sap for making beer or palm-wine, its fruit as a quite palatable

food, and its leaf fibres for thatch, hats and the like. Its stem and leaves are enjoyed as a change in diet by elephants.

The sausage tree is another large tree, up to 20 m high with a heavy, shady canopy above a sturdy stem. It grows all over the Park but is not common. Its fruit is a large, sausage-like, hard object dangling from a long stem and may be poisonous to people, but is eaten by baboons and bushpigs. Elephant and kudu like the leaves, nyala eat the flowers and monkeys enjoy the nectar. Kudu sometimes suck the pendant fruit as a child would suck a dummy.

The mopane blankets virtually the whole northern half of the Park, and a great band of these trees spreads right across the Southern African savannah, except in the lower regions near the coasts. It often occurs in shrub form: an untidy mass of stems rising to a metre or two from the ground, very bare and grey in the winter and in drought, otherwise thickly covered in broad double leaves, which consist of two identical wings, rather like a butterfly, on a single stem. In its larger form it grows into a sizeable tree of up to 18 m, usually with a single stem and a slightly spreading crown.

The mopane's saving grace is that it is nutritious. Cattle can survive well on it in times of drought and it is a staple of the elephant diet. It has other uses too. Rope is twisted from the wet bark, the timber of the larger trees is used for making furniture, and it is the exclusive environment of the mopane worm.

The flower of the common coral tree.

This opulently fat insect feeds mainly on mopane leaves and grows to a length of 5 to 10 cm. Under the right breeding conditions these worms have been known to infest a mopane so thoroughly that the tree appears to be bearing worms. The worms, however, are themselves useful: they are high in protein and are relished by a number of African tribes, eaten dried as New Yorkers might eat potato crisps or as a flavouring.

All over Southern Africa a common ornament is a necklace or bracelet of fire engine red, plump seeds about 3 cm long with contrasting black and red ends – a striking colour combination. They come from the pod mahogany, a big tree of up to 20 m in height with a very wide, spreading top of glossy green leaves and, in the right season, bunches of red flowers. The pod mahogany is common in the north of the Park but rare elsewhere. Its tough wood is impervious to insects and its leaves are yet another item on the elephant menu.

We have mentioned a small selection of the more prominent Park trees with particularly individual characteristics. Many others are not very well known or easily identified but have special uses in the same way that the marula and mobola plum are used for making liquor. Some are noted as sources of poison, for their medicinal properties, displays of flowers, or simply for their perfume.

The lavender tree is the leader in the last category. It is not very big, about 8 m at best, rather ordinary to look at and grows only in the hilly parts around Punda Maria and also south of the Sabie river. Early in the year it produces tiny white, very fragrant flowers. Its leaves are used for making a kind of tea and as a perfume – and also as a medicine to get rid of intestinal worms.

The most notable flowering trees include the red thorn with its great masses of white powder puffs, the rather shapeless weeping boer-bean, which bears thick clusters of dark red, waxy flowers rich in nectar much liked by baboons and birds; the tall and spare sjambok pod, which has a fine display of massed yellow flowers; the somewhat scraggly Transvaal gardenia bearing long, trumpet-like white flowers that quickly turn yellow with age; and the apple-leaf, a largish tree of indiscriminate, sparse shape, which distinguishes itself by producing big hanging clusters of pale

Unfortunately, most of the magnificent sycamore fig forests along the banks of the Luvuvhu river at Pafuri are beyond the reach of tourists. This photograph was taken to the west of Pafuri bridge.

mauve flowers that fall to cover the ground below in a wide, colourful carpet.

Probably the most spectacular of the flowering trees, to be found in many subtropical gardens, is the common coral tree. The big, brilliant flowers that appear before the leaves in the spring are a favourite subject for photographers with their blood red contrasting with the vivid blue of the African sky. The seeds are scarlet and black and are often used for making necklaces and other simple jewellery. What few wearers realise, is that the seeds are poisonous; however, they are so hard that nobody is inclined to try to eat them!

Although it can be seen in the Park's north around Punda Maria, the common coral tree is found mainly in the south, in the regions of Pretoriuskop and Bergen-dal – as is its less dramatic cousin the broad-leaved coral tree. Although smaller in size and full of sharp hook thorns, this tree also offers superb aloe-like crimson flowers and similar knobbly pods with black and red bead-like seeds.

Medicinal and poisonous properties
Before the advent of sulpha drugs, antibiotics and the other chemical wonders of modern medicine, people all over the world turned to nature's own remedies for their various ailments, especially in the remoter, undeveloped regions where doctors and hospitals were few and far between.

Africa was no exception. Over thousands of years the indigenous African peoples accumulated a great knowledge of the healing properties of certain plants and trees, much of which was in turn picked up by the white settlers who first came to present-day South Africa nearly 350 years ago.

Synthetic modern drugs have caused those old remedies to fall into disuse, except among black people, whose traditional herbalists still apply them. Modern science is now beginning to take a serious look at these remedies and is undertaking an enormous new field of research which will undoubtedly benefit all mankind in the future.

As crude as some of the old natural remedies might have been, they were often most effective. These are just some few examples of medicinal properties found in trees indigenous to the Park:
- The bark and leaves of the rather rare, attractive pigeonwood tree are used to make an infusion to treat bronchitis and internal parasites.
- The big, beautiful and shady Natal fig contains chemicals in its bark and leaves which are claimed to be effective against influenza, colic and even snake venom.
- While the fruit of the sourplum can be made into jelly and jam, its bark and leaves are also boiled to treat diarrhoea, eye inflammation, venereal disease and internal parasites.
- The roots and bark of the attractive bitter false-thorn are said to be effective against a variety of complaints including malaria and leprosy.
- The powdered bark and leaves of the

worm-bark false-thorn are useful in eradicating tapeworm and hookworm, as are chemicals in a wide range of other African plants.

- The dulcitol, tannin and selastrine in the bark of the common spike-thorn are used to cure amoebic dysentry and diarrhoea, among other things.
- The very useful buffalo-thorn not only has a tough and fine-textured wood but is regarded as an exceptional medicinal plant for use against stomach troubles, carbuncles and chest complaints.
- The river bushwillow seeds are effective in deworming dogs but have a curious poisonous effect on humans, causing them to hiccup for long periods.
- Chewing the fresh roots of the rare and scrawny carrot tree reputedly eases sore throats, while the bark relieves asthma.
- Kidney complaints, it is claimed, succumb to infusions of the roots of the wild firebush, a smallish, quite rare tree with unpleasant-smelling flowers.

Not all the natural properties of trees are benign, as many a victim of an African witchdoctor could testify:

- The sneezewood tree was used by farmers before the coming of iron and treated timber poles for fence posts because termites will not touch it. The oil it contains can cause prolonged sneezing but is believed to also have strong medicinal potential.
- The tamboti is a renowned African tree popular for the quality of its wood, which makes excellent furniture. It contains a poisonous latex which can cause severe eye irritation if used for firewood and serious stomach cramps if used for a barbecue. Discontented servants of hunters were reputed in the old days to deliberately put tamboti on the campfire, but it has so distinctive a smell that only the novice would fall victim to that.
- The latex of the easily identified candelabra tree and also that of the common tree *Euphorbia* is poisonous and dangerous if it gets into the eyes or is ingested.
- One of the most dangerous of all is the black bitterberry. Its small round fruits are highly toxic and produce the same symptoms as strychnine poisoning.

Flowers of the veld

South Africa is a botanist's paradise. Nowhere else on earth does there exist in one country so immense a diversity of plants, and its total extent is not yet known – new species and varieties are continually being found.

The Transvaal Lowveld, in which the Park lies, is in itself a treasure-house of plantlife which is still being explored. It offers an extraordinary range because it reaches from the hot lowlands of the Mozambique border to the Great Escarpment with its dense forest, fast-flowing waters and high altitudes. It is estimated that this zone contains more than 3 000 species of plants, of which over 2 000 have already been identified in the Park itself.

The spreading wild morning glory (above) and the brilliant impala lily (opposite).

For the Park visitor the following are some of the obvious and more impressive flowers to be seen.

Possibly the most obvious, and certainly one of the loveliest, is the impala lily, which is common to many camps and emerges from a stubby, thick-stemmed plant that looks like a miniature baobab tree. For three to four months from July it produces spectacular displays of red-pink tubular flowers which end in bursts of white petals fringed in bright, eye-dazzling red.

One of the most dramatic displays of shape and colour in the Park in spring and early summer is given by a creeper named *Combretum microphyllum*: the beautiful flame creeper. This vigorous plant climbs over trees and shrubs mainly near the banks of rivers and streams

and drapes from them in curtains of brilliant scarlet flowers – the colour coming not from the petals, which are tiny, but from the stamens. The cascades of scarlet can be metres long, mixing with the host tree's green in contrast, or drowning it out completely.

No less beautiful is *Gloriosa superba* or flame lily, most common in the northern part of the Park. These superb flowers, like brilliant flares of yellow or orange fire, have long, thin stems with blooms of six petals up to 7 cm long which rise and curl back above a cradle of long, slender stamens. They appear from November to March and many people regard them as the floral queen of the Park.

The Barberton daisy is known all over the world and, because of hybridisation, comes in many colours. Here, in its natural state, it is a glowing red, star-shaped flower which opens on a long stem above a crown of ground-hugging green leaves. After the first spring rains they form carpets of colour on flat stretches of veld.

Widespread in the Park and all over the Lowveld is the tall, glorious flowering shrub nicknamed "Pride of De Kaap" (after the great De Kaap valley south-west of the Park) or the "Flame of the Flats". It is *Bauhinia galpinii* which bears large bunches of salmon to rich red flowers with slender stemmed petals contrasting against the shrub's rich, green foliage. In the Punda Maria and Berg-en-dal areas, they grow wild in abundance, preferring the forest and thicker bush of the riversides.

Aloes are to be seen all over the Park – their tall flower stalks rising from clusters of thick, fleshy, spiny leaves. There are a great many kinds of aloe in Southern Africa, some so specialised they exist only in a single valley or on one hillside, and several of the commoner varieties inhabit the Park. Thorns and bitter juice make their leaves unpalatable to browsing animals, but their clusters of red or orange blossoms on top of their long stalks are a favourite feeding place for hovering sunbirds.

The Park's quiet pools and dams are rich in water lilies, whose beautiful flowers open with the dawn to the sun and close in the evenings. Loveliest, in the eyes of many, are the *Nymphaea* species, whose blooms range in colour from pure white to pale blue.

Spring collection . . .

The Lowveld is blessed with hundreds of species of wild flowers that bloom mainly after spring and summer rains. Some of the more common varieties are shown on these pages – all of them photographed within the bounds of camps and picnic sites. They are: **1.** Wild hibiscus. **2.** Blouselblommetjie. **3.** Wild forget-me-not. **4.** Barberton daisy. **5.** Devil's thorn. **6.** Crossandra. **7.** Summer impala lily. **8.** Gladiolus varius. **9.** Bloodflower. **10.** Pride of De Kaap. **11.** Bobbejaanstert.
Note: the botanical name is given where there is no listed common name.

Gills and fins

Almost every casual visitor to the Kruger National Park counts his experience of it in a series of distinct stages. First is that extraordinary, indefinable feeling of arriving in another world, strange yet familiar, as if the act of entering its gates takes one back to a dim distant time which stirs ancestral memories. The senses are immediately alert to the sudden absence of the modern world outside and attune to the Park's own scents, sights and sounds – those of original Africa.

And then in swift succession come the absorption and excitement of the big actors on the Park's stage – the elephant, lion, buffalo, antelope, crocodile, hippo and all the others of high profile, followed by the birds and smaller creatures in endless panorama. And then comes their backdrop of flowers, trees and rivers.

There is one facet of Park life which is no less important in its ecological structure than any other but is very rarely seen because no facility has yet been created to make it visible, and because it is under water: fish.

Of more than 100 species and subspecies of fish indigenous to South Africa's fresh water, 49 are found in the Park. Most of these are such small, mundane or highly localised kinds that they are beyond reach of the visitor, but some are of particular interest and all slot into the natural cycle of eating and being eaten.

Two in particular are remarkable. They are the strange hibernating lungfish (see p. 35) and a small, rare fellow which has potential importance for all of malaria-beset Africa because a prime part of its diet is mosquito larvae: the killifish (see p. 35).

Some of the more common notables are mentioned here.

The tiger fish is the freshwater king of Africa, a voracious hunter which reaches nearly 6 kg in the Park rivers but considerably more in larger bodies of water, such as the Kariba and Cahora Bassa dams and the Zambesi river. The fish is silver with a red tail and rows of dark lateral stripes, and is equipped with teeth so formidable and jaws so strong that anglers have to use steel traces between hook and line. They are renowned fighters when hooked, leaping clear of the water between deep plunges, but the larger ones tend to be more sluggish. They are fast and will eat anything alive in the water small enough to conquer and, in turn, are eaten by crocodiles and fish eagles.

One of the problems the Park authorities have found in recent years is that tiger fish have become rare higher up in the rivers above dams and weirs, which they cannot easily negotiate in their annual migrations from their breeding grounds in the coastal swamps and lagoons. Steps are now being taken to provide "fish ladders" alongside dam walls, akin to those built for migrating salmon in some North American rivers.

Another widespread denizen of Park waters is the common catfish whose flat, long-whiskered head forms roughly one-third of its slippery body which is surmounted by a low dorsal fin and weighs up to 25 kg (55 lb). Catfish, also known as barbel, are in the water what the impala are on the plains – a prime food source for crocodiles and other predators. They themselves eat almost anything, including fallen fruit.

They vary somewhat in colour according to the water and generally feed close to the bottom. When the Sabie river is low they can be seen clearly in the water from the patio of the Skukuza camp, moving gently to keep station in the slow stream.

Hardy creatures, these fish are very adaptable and can live in water too muddy and foul to sustain any other fish. They have the unusual feature of an accessory breathing organ which enables them to breathe fresh air in addition to filtering oxygen from the water through gills. In dry seasons catfish are often trapped in ever-shallower pools as the water recedes, sometimes in large numbers, and these then become cafeterias for all kinds of other animals including crocodiles, leguaans, vultures, marabou storks, hyenas, jackals and even lions. The common catfish is one of several kinds of catfish in Park waters.

The brown squeaker is so named because when taken out of the water and handled it emits grunts like a small pig. Another oddity is that the spines of its dorsal and pectoral fins are covered with toxic mucus. If the spines cause a wound it bleeds profusely and can be very painful. They are bottom feeders that like stiller water.

Several kinds of eel are found in Park rivers and, like all other eels, they come from far out in the Indian Ocean and go back there to breed and die. They too are predators, fast-moving and strong.

Several marine fish have been found in the Park rivers. The most notable was a single Zambezi shark caught in shallow water in July 1950, at the confluence of the Luvuvhu and Limpopo rivers at Pafuri, about 400 km upstream from its natural habitat – the sea. Rated as one of the most voracious of sharks and one of the chief culprits in shark attacks on bathers on South Africa's east coast, it is peculiarly adaptable to fresh water and had previously been found well inland along rivers in Mozambique. Zambezi sharks are believed to reach a maximum length of between 2 and 3 m.

Another ocean fish found this far inland in the Park is the river bream, one of which was caught near Crocodile Bridge in 1970. Others at home in either saline or fresh water are the red-fin tilapia and the common catfish.

* * *

That, then, is an encapsulation of the Kruger National Park by its most salient features – necessarily a superficial one, so great is the compass and so huge the accumulated knowledge of its life. Many people have made this relatively small segment of Africa their lifetime work and yet none knows it in totality, for this is beyond the scope of a single human mind.

Further aspects of the Park will emerge in later pages of this book: its fascinating history, its personalities and its creatures, as well as what it has to offer the visitor; but there is no pretence at its being comprehensive. All the authors have attempted to do is to give a sampling of the wealth of experiences, challenges and, above all, of the enormous balm for mind and soul that lies within its boundaries for those who have the desire to reach and find. Here man can shed care, prejudice and status and let nature lead him back through the aeons to bask in the glow of its pristine tranquillity.

TWO UNDERWATER WONDERS

After reproduction, survival is the most powerful motivating force in all living creatures; without it there can be no continuation of the species. All things are equipped with special skills or tools to help ensure their survival . . . from the ability to produce millions of offspring, like termites, to highly effective deterrents, like the skunk's scent glands, to simple brute strength, as in the case of the lion and the elephant.

Two of the most remarkable adaptations for survival in the world exist in the Park and by a curious coincidence, the search for one also led scientists to the other. They are the killifish and the lungfish.

Killifish is an odd name for an unusual creature. It is a small fish found previously in the Mozambique coastal lowlands and in Natal but in the Park only as recently as in the 1960s.

There are two kinds, the spotted or Tongaland killifish and the crescent-tailed killifish, the former no longer than 9 cm and the latter 6 cm.

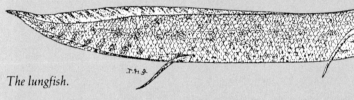

The lungfish.

The chief warden, Dr Pienaar, who first found the spotted killifish in the Park in 1959, writes that they have the potential of being a considerable boon to mankind in Africa, where malaria is one of the most widespread and enervating of diseases. This is because a major part of their diet comprises mosquito larvae and the *Anopheles* mosquito, of course, is the carrier of malaria.

"They are such voracious mosquito larvae carnivora," Dr Pienaar states, "that they have been artificially introduced in swamp lands to aid mosquito control in countries far removed from their natural habitat. Results achieved in this respect look very promising."

The spotted killifish is a brightly coloured creature with a blue-green top shading to pale gold below in the male, whose fins are trimmed with mauve or indigo edges and magenta spots. The fins of the females are olive with brown spots and edged in blue.

The crescent-tailed killifish is even more spectacular. Males are an iridescent blue-green to green-gold with red throats and broad red and blue bands on the tails. Females are olive to brown with a green sheen.

In the Park both inhabit rather shallow ponds, seldom deeper than 1 m, living in the reeds and weeds and emerging only to feed in the adjacent open water. They live for less than a year, growing so fast that they can breed within two to three months. This is necessary not only because of their short life span, but because in dry seasons such pools very often dry up completely. The killifish has remarkable protective devices against this to ensure the continuation of the species.

The eggs they lay sink into the mud at the bottom of the pool and stay there until the conditions are right for hatching, which is not just any rainy season. According to Dr Pienaar, it seems that the eggs will not hatch until the salinity of the water and its acid factor are at levels which indicate that the pool will not dry up before the next life cycle is completed. If the pool is not full enough – if only the top layer of mud has been made wet – the eggs stay unhatched; likewise if the water's temperature is too high. If there is too much rain and floodwaters go through the pool, the salinity, temperature and acid factor will be such that the eggs will not hatch.

And as a final touch of ingenuity, nature has so arranged things that not all the eggs laid during any season will hatch the first time conditions are right. Some will stay in the mud for another year or two, and then hatch.

The little killifish is one of the most dramatic examples of adaptation for survival through countless aeons.

Research workers making a detailed study of killifish recently discovered, quite by accident, that among the Park's denizens is one of the oldest known creatures indigenous to Africa – the extraordinary living "fossil" known as the lungfish.

Dr Pienaar recalls that the discovery was made when two zoologists from the University of the North in South Africa visited the Park to net killifish for study purposes.

At a water-filled pan close to the Mozambique border between the Satara and Olifants camps, where the killifish was originally discovered, the scientists cast their nets.

To their astonishment the first thing they pulled up was a young lungfish – for which Park staff had searched unsuccessfully for 25 years, knowing it to have been found not much farther north near the Limpopo river.

Rejoicing in the name of *Protopterus annectens brieni*, the lungfish is known from fossils found in South Africa's Karoo, a treasure house of the remains of prehistoric creatures that existed there some 150 million years ago. They are, in a sense, live fossils and differ from other fish in that each has an air-containing sac which supplements its breathing through gills. Zoologists regard the lungfish as a creature representative of the period when vertebrates evolved from life in water to life on land.

A curious, eel-like creature with feathery fins, which can reach 1 m in length, the lungfish has the remarkable facility of aestivating when the water it lives in dries up. When the water level drops it digs a bottle-shaped hole in the silty bottom and curls up inside. It then excretes a thick mucus which, when dry, covers the fish with a hard plastic-like coating preventing it from dehydrating. It breathes through holes in the coating which form at the nostrils.

Hibernating lungfish have been dug out of dry beds and kept for four years, Dr Pienaar states. When returned to water they soon revive and begin behaving normally again. They have to surface regularly to breathe through their nostrils.

Fossils of lungfish have been found worldwide and indicate that they lived as long as 300 million years ago.

When Africa cooled and its crust ceased heaving it formed, far down in the south-east corner, a narrow, undulating plateau like a shallow step between the higher hinterland and the soggy lowland of the Indian Ocean coast.

Less than 1 000 km long and rarely more than 100 km wide, it lies north-south between longitude 30 and 33 east and latitude 20 and 30 south; from the south-east corner of Zimbabwe down the easternmost edge of South Africa's Transvaal province; through the tiny, landbound Kingdom of Swaziland and into the northern tip of Natal province, where it peters out in the homeland of the mighty Zulu nation.

There is nothing else quite like it on the African continent although there are similarities with the bed of that giant geological slash down Africa, the Great Rift Valley, of which some people speculate this might be the southernmost extension.

It is commonly called the Lowveld ("Low country", pronounced Lowfelt) and it is most clearly defined in the Transvaal. It is bounded to the west by the spectacular, jagged ramparts of the Great Escarpment plunging over 1 500 m from the vast inland plateau above. To the east it is bordered by the long, low range of tilted rock formation named the Lebombo mountains, beyond which lies Mozambique where the land slopes down to the miasmic deltas and marshes, lagoons and brilliant white beaches of the Indian Ocean.

A complete cross-section of this Lowveld plateau as it was before the advent of modern man has been preserved by the creation of the Kruger National Park.

It has taken over a century for it to evolve from the brainchild of President Paul Kruger to its present remarkable status, and the process has preserved much more than wildlife. Archaeologists, anthropologists and historians have found it to be a rich field for their research, taking them far back in time.

Mankind, they are discovering, has come and gone in the Lowveld for up to 500 000 years, flourishing and waning, being pushed aside by newcomers who mysteriously disappear, to be replaced centuries later by yet others pursuing their own goals.

The researchers are beginning to uncover a vast historical tapestry of peoples and migrations and wars and characters larger than life, spilling over right into this century, as dramatic as any other in Africa. It will take them many years to reveal it fully, but enough is visible to whet the historical appetite.

Early man, the hunter

Archaeologists of the University of Pretoria began to take a serious interest in the Kruger National Park in the early 1970s, and soon found ample evidence that prehistoric man – *Homo erectus* of the Acheulean Period between 500 000 and 100 000 years ago – had lived there, leaving behind a variety of hand axes, cleavers and other implements shaped from stone.

This was not unduly surprising because very ancient hominids had been discovered elsewhere in South Africa half a century earlier, such as the fossil skull of a child who died one to two million years ago which was found in the Taung quarry near a small railway siding on the edge of the Kalahari desert in the northern Cape Province.

No skeletal remains of Stone Age man – no jawbones, pieces of skull or any other physical relics – have so far been found in the Park, although cultural remains of successions of peoples living in that area during the Middle Stone Age from 100 000 to 30 000 years ago have been identified. Following them were the Later Stone Age people of 30 000 to 20 000 years ago and their Bushman descendants, Iron Age man of about 1 500 years ago and finally the European explorers and settlers.

The University of Pretoria archaeologists have discovered more than 300 sites once inhabited by people of these various ages and have started a programme of intensive investigation, some of which has yielded further places of historical interest for visitors to the Park.

Among the most intriguing of these vanished Park dwellers are the diminutive Bushmen folk (San) who apparently evolved in Southern Africa

*Above: Some of the most vivid examples of rock art can be seen on Wilderness Trails. **Facing page:** This water leguaan in casual mood was seen on the Mphongolo loop north of Shingwedzi camp. It had probably raided a bird's nest.*

and roamed most of it until advancing Bantu migrations from the north, and then European migration from the south, pushed them into the vastness of Botswana and Namibia – where they are found today.

Archaeologically the Bushmen belong to the Later Stone Age (their stone-chipping skills are still being practised by a single, remote tribe in Namibia) and their former presence in the Lowveld was detected early in the Park's history in the form of paintings on the rock walls of caves and overhangs which they used as shelters.

So far nearly 100 sites have been discovered in the Park with Bushman paintings – many crude, some finely drawn – depicting animals then common in the Lowveld such as eland, tsessebe, kudu, roan antelope, elephant and giraffe, as well as stylised figures of people. The paintings are usually in red but also in yellow, black or white or combinations of these.

Experts are still puzzling over the meaning and purpose of this form of art, as it is completely forgotten by present-day Bushmen. The pictures vary widely in content, from single figures to complex friezes of human forms and animals. Many of the people depicted are obviously Bushmen from their physical shape, while others appear to represent the Bantu newcomers who eventually pushed them out.

A few of the rock paintings in the Kruger National Park show people hunting or actually shooting at animals with bow and arrow. Such pictures found elsewhere have led Bushman-ologists to speculate that painting animals on rock faces was a ritual developed to bring success in the hunt. But other speculation has it that the painting could have been no more than a leisure pastime; a form of record-keeping; a way of marking certain areas as suitable for certain kinds of prey; part of some unknown other form of ritual; or merely, in its cruder form, the work of children at play. The speculation is intense and inconclusive.

Excavations in the shelters where paintings have been found have brought to light various other Bushman artefacts including pieces of spears and arrows, stone implements and ostrich shell beads. Archaeo-logically, many of the sites are relatively recent and the last Bushmen are believed to have disappeared within the past few centuries.

Precisely why they vanished is unclear, but the probability is that their demise began with the first influx of Bantu peoples from the north. The experts are unravelling the long drawn out trekkings of the Bantu by the ingenious method of studying the pottery they have left behind. Each tribe, or the culture to which it belongs, can be fairly accurately identified by the "fingerprint" of its pottery – its shape and style and particularly the decorative patterns on it. Given a reasonably large piece of the lip and body of an earthenware pot dug up at some ancient dwelling site, archaeologists can fairly accurately reconstruct the whole pot.

In a recent exploratory survey in the Kruger National Park a University of Pretoria archaeology lecturer, Mr André Meyer, identified more than a dozen Iron Age cultural groups or "industries" dating as far back as the year AD 200. The first of these the researchers entitled the Silver Leaves industry (from the name of the farm where the find was made), which represented the coming of the Iron Age to the Lowveld. By correlating the finds of other archaeologists, it has been possible to backtrack the trail of broken pottery left by the Silver Leaves people in their migration.

This apparently began in what is now the southern Kenya and northern Tanzania region, from where they slowly moved south along the east coast of Africa as far as modern-day Maputo, capital of Mozambique, in Delagoa Bay. Numbers of them turned inland north of Maputo and eventually moved along the Letaba river, obviously because of its fresh water and the animals that concentrate close to rivers. Some settled in what is today the Park, while the rest continued inland along the river courses. What ultimately became of them is unknown. They simply vanished, like so many other tantalising factors in the Lowveld's history.

Several other cultural groups migrated through after the Silver Leaves people up to about AD 900, pausing here and there in the game-rich wilderness, leaving their mark in broken pots, making new ones, settling down for brief periods and then travelling on.

The diminutive Bushmen once roamed throughout Southern Africa, leaving behind them their fascinating rock art as they moved on. They were persecuted by settlers, both black and white, to the brink of extinction and only a few thousand now survive in the Kalahari Desert and Namibia.

Research is building up the glimmerings of a picture of these semi-nomadic farming and hunting communities, the first of which appeared to subsist almost entirely on wildlife and what fruits the bush could offer. In the period AD 700-900 there was a marked increase in the number of settlements and, Mr Meyer has found, the settlers increasingly depended upon livestock such as cattle, goats or sheep which they brought with them.

Here and there, he and his assistants have found what could be termed the first signs of some foreign contact with the Lowveld: trade beads, which could well have been brought from East Africa, where the Arabs were already long-established traders, or which the inhabitants could have traded for ivory they hunted for in the Lowveld.

Something must have happened in the Middle Iron Age which deterred the migrating groups, because few traces of settlement between AD 900 and 1600 have so far been found in the present Park area. Whether the deterrent was disease, climatic conditions or some other factor has yet to be established.

From the late 1600s they appeared again, some along the old east coast route and others from the north through today's Malawi and Zimbabwe.

Their life and times in the Park area are now being traced by excavation and research at 12 of the approximately 300 Iron Age sites found by the University of Pretoria archaeological department, under the guidance of Professor Hannes Eloff and Mr Meyer.

The migrants brought with them from East and Central Africa the skills of making iron and settled where they found iron ore. The same iron sources were also used by the Bushmen for the red pigment in their rock art. Remnants of iron-smelting furnaces, implements and pottery have sometimes been found at the same sites where the Bushmen decorated their sheltering rock walls.

"We got the impression that the earliest Iron Age people lived in huts made from sticks and grass with floors raised above the ground," said Mr Meyer. "Apparently they swept rubble in under the huts." Today that rubble provides the clues to the past.

The wealth of archaeological discovery now being made is adding yet another facet to the rich attractions of the Kruger

National Park. A new camp, the Bushman Trail camp, has been built in a selected area from which visitors accompanied by a ranger can make daily hikes to see ancient artefacts and Bushman paintings, as well as study the Park's wildlife.

About 50 rock art sites are within reach of the camp, most of them situated low down on hillsides, close to water and often very close to well-used game paths. It seems that the Bushmen cunningly chose these sites so that they would not have to carry water far, and because game was always plentiful and easy to shoot from close range, their light bows not being able to send an arrow far.

Another new attraction is Masorini, a quite recent Iron Age site on a hillside near the Phalaborwa entrance to the Park. This was inhabited, states Professor Eloff, by a group of people belonging to the baPhalaborwa tribe who made a living by making and selling ironware, and so developed a specialised economy and a fairly advanced technology.

The industry was sophisticated for its time, involving the shallow mining of iron ore from a number of places in the vicinity, crude but effective smelting at temperatures between 700 and 1 000 °C, and the employment of hundreds of people to make charcoal for the smelters and to work the iron into spears, hoes and other implements.

When Masorini was discovered, it comprised little more than a few stone walls, grinding stones, shards of pottery, the remains of a nineteenth century smelting furnace and stone implements

A rare photograph of a smelter of a kind used by tribespeople for centuries in the Lowveld. The three workers are operating bellows to pump air into the simple furnace containing a mixture of iron ore and charcoal.

dating back to the Stone Age. The archaeologists decided to restore it as accurately as possible.

Excavations revealed hut floors and other remains which told how the former inhabitants had lived, their way of commerce and what their domestic life was like. This information was supplemented by the word-of-mouth history provided by the chief and elders of the baPhalaborwa tribe living nearby, outside the Park.

Today Masorini lives again, but without inhabitants. It has been rebuilt from scratch – a collection of huts blending pleasantly with the big boulders on the hillside, complete with granaries, a reconstructed replica of a smelting furnace, all the usual household articles and implements of the past, and a magnificent view from the top of the hill.

"Masorini is unique in the sense that it is a prehistoric dwelling site which was literally rebuilt from the foundations," says Professor Eloff. "The restoration was not merely a flight of the imagination but the result of extensive research.

"The living people are no longer there, but visitors can imagine the pulsating life of this village. It is as if the actors have left the stage but the set remains as a reminder – in this case as a reminder of the life-style of a group of people whose traditional customs were not influenced by a foreign culture."

Explorers, pioneers and fortune-seekers

Europe's age of exploration and conquest began to embrace this part of Africa even as the latter waves of Bantu migrants moved southwards into the pristine wilderness of the Lowveld, long before Masorini became an iron-working centre.

Other foreigners had made their mark in East Africa for hundreds of years before the advent of the Europeans, but their influence somehow never extended much further south than the present-day Mozambique port of Beira or into the southern hinterland. These people were the Arabs, Persians and Indians who had set up regular trade routes with East Africa several centuries before the birth of Christ, riding the monsoon winds between Arabia and Africa in their dhows. Their brisk presence on what is now the Kenya coast was recorded by a Greek explorer in 60 BC.

Drawn by ivory, gold and other riches of Africa, including black slaves, the Arabs set up city-states down Africa's eastern seaboard at places such as Gedi, Malindi, Mombasa, Zanzibar, Pemba, Kilwa, Mozambique Island and, just south of Beira, at Sofala – at the steamy mouth of a small river where the monsoon winds peter out.

Their trading took them far inland into what is today Zimbabwe, to the crude gold workings of the kingdom of Monomotapa, legendary source of the riches of King Solomon and the Queen of Sheba. Their journeys would have

where the port of Maputo is now sited, and three years later seized Sofala from the Arabs but abandoned it in 1508. Today the original Sofala has vanished, enveloped by the tropics and washed away by floods.

Competition between the Portuguese and Arabs at that time led to a long succession of conquest and reconquest as each fought for supremacy. The first European to penetrate the interior was a Portuguese ship's carpenter, Antonio Fernandes, who in 1514 travelled far and wide in Zimbabwe and encountered the powerful chief, Monomotapa.

Thereafter followed centuries of Portuguese penetration and conquest into what is now Mozambique and eastern Zimbabwe. They had the field to themselves until other European nations' interests in Africa's riches – for which the Dutch and British began to vie – were aroused.

The honour of being the first European to reach the Transvaal Lowveld went to a Dutchman. The reports of Monomotapa's gold persuaded the Dutch East India Company to occupy Delagoa Bay – then inhabited only by Africans – in 1719, and six years later an expedition was sent inland under the command of one Francois de Kuiper.

It was an ill-starred trip, however. They crossed the Lebombo mountains, according to De Kuiper's diary, and had travelled north-west for only another 30 km or so when hostile blacks attacked them at a place called Gomondwane (between the present Crocodile Bridge

Their cousins, the Shangaans, struck from the east coast. The blacks who remained in the Lowveld lived largely in hiding and many were refugees. To add to their troubles they had to suffer the bilharzia disease which still contaminates all east-flowing rivers, endemic malaria, and the dreaded tsetse fly which infects cattle with "nagana" and humans with sleeping sickness.

In the nineteenth century yet another force was added to the rising tensions in the Lowveld; the South African Dutch, or Boers – the forefathers of today's Afrikaners.

These indomitable people gradually moved from the original Dutch settlement in the Cape into the interior of what is today South Africa, from the seventeenth century onwards. In the fourth decade of the nineteenth century this movement became a flood as the Voortrekkers (pioneers) poured northward to escape creeping British hegemony. By the time they had ensconced themselves in what is now the Transvaal, the Lowveld had been partly explored by ivory hunters, slave raiders and a few other hardy adventurers who had braved the fighting and pestilence to blaze paths across it.

The first organised expeditions of white men known to have crossed the Lowveld were those led by the Voortrekker leaders Hans van Rensburg and Louis Trichardt in 1836, both seeking routes to the Indian Ocean and Delagoa Bay, then little more than an outpost occupied by Portugal, but also being eyed by Britain.

Van Rensburg and his group took an easterly route towards the Limpopo river in their ox-drawn *kakebeen* (jawbone) wagons to meet their deaths. In August that year, at a place called Combomune, they were massacred by the warriors of Chief Manukosi. Manukosi (or Soshangaan) was the founder of the Shangaan nation – another who defected from the tyranny of Shaka, the Zulu king.

Trichardt was hardly more fortunate. Indeed, with nine wagons, 10 men and 39 women and children, he reached Delagoa Bay, but most of them died of malaria contracted on the journey.

The Transvaal Republic government set up by the Boers made repeated efforts thereafter to find a good route to Delagoa Bay, the nearest seaport, but so hazardous was the trek through the

The first European to penetrate the interior was a Portuguese ship's carpenter...

taken them across the northernmost end of the Lowveld but never, it seems, as far south as the present Kruger National Park – which, ironically, is close to far richer goldfields.

The Arabs and Indians monopolised this contact until the coming of one of history's great discoverers: Portugal's Vasco da Gama, who circumnavigated the Cape of Good Hope in 1497 and, on his way in search of a sea route to India, stopped at Mozambique Island.

His finds, and the Arab gold trade, quickened Portuguese interest and in 1502 one of their squadrons visited the fine natural harbour at Delagoa Bay,

and Lower Sabie camp). The Dutch hastily retreated to the coast.

The Lowveld was left to itself for more than a century after that by the Europeans – but not by blacks. Population pressures and tribal conflicts were building up a head of steam following growing waves of migration which saw major black settlement west and south of the Lowveld and the emergence of powerful tribes who still wield heavy influence in Southern African affairs: Zulus, Swazis and Shangaans.

Zulu and Swazi impis (regiments) came storming up from the south, fighting each other or simply plundering.

An historic glimpse of the old wagon road as it approaches Ship mountain near Pretoriuskop.

Lowveld that in 1847 the government forbade another attempt.

At about this time, however, one of the Lowveld's early characters appeared on the scene: a Portuguese named João Albasini, who became the first European to actually settle in the area, at Ngomeni near the confluence of the Phabeni creek with the Sabie river, where the ruins of his house can still be seen.

Juwawa, as the blacks called him – a corruption of the pronunciation of João – had some idea of annexing the Lowveld for Portugal and calling his new colony Santa Luiz, but instead turned to helping the Boers find the route they sought for their wagons. Such was Albasini's influence in the area that blacks all along the Sabie river regarded themselves as being under his protection and worked for him as porters, carrying trade goods through the malaria and tsetse belt to the interior.

The Boers' new wagon road was at first used sporadically, and until quite recently all that was known of the route was that it started on the Highveld plateau near Ohrigstad, descended the Great Escarpment to Pretoriuskop and then went on somewhere through the southern sector of the Park. Recently the exact route was rediscovered, thanks to senior ranger Thys Mostert and his assistant rangers who devoted hundreds of tedious hours to tramping through the bush searching for clues such as old discarded food cans, rusty bullet casings, burial sites and even likely river crossings through which the wagons may have passed. Methodically Mostert remapped the old road. From Pretoriuskop it travels south-east just north of the present (Voortrekker) tourist road to Ship mountain and then roughly in a straight line south of the new Jock of the Bushveld camp to the Park's border between Crocodile Bridge and Malelane.

The importance of Delagoa Bay, and a road to it, soared dramatically in 1870: one Edward Button found gold at Spitskop near Lydenburg up on the Highveld.

Gold fever had reached South Africa decades before and there were many finds but few of any value. The insular Transvaal Republicans took a jaundiced view of the metal, not wanting an inrush of fortune-hunting foreigners. As early as 1853, Capetonian Pieter Jacob Marais, who had been in the California gold rush, found gold on the Jukskei river far to the west, tantalisingly close to the vast Witwatersrand deposits later to be discovered.

The government offered him a large reward if he could find more – and threatened him with death if he told any foreign government about it. He left in disgust.

Button's find at Spitskop was followed a year later by others at Mac-Mac on the Great Escarpment (so named because of the many Scots who mined there) in 1871, at Lydenburg in 1873, at Pilgrim's Rest two months later, and in the Sheba Valley outside Barberton in 1884 – the biggest and richest, which led to a rush matching that of the Klondike.

Hordes of fortune-seekers streamed into the Eastern Transvaal and Lowveld – prospectors, miners, crooks, buyers, assayers, traders, pubkeepers, good-time girls and simple hunters and adventurers from all over the world. South Africans of all sorts rubbed shoulders with Americans, Australians, Canadians, Poles, Hungarians, Britons, Irishmen, Germans . . . all living in a motley collection of rough mining camps and tin towns scattered through the rugged valleys and forests of the Great Escarpment.

Life was harsh and extrovert: whisky, disease and violence took a steady toll of lives. Coaches carrying gold from the scores of shafts and adits in the bush were regularly held up and robbed. Many men carried rifles or pistols and dealt summarily with thieves.

Delagoa Bay and the grubby Portuguese port settlement there of Lourenco

Marques became the main import link for all the mining equipment which the diggers were unable to make for themselves by Heath Robinson techniques. This gave birth to the "transport rider" era of intrepid men who lived by ferrying machinery and basic foodstuffs inland from Lourenco Marques in their big flatbed wagons drawn by teams of plodding oxen.

For fresh meat these men, the miners and the bustling communities which lived off the mines turned instinctively to the wildlife abounding in the Lowveld. Hunting for the pot became as natural as going to the supermarket today and much of it was grossly wasteful: shooting a buck for the hindquarters only and leaving the rest to the hyenas and jackals.

'A great deal of this killing took place inside the boundaries of the present Kruger National Park, for the main wagon roads traversed it. To add to the slaughter came those not interested in gold but in hunting for profit – the meat providers to the camps, the ivory hunters and also, from the Highveld, the Boers who had lived off the land for generations and regularly came down to the Lowveld to stock up with meat, hides and other animal products such as fat for making soap.

The Eastern Transvaal gold boom

ALBASINI, THE PATHFINDER

He could well have become a privateer, a South Seas pearl trader, a fur trapper in the Arctic, a mercenary in the Orient, or any one of those many other daredevil callings which beckoned the men of Europe in the early 1800s. But for some unknown reason the call to João Albasini came from Africa, then still challengingly wild, largely unexplored and wreathed in tantalising stories of great mysteries and greater riches.

Italian born and of Portuguese nationality, he somehow found his way to the miasmic, enervatingly humid and pestilential little port of Lourenco Marques, capital of the south-east African colony of Mozambique which Lisbon claimed but over which it was still dithering to establish its suzerainty.

The Portuguese had earlier hastily established the little town in Delagoa Bay, on the north bank of a sluggish estuary formed by several rivers, to stake their claim to the region when the British and Dutch showed interest in it. It was no health resort. Surrounding marshlands fed by tropical storms and the waters of many rivers emptying from inland bred hordes of malarial mosquitoes from which the locals took refuge in crude homes and in *aguardente*, their powerful brandy.

Hordes of natives who resented the heavy-handed and uninvited presence of the Portuguese regularly pestered them and forced them to take refuge in a small, thick-walled fortress of stone beside the estuary.

It was at this insalubrious outpost that Albasini arrived early in the 1800s – precisely when is not known, but probably in the 1820s or 1830s. Apart from its role as a flag-flying station for Portugal, Lourenco Marques happened to be also the starting point of a very old trading route from the coast to the African interior, possibly used by the Arabs who came down the east coast in their dhows.

Albasini must have been a strong man both physically and in personality: one had to be to survive the rigours of Africa then. He set himself up as a trader and elephant hunter and became a personage of some note in Lourenco Marques. He was destined to play a significant role in the

João Albasini in the uniform of a Portuguese vice-consul.

launching of the little port's future as the main gateway to the sea for the emerging Transvaal Republic. Even today a great deal of the imports and exports of the Transvaal province of South Africa still go that way.

Albasini was in Lourenco Marques in 1838 when a famous Voortrekker arrived there by oxwagon at the end of a daunting journey to find a way from the Transvaal hinterland to the sea. It was Louis Trichardt, whom Albasini met before he and most of his party succumbed within a short time to malaria and were buried in the town – where a memorial now marks their venture.

He was there again six years later when there arrived another great Afrikaner leader and pioneer from the Transvaal Republic: Hendrik Potgieter and his entourage. A year later Potgieter founded a frontier town on the Transvaal Highveld near the Great Escarpment which sweeps down to the Lowveld and today's Kruger National Park, and named it Andries-Ohrigstad – now known simply as Ohrigstad, a hamlet north of Lydenburg. Both Ohrigstad and Lydenburg were to become known around the world in the gold rush of the late 1800s.

By this time Albasini, in his peregrinations, had set up a small network of trading routes reaching inland as far as the Lowveld and a trading post on a site just north of today's Pretoriuskop camp in the Park. It was Albasini, in fact, who buried another Afrikaner who died on the search for a way to the coast: Willem Pretorius, after whom the camp and nearby hill are named.

By agreement with these doughty trekkers, Albasini transported goods for them from Lourenco Marques to this trading post, from where they fetched them with wagons which came down the Escarpment. The reason for this was that curse of Africa, the tsetse fly.

The fly vanished from the Lowveld after the outbreak of rinderpest in 1896, but then it was fatal for cattle to travel through the Lowveld. Horses too were vulnerable, because of African horse sickness in that zone.

The arrangement was for Albasini to use black porters to

lasted only a few years and by 1890 even the spectacular Barberton discoveries had dwindled to almost nothing, except for a very few mines like Sheba. By then, however, the pioneer Struben brothers and other prospectors had uncovered the huge gold deposits of the Witwatersrand hundreds of kilometres westwards, and a horde of fortune seekers swooped there like vultures on a new kill.

The importance of the wagon road to Delagoa Bay continued until, in 1894, the last spike was driven in the new Pretoria-Lourenco Marques railway commissioned by the Transvaal Government, and the wagons and the oxen faded from the scene.

Although everybody in those days took wildlife for granted as a natural larder or regarded it simply as a menace to the farmers beginning to settle the western edge of the Lowveld with cattle and sheep, there were wise heads who foresaw that if the killing continued at the current rate there would soon be no game left.

Inveterate hunters though they were, having lived off game during their generations of trekking north to escape from British rule and find a country of their own, the Boers were not insensitive to the danger. Not long before, in 1860,

carry his goods all the way from Lourenco Marques across the Lebombo mountains to his post, and then for the Transvaalers to fetch the goods by wagon, travelling fast at night so as to be away before daybreak and the onset of the tsetse.

The post expanded into a shop and homestead on the west bank of the Phabeni creek (north of today's Pretoriuskop camp). It was a mud-brick building with a peaked roof of reed and grass thatch, an open oven outside, door frames of selected timber and, probably, a floor of hard-packed mud or clay. There Albasini made himself comfortably at home, leading water by furrow from the creek to irrigate his domestic crops of corn and vegetables and a small orchard.

He traded the land from Magashula, a sub-chief of the local Sotho-speaking Kutswe tribe, for 22 head of cattle and it soon became an oasis of semi-civilisation in the middle of the wilderness. The Park authorities rediscovered and cleared the site and found the ruins of the old homestead-shop. Its partly rebuilt walls are now one of the Park's historical attractions and can be reached by a short drive from Pretoriuskop. A Spanish reed bush which Albasini planted to provide runners for his beans can still be seen on the banks of the Phabeni creek nearby.

Clearly an entrepreneur of some force, Albasini gathered about him a sizeable retinue of local blacks and appointed his two personal headmen, Manungu and Jozikhulu, to run other outposts for him on the Voortrekker wagon road between Ohrigstad and Lourenco Marques. Manungu's outpost was beside what is today Manungu's peak, just south of Pretoriuskop, while Jozikhulu ran his establishment further east and south near the well-known Park landmark, Ship mountain.

The Italian-Portuguese Albasini thus became the first white settler in this magnificent, animal-rich corner of Africa. It was a rough but good life for the adventuresome, with plenty to hunt, many locals willing to act as servants for the rewards of livestock and food, and virtually no laws to bother one except the unwritten code by which men helped one another.

But, surprisingly, Albasini stayed only two years at Magashulaskraal, as he named his trading post. In 1847, attracted by the growing settlement of the Transvaalers up on the Highveld, he bought a farm near Ohrigstad and opened a shop there. When three years later a plague of malaria forced the evacuation of Ohrigstad, he moved to the higher land of Lydenburg, set up shop again and married Gertina Maria Petronella Janse van Rensburg,

daughter of two local pillars of the community.

In all this time Albasini kept on much of his black retinue and his life took on the overtones of a country squire, albeit at a rather rough and ready African level. He was a restless soul and three years after settling at Lydenburg he packed up to move again – this time with seven wagons and his retinue – to a place considerably further north, then named Zoutpansbergtown (after the Soutpansberg range of mountains in the northern Transvaal) and later Schoemansdal. This old Voortrekker town was situated in a valley 13 km west of today's Louis Trichardt and was later abandoned because of constant raids by Venda tribesmen – and also serious malaria epidemics. The site is now being excavated by a team of archaeologists from the University of Pretoria and will eventually be rebuilt as an historic monument.

Albasini stayed there for four years before moving once more, now back into the Lowveld depths to a farm on the Luvuvhu river which in part now forms the north-western border of the far north of the Park near Punda Maria camp.

There he reached the peak of his colourful career. He built himself a fortress-like home on his farm, Goedewensch (literally, "Good Wish"), and lived surrounded by his black retinue in almost feudal splendour and power. He trained regiments of local black people and formed his own small private army while continuing to hunt elephant and trade with the locals.

Albasini achieved a curious dual official status. In 1858 he was appointed the Portuguese vice-consul in the Transvaal Republic, and the following year was appointed by the Republic itself as superintendent of the native tribes in a large region east, north-east and south-east of Schoemansdal, with the specific task of collecting head taxes from them. His trained regiments, or *impis* in the African language, became known as the *gouvernementsvolk* or "people of the government" and he used them to quell rebellious tribes in his domain.

In 1867 tribes in the Schoemansdal vicinity rose up against the Republican government and Albasini, with a few whites and his faithful black followers, stood firm when other whites had to hastily evacuate Schoemansdal. As a result of his stand the Soutpansberg district stayed in the control of the Republic.

Albasini died in 1888 and was buried on Goedewensch farm. The Albasini dam on the Luvuvhu river was named after him.

they had seen or heard of a single hunt organised in the Orange Free State for England's Prince Alfred in which more than 1 000 head of game were wantonly shot. Small wonder the huge herds of springbok and other antelope which once roamed the Free State had all but vanished.

"The wildlife of the Transvaal Lowveld was inexorably drawn into this whirlpool of destruction," the present warden of the Kruger National Park, Dr U. de V. "Tol" Pienaar wrote recently.

One key man who clearly visualised a Transvaal without wild animals was himself once a professional and sporting hunter of great renown: President Paul Kruger of the Transvaal republic (*Zuid-Afrikaansche Republiek*), the patriarchal giant who was among the leaders of his people's dogged drive for independence in their own land.

One of Kruger's first undertakings after he became President was to try to conserve wildlife. It was an extraordinary move considering the problems besetting the beleaguered young Republic, which was so poor it subsisted largely on a bank loan and had to pay one top civil servant in postage stamps. It was also revelationary that this man whose livelihood had depended so much on hunting should suddenly wish to curb it – but it is axiomatic that the first people to discern damage to the ecology are the hunters themselves.

In 1884 the long-bearded President solemnly proposed to a meeting of the *Volksraad* (People's Council, or Parliament) that a tract of land be delineated where hunting would be totally prohibited.

"It is becoming advisable," he told the *Volksraad* members, "to set aside some kind of sanctuary in which game may find a refuge. Continued shooting by farmers, poachers, hunters and others would destroy a valuable national asset," he asserted.

The *Volksraad* is said to have been stunned. Never before had anyone suggested that their complete freedom to shoot what they would, where they would, should be curtailed. Never had the thought arisen among most of them that the reservoir of wild animals was anything but bottomless.

President Kruger's proposal did not gain immediate acceptance, but he had planted a seed. Debate on conservation

A MAN OF VISION

Stephanus Johannes Paulus Kruger was born at Cradock in today's Cape Province of South Africa on 10 October 1825, of Prussian ancestry. His parents were fairly typical of that time: they owned livestock but not land and so were forced to be continually on the move, living in wagons between temporary homes while they sought grazing.

Like other youngsters, Paul Kruger had to tend the stock from an early age and a horse and a gun became part of his everyday life.

The Kruger family migrated with thousands of others northwards from the Cape, crossing into the Orange Free State, trekking into Natal, then into the Transvaal where they at last began to settle. At the age of 16, Paul obtained his own farm in the western Transvaal at Rustenburg and in the following year, 1842, he married.

A country largely in name but not in fibre, with ill-defined borders and populated by highly self-reliant, almost excessively independent Boers, the Transvaal was then a frontier land in which the white migrants competed for territory with the black migrants and, in the end, gained superiority because of their guns and their organisation.

During this time the young Paul gained a reputation for paying less attention to his livestock than to hunting for food and for profit, but he became increasingly involved in the political life of the new Republic.

In 1846 his wife and first child died of malaria. He married again a year later and sired 16 children, some of whom died in infancy. In the following years he became renowned for his bravery in conflicts with unruly tribes and, in 1858, for his skill as a negotiator in assisting in a peace treaty between the Orange Free State and the legendary king Moshoeshoe (pronounced Moshweshwe) of the Basuto people.

His political progress led him to stand for the presidency of the Transvaal in 1877 but in that year Britain bluntly annexed the Republic, whose government, the *Volksraad*, elected Paul Kruger to the vice-presidency.

For years he wheedled and argued with the British for the Transvaal's independence, until his government exasperatedly redeclared it a republic in 1880. This triggered the first Anglo-Boer War, which the Boers swiftly won.

After an armistice which placed British conditions on the resurrected state, Kruger became president in 1883. Soon afterwards gold was discovered on the Witwatersrand and the influence of the *uitlanders* (foreigners) was already seriously perturbing the Transvaal government; these two events brought things to a head. Against a background of *uitlander* dissatisfaction, the megalomanic entrepreneur, Cecil Rhodes, tried to topple the Republic government with the Jameson Raid in December 1895, which Kruger's forces quashed.

On 11 October 1899, the second Anglo-Boer War (the South African War) began – a long and bloody confrontation in which the Boers put the word "commando" into the English language, initiated guerrilla warfare and nearly brought Britain to defeat, until overwhelmed by sheer numbers and the infamous "scorched earth" and "concentration camp" policies introduced by Lord Kitchener.

Kruger did not see the end of this. By then too old to take to horseback in guerrilla warfare, he was ordered by his government to go to Europe to seek international support. There he died in exile at Clarens on Lake Geneva on 14 July 1904.

His last sight of his beloved Transvaal as he crossed from it into Mozambique by train on the way to Lourenco Marques, was of his most lasting creation – the Sabie Game Reserve, forerunner of the Kruger National Park.

Left: Paul Kruger – President of the Transvaal Republic for almost 20 years.

smouldered for another five years, until 1889, when again he stood up in the *Volksraad* and this time proposed that two specific areas be proclaimed as game reserves, where no killing would be allowed; a portion of the Soutpansberg district (today's Shingwedzi area) and the district of Pongola on the Swaziland-Natal border.

It took Kruger another four years to win approval for a reserve . . .

Fierce debate erupted, but this time the idea was accepted in principle by the *Volksraad*. It took another five years, however, before the slow-turning machinery of government could put it partly into effect.

On 13 June 1894, the Pongola reserve was proclaimed but it did not prove viable as a game reserve because very few animals inhabited it permanently: most migrated and in later years it was deproclaimed. The Pongola was the first such reserve created by the Boers, who were much maligned as random killers of game, and testified to their acceptance that, as Dr "Tol" Pienaar put it, "the conservation of his environment and fellow creatures on earth is a supreme duty of man."

It took Kruger another four years to win approval for a reserve in the area that he regarded as most important – the Lowveld. Arguing persuasively in the *Volksraad* he gained increasing support, including that of Mr R. Kelsey Loveday, member for Barberton, and on 26 March 1898, he was able to proclaim the *Goevernements Wildtuin* (Government game reserve) between the Sabie and Crocodile rivers, an area of about 4 600 km² (1 800 sq. miles). This was the Sabie Game Reserve, the nucleus from which grew today's Kruger National Park.

It was a singular achievement in the context of the rising disorder in the Transvaal caused by political tensions, the gold fever, clashes with black tribes, the confrontation with Britain and the clamour of grievances from the *uitlanders* (foreigners). Had Kruger delayed it today's Park might never have been, because 18 months later the second Anglo-Boer War began.

The Sabie Game Reserve, however, existed on paper for that time, though in little else. A Sergeant Izak Holtzhausen of the *Zuid-Afrikaansche Republiek* Police, stationed at Komatipoort on the Mozambique border south of the Reserve, was allotted the task of looking after this large chunk of bush.

Before the Reserve's borders could be properly demarcated, the war began and the Lowveld became part of the general battlefield in this fight between the British juggernaut and the swift-moving Boer forces, who skirmished on horseback and lived off the land. Wild animals again simply became food for both sides in the conflict.

The war is another story, but it did produce a further conservationist from a strange quarter, a Major Greenhill-Gardyne, adjutant of an irregular force operating along the southern Lowveld/ Mozambique border named Steinaecker's Horse – certainly one of the most curious military units ever to operate in Southern Africa – under the command of the infamous Colonel Ludwig Steinaecker (see p. 48).

From the first days of his arrival in the Lowveld to join Steinaecker's Horse – of which he was frequently acting commander during Steinaecker's absences and, in fact, the driving force – Major Greenhill-Gardyne of the Gordon Highlanders was quick to see the need for conservation. He had become an experienced hunter and keen naturalist during shikars in India.

The local unsophisticated black tribespeople made widespread use of snares and traps of all kinds: deadfalls, weighted falling spears, hunting dogs and other means to kill wild animals for their

meat, hides and horns – as, of course, their forefathers had done for many generations.

On top of this came the soldiers, both the Boer guerrillas – experienced hunters to a man who could live off the wild as naturally as leopards do – and the British "Tommies", who had no choice but to shoot for the pot to supplement their army rations.

Greenhill-Gardyne did a great deal to curb wanton killing, although it did not always endear him to the troops or to

the Lowveld blacks. In his last year in the Lowveld especially, he did much to introduce basic conservation measures and some awareness of the fact that the game was being depleted with dangerous speed.

His most important contribution was a report he wrote just after the fighting stopped in which he made a number of proposals on how to save the Lowveld as a wildlife area, and on conservation in general. This helped to provide a blueprint for later policy in the Reserve, which was yet to be officially born under the aegis of the new British colonial regime in the Transvaal.

Greenhill-Gardyne, like Steinaecker, also left the Lowveld after the war, but where he went is not known. Quite coincidentally his conservation efforts were to receive backing from an unexpected quarter, of which he was probably not even aware.

Just before the outbreak of war a then renowned hunter and naturalist, Abel Chapman, happened to have been on a hunt north of the Sabie river with a Lowveld resident by the name of J. C. Ingle as his guide. So impressed was Chapman by the Lowveld's potential as a game reserve that in 1900 he wrote a detailed report proposing that the entire area be proclaimed as one – from the Limpopo river in the north to the Crocodile river in the south, and extending from the Lebombo mountains in the east to the Drakensberg escarpment in the west.

His report was adopted by the International Convention for the Preservation of Wild Animals, and a copy reached the British Government. The Governor of the new Transvaal colony was Lord

"Go down there and make yourself thoroughly disagreeable to everyone."

Alfred Milner, himself a conservationist. Armed with Chapman's report plus Greenhill-Gardyne's, and with strong prompting from the same Loveday of Barberton who had originally backed President Kruger's proposal, plus growing support from other Lowveld residents, Lord Milner reproclaimed the Sabie Game Reserve when hostilities ceased. So closely did the reproclamation follow the now exiled President's ideas that Kruger's original proclamation was used almost word for word.

A man and his Eden

Then there came upon the scene the man who did more than any other to bring President Kruger's vision to fruition and build the Sabie Game Reserve into the huge, unique Kruger National Park of today. He was Major James Stevenson-Hamilton, a Scot born on 2 October 1867, who was educated at Rugby School, entered Sandhurst Military College and rose to become Laird of Fairholm in Lanarkshire and an officer of the 6th (Inniskilling) Dragoon Guards.

Like many Britons around the turn of the century, Stevenson-Hamilton was lured by the challenges of sport and adventure in the then still mostly dark Africa, where the powers of Europe and the great trading companies were opening up new territories and competing for profit and influence.

For the last few years of the nineteenth century, Stevenson-Hamilton had been exploring and hunting big game in the vast and remote wilderness of Northern Rhodesia, now Zambia. After the Anglo-Boer War the short, stocky Scot happened to be in the booming gold-mining city of Johannesburg, where he met the colonial government's newly appointed Commissioner for Native Affairs, Sir Godfrey Lagden, also a keen naturalist. Responsibility for the reproclaimed Sabie Game Reserve had been given to Sir Godfrey and he was seeking someone to go out there and run the place. Stevenson-Hamilton appeared to have been sent by Fate and on the first day of July 1902 was appointed to the job of "head ranger".

The adventure-seeking major had never been to the Lowveld, which people described to him variously as "a white man's grave" or "a game paradise" and (as he wrote in his biography) he "looked forward to the two years for which I was to be seconded from my regiment merely as a pleasant interlude, to be spent in interesting organisation, with unrivalled opportunities for studying wild life at first hand."

On a large-scale map on a wall in the Rand Club, he saw the Reserve as a blank space crossed by a few dotted lines indicating unsurveyed watercourses. His instructions were equally vague: "Go down there," Sir Godfrey Lagden told him, "and make yourself thoroughly

disagreeable to everyone."

He made his way to Lydenburg and set out from there with a wagon drawn by six oxen, three riding horses and three helpers: a Cape man named Nicholas, a Basuto horse attendant and an elderly black man nicknamed "Toothless Jack".

They followed the wagon road down the Escarpment and Stevenson-Hamilton made his way to the outpost of Nelspruit, to Barberton and to Komatipoort to find out from local officials and others as much as he could about the Lowveld and the new Reserve.

His first task was to recruit men who

Major James Stevenson-Hamilton in the uniform of an officer of the 6th (Inniskilling) Dragoon Guards.

could help him establish the Sabie Game Reserve and here he profited from the presence of Steinaecker's Horse, now nearing the end of its service. Major Greenhill-Gardyne told him of a Captain Gray stationed at the Steinaecker's Horse outpost at Gomondwane northwest of Komatipoort, inside the Reserve. Stevenson-Hamilton trekked up there and found him at Lower Sabie, a camp still further north on the river. This was "Gaza" Gray, a "hard-bitten looking man of about 50" who had come from the Eastern Cape and worked for a long

time in the Lowveld as a recruiter of labourers for the gold mines.

Gray became the first ranger in the Reserve, agreeing to act in an honorary capacity provided he could stay on at Lower Sabie. It was he who pointed out to the new head ranger that, although there was still a fair quantity of wildlife in the Reserve's eastern part, it was rapidly being diminished by tribespeople who had moved into the area from adjacent Mozambique and other areas where game had been virtually wiped out since the start of the Anglo-Boer War.

Largely because of this, Stevenson-Hamilton decided to act on his conviction that under the existing circumstances wild animals and people could not live together, and to move people out of the Reserve.

The tribespeople were told to return to their former areas, back into the domains of their tribal chiefs, before they planted their next crops. In compensation they were released from paying taxes for a year, and were told that if they had to travel through the Reserve, they must follow certain special routes.

Stevenson-Hamilton set up his own headquarters at Sabie Bridge, where the river was crossed by a railway line, which in later years was to link the Selati goldfields further north-west, in the Lowveld beyond the Reserve, to the main Pretoria-Lourenco Marques line just west of Komatipoort.

Sabie Bridge is today Skukuza, the headquarters camp of the Kruger National Park, after the name given to Stevenson-Hamilton by local blacks meaning "the man who sweeps clean" – referring to his "sweeping" them from the Reserve.

Greenhill-Gardyne introduced Stevenson-Hamilton to his next ranger recruit in Komatipoort: a tall, lean, black-moustached man of 27 named Harry Christopher Wolhuter, who was to work with him for over 40 years and become one of the most famous rangers in Africa, thanks to a single incident of incredible courage and fortitude (see p. 107).

Wolhuter had come north as a youngster with his farming parents from the Cape, his birthplace, and grew up in the Lowveld. He too served with Steinaecker's Horse in the war and was stationed at Olifants river north of the

THE BUSHVELD BISMARCK

Colonel Ludwig Steinaecker was something of a mystery, a flamboyant little buccaneer of a man who might well, a century or two earlier, have become a Henry Morgan style pirate. He spoke fractured English with a strong Bavarian accent and claimed to have been an officer in the Prussian Guard which came to South Africa on some mysterious German bid to gain a foothold.

When the Anglo-Boer War began, he is said to have enlisted in the Imperial Guides and to have been made a cook. Tiring of this, he put a proposition to the British command that he gather a small force in Natal, march north and blow up the railway bridge at Komatipoort, to block the Boers from moving their artillery.

Headquarters liked the idea and gave him the temporary rank of major and the authority to raise a small band. He took them north but too slowly, for by the time he reached Komatipoort the Boers were already there and with it went the chance of blowing the bridge.

Frustrated, Major Steinaecker looked about for something else to blow up to justify his trip. Some 40 km west of Komatipoort, at Malelane, he found an unguarded railway culvert. This he destroyed – causing no damage to the Boers, but holding up the British forces for two days . . .

Somehow he not only talked his way out of this blunder but further persuaded the British commander to let him raise a larger force to patrol and guard the Mozambique frontier, the Portuguese being neutral in this conflict and anyway unable to stop the Boers making free with their territory.

So was born Steinaecker's Horse, a most irregular collection of about 300 men recruited largely from Lowveld residents of British origin who were paid 10 shillings a day plus pickles, fresh milk and whisky to "keep off the fever".

Steinaecker – now a colonel, but whether by his own or British promotion is uncertain – set up his headquarters at Komatipoort and established a string of outposts for nearly 200 km along the Mozambique border from Swaziland northwards. When the British pulled out most of their regulars for war service elsewhere, Steinaecker set himself up as king of the region, taking most of the native cattle into his care (for milk) and wielding martial law.

If war ever touched his area it was very rare. The colonel was able to indulge himself in frequent visits to the flesh-pots of Lourenco Marques and even faraway Durban, while his men lived comfortably in their outposts, shooting game for food and receiving their regular rations of whisky and pickles.

A thin, small man of little over 1,5 m (just over 5 ft), the colonel adorned himself with imperious pomposity and a uniform to match. Major James Stevenson-Hamilton, the man who later did most to bring the Kruger National Park to life, described him thus:

"The first things about him that caught the eye were his tremendous moustachios, which were pendant quite ten inches on each side of his rather thin and bony face and were, so to say, balanced by a long goatee or imperial which adorned his chin. He possessed bushy eyebrows, a large hooked nose and aggressive dark eyes.

"His uniform, which I believe was entirely of his own design, consisted of a staff cap, deeply encrusted with silver lace, and encircled by a broad green band. He wore a double-breasted khaki frock-coat, padded and wasp-waisted, which reached to his knees. Adorned with rows of large silver buttons, and on his shoulders, sat huge and solid epaulettes of the same material.

"A pair of long soft brown leather boots, the heels adorned with box spurs, met his neatly made riding-breeches at the thighs. Perhaps the most striking item of equipment was the enormous sword, which might have fitted a cuirassier of the guard, and was supported by a massive belt with two large fringed silver laced tassels hiding all the upper part of the weapon."

The men of Steinaecker's Horse seemed to have had more troubles with wild animals than with the Boers. In 1900 about 20 were sent to set up a post in the Game Reserve at Sabie Bridge, where the railway running north from the Mozambique line crossed the Sabie river.

Until they could complete a thornbush stockade or *zeriba* in which to sleep at night, they camped around a large fire with guards posted to protect the sleepers and keep the fire fuelled. There was no doubt this was necessary: several blacks in that area had disappeared and horses had been taken from the camp – the work of lions.

One night a man named Smart was on guard and saw that the campfire was burning low. He walked a few yards to fetch a fresh log and was immediately taken by a lion. His yells awakened the others who loosed a ragged fusillade of shots in the dark, which frightened off the animal. Smart was brought into the camp badly torn and bitten. He died the next day.

That same night the lion came back and took one of the party's black members, without anybody noticing. They missed him the next morning and found his partly eaten remains in the bush nearby. Obviously there was a man-eating lion in the neighbourhood. Then the party had several other narrow escapes, including one when the lion reached into a railroad cattle truck in which they were sheltering one night and lashed out with its huge paw, inches from the leg of one of them.

Another black man and several more horses were taken. The group lived in constant fear, until finally the lion was wounded by a trap gun and then finished off by an officer when it charged him. It was found to have a bullet wound in the flank, which is why it could not hunt its normal prey and had turned on people. Ironically, one of the party recognised the old Martini-Henry bullet as coming from a rifle with which he had wounded a lion two years before.

Steinaecker's Horse, its whisky-and-pickles life in the game-rich Lowveld (a wartime idyll which involved virtually no fighting) and his own flamboyant command were the last high points in the lifetime of this curious little man. When the war ended he tried to keep his unit going, as did a number of its members who liked the bush life, but his ambition brought about its end rather abruptly.

Steinaecker heard that a number of British regiments still stationed in South Africa were sending contingents to London to take part in the parade at the coronation of King Edward VII. To his indignation nobody had invited Steinaecker's Horse to send representatives.

An artist's impression of Colonel Ludwig Steinaecker.

So he selected a group of his men and talked a shipping company into taking them, and himself, to London. Surely, he reckoned, this famous fighting force which had helped to stave off the Boers in darkest Africa would be welcomed with fanfare. Unfortunately, it was not.

Whitehall was astonished to discover the presence of a small, moustachioed German of ferocious demeanour, dressed in a stunningly bizarre uniform, demanding a place in the big procession. Someone started checking the military records and discovered that at best, Steinaecker's Horse was a locally generated irregular force with an oddly mysterious history. Wheels turned rapidly and the tiny colonel was quickly deprived of his command and rank and told to go home.

That he did, but once back in Komatipoort he refused to quit his command and a long and saddening struggle ensued between Steinaecker and the authorities. Major James Stevenson-Hamilton was asked to look into the matter, and his report led to the final disbandment of Steinaecker's Horse. Steinaecker himself stayed in Komatipoort as long as he could and was eventually evicted from his once-official residence there by his former secretary, Jules Diespecker.

By now a pathetic figure, the little German was seen about Komatipoort for a time, still resplendent in his beloved uniform and sword – until he disappeared.

Author Alan Cattrick tells of Steinaecker's sad end in his book on renowned South African characters, *Spoor of Blood*, as it was told to him by Harry Wolhuter who was a member of Steinaecker's Horse before he joined the Sabie Game Reserve as a ranger.

When it finally sank into Steinaecker's mind that his cause in Komatipoort was lost and his military career finished, he moved up to the beautiful village of Pilgrim's Rest, at the top of the escarpment above the Lowveld, and began farming cotton.

This failed and Steinaecker lost everything, says Cattrick. Then another ex-member of Steinaecker's Horse, J.C. Travers, took pity on him and invited him to stay with his family in the Pilgrim's Rest area. Steinaecker lived with them for a number of years, until the start of World War I. Then, apparently, he lost his mind. He acquired several pistols and plotted to murder Travers and his wife, but was disarmed in the nick of time.

Travers called the police from nearby Bushbuckridge and Steinaecker was taken into custody. On the drive back to the police station, says Cattrick, the prisoner asked his escort to stop and then swallowed a poison he must have been carrying. Steinaecker died within minutes and was buried right there.

"For all his faults," writes Cattrick, "he deserved a better end than this and certainly his name will never be forgotten in the Lowveld where he was once monarch of all he surveyed."

Strange, erratic character though he was, Colonel Ludwig Steinaecker did serve the purpose of considerably curbing hunting and poaching in the Lowveld. This was chiefly due to his adjutant, Major Greenhill-Gardyne of the Gordon Highlanders.

(*Note:* Most references, including works on South African history, use the spelling "Steinacker" for the Colonel's name. Modern historians, however, state that the correct spelling is "Steinaecker".)

Sabie when he was offered the post of ranger. He accepted after some initial hesitation and became one of the key figures in the Reserve's development and growth.

Having acquired two rangers and sent Wolhuter off to look after the northern and western parts of the Reserve (based at Pretoriuskop), Stevenson-Hamilton set about surveying his new territory on horseback, with his oxwagon following.

He came across a fair variety of wild animals, but not in large numbers and all of them shy – wildebeest, impala,

The warden's headquarters at Sabie Bridge, now Skukuza.

waterbuck, reedbuck, warthog, wild dog, jackal and occasionally a lion. Where there were tribal villages, particularly along rivers, there was no game.

Camping out on trek he had encounters with lions. On one of his first trips through the Reserve he bivouacked near a river and tethered his three horses and two pack mules to a rope pegged to the ground for the night. He was about to go to sleep when there was an uproar from the line.

Charging out of his tiny tent, he found Toothless Jack hurling burning brands from the campfire into the dark while another of his men tried to calm the horses and mules. After Stevenson-Hamilton had fired a few shots into the bush Jack told him that when he put a fresh log on the fire, the flames sprang high and in the burst of light he had seen a lion less than four paces away. Jack had tossed a burning stick at it and it had vanished into the dark. In the morning they found that a second lion had been within 4 m of the horses when Jack discovered the first. Thereafter, whenever he travelled in lion country, Stevenson-Hamilton always enclosed his transport animals in a *zeriba* or shelter made of thorn tree branches.

In 1903 the colonial authorities decided to give further life to President Kruger's vision by proclaiming yet another large parcel of land the exiled leader had wanted preserved. This was the Shingwedzi – a broad expanse of mopane forest and savannah which lies between the Letaba river to the south and the Limpopo river to the north (the border with Southern Rhodesia as it was then), and stretches roughly 60 km westwards from the Mozambique border.

Quite unlike the Sabie in character, it is rich in birdlife and when it was proclaimed was largely a trackless wilderness populated by a few tribespeople – so remote that one of Stevenson-Hamilton's workers quit soon after going in there with him, declaring this was no place fit for human beings.

Having the Shingwedzi too placed under his control, however, presented Stevenson-Hamilton with a new problem. Sabie and the larger Shingwedzi were separated by a great chunk of land between the Olifants and Letaba rivers about 32 km wide which was a proclaimed mining area.

It was not until the main boundaries of the Kruger National Park were finally defined in 1926 that Stevenson-Hamilton was able to get this thorn out of his side, but meanwhile he negotiated successfully with the mining and land companies to gain game protection rights over some of the land there, thus linking the Sabie and Shingwedzi Reserves.

At about the same time he was given the powers of a Justice of the Peace so that he could deal with poaching and other infringements on the spot, instead of having to travel for days or weeks to the nearest magistrate. He took over the

Disease was a major problem. A few years earlier, just before the turn of the century, the dreaded rinderpest had swept down from East and Central Africa wiping out livestock and wildlife and ravaging the numbers in the Lowveld. By wiping out most of the animals, it also wiped out the tsetse fly in the Lowveld, which has been free of this major African pest ever since.

Fear of rinderpest, and then the advent of East Coast cattle fever, made many of the still rough-and-ready farmers living outside the Reserve oppose its existence vigorously. The wild animals, they argued, were reservoirs of disease or at least carriers of it. And, they argued, all lions should be shot out because they were a menace to livestock.

Undeterred, Stevenson-Hamilton went ahead with his plans, still in the formative period, to firmly stamp his authority over the whole area. To Shingwedzi he sent as a ranger one of the most colourful and eccentric characters in Lowveld history – Major A. A. Fraser.

Major Fraser, recently of the Bedfordshire Regiment and well experienced in hunting in India where he served for 25 years, was a powerful man of nearly 2 m (well over 6 ft) with a great red beard, total impatience with administration and paperwork, superb marksmanship with rifle or shotgun, and a seemingly bottomless capacity for whisky.

Of Fraser, Stevenson-Hamilton wrote: "I have seldom met anyone more observant of facts relating to wildlife, or with a memory more retentive of what he had noticed." He added: "He was, in fact, a sportsman in the higher sense of the word, and a born gamekeeper – essen-

. . . the magistrate felt himself wilting under the steady stare from the fierce grey eyes . . .

responsibility of patrolling the Mozambique border, established friendly relations with the Portuguese authorities on the other side, and employed more rangers.

By 1905 Stevenson-Hamilton's domain covered more than 30 000 km² (about 11 500 square miles – considerably larger than today's Park) between the Crocodile and Limpopo rivers. Running it was no easy task for him and his small band of five rangers and a few score black policemen.

tially of the Highland variety, be it said, for no gillie, even of the most hardened type, could have excelled him in the absorption of unlimited quantities of Scotch whisky without the slightest visible effect."

Fraser cared nothing for desk work or, it appears, for paper in any form except that of the English country life magazine *Field*. This he avidly read every week.

Stevenson-Hamilton first posted Fraser to the brief-lived Pongola Reserve south of Swaziland, but moved him to an outpost named Malunzane, roughly halfway between the Letaba and Shingwedzi rivers, soon after the big Shingwedzi area was made a reserve.

There Fraser stayed for the next 16 years, apart from occasional visits south, and became a renowned and respected figure.

In 1917, when Stevenson-Hamilton went off to serve in the First World War, Major Fraser was brought south to Sabie Bridge to act as warden for the whole Reserve – an episode which made the hair of bureaucrats turn grey.

Such contempt did he have for red tape and official documents that he simply ignored them. If he had to fill out papers to draw employees' wages or rations, he preferred to pay out of his own pocket. Having judicial powers as acting warden, he held court and issued sentences but never bothered to keep records, so that the inspecting magistrate had to throw out just about all the cases.

The result was chaos, which later took months to sort out. When Fraser moved to Sabie Bridge he took with him his pack of 25 dogs, answered no letters and so neglected his administrative duties that when officials visited to find out what was going on, the office was so congested with cobwebs that labourers had to cut a way in.

Stevenson-Hamilton tells the story that when Fraser was in Shingwedzi he collected fines for a year, then visited the Pietersburg magistrate to hand over the cash. He dumped a bag of sovereigns on the astonished magistrate's desk, but with no supporting papers, receipts or other documents.

When the magistrate said he could not accept the money that way, Fraser reacted angrily. The big red beard began to bristle, the magistrate felt himself wilting under the steady stare from the fierce grey eyes, in which a dangerous light was slowly gathering.

"'Papers! Receipts! If you don't think, sir, my word is good enough, you had better tell me so! I don't care a curse what you do with the money; I have not even counted it; you can do that yourself and make out your own damned receipts!'"

Fraser stalked out and there is no record of what happened to the cash.

Ranger Harry Wolhuter visited Fraser one night at Sabie Bridge and was put

Major A. A. Fraser surrounded by some of the 25 dogs he brought with him to the Reserve when he became a ranger.

up in the spare room with a bed, two blankets and a pillow. During the night he became cold and sought out Fraser to borrow another blanket.

"It was still quite dark and by the light of a candle he could make out, at the farther side of the room, what looked like a great dark pile from which emanated a variety of grunts and snores," Stevenson-Hamilton wrote later.

"Approaching closer, the pile disintegrated and revealed itself as consisting of the 25 large dogs, which were sleeping on and around their master who, fully dressed, was extended on his back on the bare floor, snoring quite happily. He explained that having no other bedclothes than those he had lent to Wolhuter, he had spent thus such part of the night as he devoted to slumber, adding that it was not an unusual practice with him, even when the ordinary amenities were available, as the dogs kept him much warmer than any blankets could ever do."

Not that Fraser slept much at night. His practice was to get up very early, fish or hunt warthogs and go to bed after a late breakfast. He would rise again at about 6 p.m. and spend most of the night working on his guns, mending

clothing and reading *Field* before getting a little more pre-dawn sleep.

For all his quirks, Fraser was, however, a highly efficient ranger who for years controlled virtually single-handed a huge and difficult region beset with problems like snaring by poachers, elephant hunting by ivory traders and disease.

Many of the difficulties that Stevenson-Hamilton and Fraser had to deal with in the Shingwedzi Reserve came from a curious and fascinating little enclave which has carved itself a permanent niche in Southern African history: Crooks' Corner in the far north-eastern corner of the Transvaal, an inaccessible triangle where three international borders meet, which became a refuge for every scoundrel – and poacher – in the area (see p. 52).

For several more years Stevenson-Hamilton and his small but formidable band of rangers continued slowly to expand their control over their huge and unwieldy charge, in spite of the anti-conservation clamour of neighbouring farmers.

But towards 1912 the political storm clouds began to gather and they loomed for more than 14 years, until the Sabie,

Shingwedzi and the territory between were finally consolidated into a single national park.

The first troubles came from the land and mining companies which had granted game control rights to the Reserve over an area between the Olifants and Letaba rivers for five years, which had been extended for another five. Now, with the Lowveld becoming more settled and prosperous, they wanted to make use of the land, sell it, or somehow profit from it.

Sheep farmers demanded – and were granted by the government – the right to bring their flocks down to graze in winter in the Pretoriuskop area of the Reserve, for a fee. This included the right to protect their flocks, so they shot any predators on sight and generally caused considerable disruption – until Stevenson-Hamilton blocked them out the next winter by simply doubling the grazing fees.

One day in Komatipoort, while Stevenson-Hamilton sat drinking beer on the front veranda of Hanneman's Hotel with the innkeeper, his thoughts were turned to tourists. "When is this Reserve of yours going to be thrown open for shooting?" he was asked.

"I hope never," Stevenson-Hamilton replied.

"What!" exclaimed the innkeeper in astonishment. "Do you mean to tell me that the government is going to spend thousands of pounds every year just to keep game? The public won't stand for it! Some day this Lowveld will be opened up and then where will you be?"

His reaction gave Stevenson-Hamilton food for thought.

"After all," he wrote later, "what goal were we striving for? I had lazily accepted the idea that some day, I hoped a long way ahead, parts at least of the Reserve when fully stocked, would be hired out for some kind of controlled shooting.

"On the other hand it did seem a waste of time and money if all the labour and care of the last three years, not to speak of the £15 000 or so which had been spent, was merely to provide hides and biltong [dried, spiced meat] for that section of the local public, with the result that in two or three years we should be back where we started."

In 1912 he put a plan to the British colonial secretary, Sir Patrick Duncan, later South Africa's first Governor-General, and to General Jan Smuts, then Minister of the Interior, Finance and Defence in the new Union of South Africa, for the Reserves to be made a national park. He had strong backing from the Game Protection Society of South Africa, later to become the Wildlife Society of Southern Africa.

But for it to be a national park it would need public support, and to gain public support he would have to lure visitors, Stevenson-Hamilton realised.

World War I intervened, and Stevenson-Hamilton and some of his rangers who were young enough went off to fight. One of those who did not come back was a vigorous man who joined the Reserve staff in 1904 at the age of 21

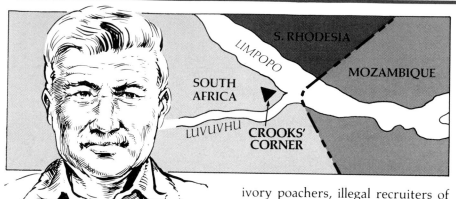

Stephanus Barnard, known as Bvekenya.

THE HAVEN OF SCOUNDRELS

Right up in the far north was a small piece of no man's Africa, triangular in shape, sandwiched between the Luvuvhu (then better known as the Pafuri) and Limpopo rivers. Since 1913 the Luvuvhu formed the northern boundary of the Reserve up to where it joined the Limpopo, and the Limpopo, of course, was the boundary between the Transvaal and Southern Rhodesia (now Zimbabwe). The Mozambique border also began at that confluence.

Thus the triangle of land, while it lay in the most north-easterly corner of the Transvaal, was very close to two other countries and a large game reserve. It was also extremely remote, accessible only over long, arduous wagon roads through thick bush, in very hot conditions.

It was the natural refuge for all kinds of people who had no great wish to look into the eyes of the law, who might suddenly feel the urge or the necessity to flee across an international border and who liked the hard but never dull life of the wilderness.

Inevitably, the place was nick-named Crooks' Corner, home of ivory poachers, illegal recruiters of black labour, gun-runners and dozens of others who had chosen the free life of the buccaneer.

South African author Tom Bulpin describes the place in his book, *The Ivory Trail,* the intriguing story of a man named Stephanus Cecil Rutgert Barnard, named *Bvekenya* by the blacks – "The one who swaggers as he walks".

Up at Crooks' Corner a loquacious Irish bush man named William Pye told Bvekenya that if he got into trouble while poaching ivory, he should head straight back for this place.

"If you want to know why so many of us live here in the bush, I'll tell you," Pye said. "Crooks' Corner, they call this. Well, at the corner, where the rivers meet, there's a wonderfully handy beacon. East of it is Portuguese land; north of it is Rhodesia; west of it is this new-fangled Union of South Africa; and south of it is what they want to call a game reserve.

"If you ever get into trouble just remember that beacon. That's why most of us live here. Whoever comes for you, you can always be on the other side in someone else's territory; and if they all come at once, you can always sit on the beacon top and let them fight over who is to pinch you."

and, like so many others, left his stamp on the Kruger National Park history: G. R. Healy.

During the war years conditions in the Reserves declined sharply. With less than half the usual staff running them, poaching increased so fast that organised parties of hunters actually camped inside the Reserves for weeks and months, trapping and shooting game to such an extent that the numbers of wild animals declined steeply. Stevenson-Hamilton returned to find a grim situation.

The change was due to the advancement of civilisation riding down the railroads – the coming of more and more settlers and of big business.

"The old free and easy days had already gone," wrote Stevenson-Hamilton, "those days when a party of prospectors would subscribe to buy among them a horse which had the fascinating habit of drinking beer out of a bucket; or designing to convey on a mule trolley a defunct comrade to the place of interment, would so well drown grief before and during the journey as to arrive at their destination without the central figure."

During the war, from which Stevenson-Hamilton returned a colonel, a game commission visited the Reserves and said the government should aim towards creating a "great national park" – but then weakened their case by saying parts of this should be open to farmers for winter grazing.

The years right after the war were bad ones for Stevenson-Hamilton and his objectives. Companies which owned private land inside the Sabie Game Reserve began to demand their land back so that they could develop or sell it. The protagonists of a national park gloomily concluded that it would be too costly to buy the land from them for the Reserve.

A coal syndicate had acquired a concession along the Selati railway running north-west through the Reserve from Komatipoort and marked out sites for prospecting shafts. Farmers were agitating to be allowed deeper into the Reserve for winter grazing. Someone had a scheme to mine gold along the Luvuvhu river in the north.

The biggest of the land companies had launched an anti-Reserve campaign in the Press, accusing it of being a breeding ground for lions which menaced farming. The campaign was so strong that one newspaper (contrary to most) acidly stated: "This so-called game reserve is merely a refuge for dangerous wild animals, a focus of disease, and should be swept away."

The South African Railways administration entered the fray against the Reserve. It had taken charge of the Selati line, which linked Komatipoort through the Sabie to the goldfields beyond the Reserve in the north-west Lowveld. This line was started in the late 1890s by the old Transvaal republic government, but construction stopped when the contracting company went bankrupt. Building began again in 1909 and was finished in 1912.

In 1922 the administration demanded that a corridor of land bracketing the railroad be deproclaimed as a reserve to be used for growing timber and possibly for farming. The authorities believed that this would improve its business, railways being in existence to open up new territory.

In the avenues of government in Pretoria and in the public view, the pendulum began to swing against conservation in general and the Sabie and Shingwedzi Reserves in particular. Characteristically, Stevenson-Hamilton dug in his heels and prepared to fight.

At a meeting in Pretoria he was told bluntly by the Department of Lands that they wanted the entire Sabie Reserve abolished and used for ranching, sheep grazing or cotton and citrus farming.

Then, two singular events occurred which changed the tide, both of them initiated by Stevenson-Hamilton.

The Transvaal Consolidated Land Company was formed to buy up most of the privately owned land in the Sabie Game Reserve and was determined to stock it with cattle – in spite of the horror the company executives expressed at the number of lions that roamed the area. By the end of the year they had about 800 head of cattle a mere 7 km from the Sabie Bridge camp.

Stevenson-Hamilton's worry was that the government, faced with a *fait accompli* and the high cost of buying back the land for the Reserve, would simply give up and deproclaim the area. Fortunately he had an ally from an unexpected quarter: a Mr Crosby, who was sent by the land company to manage their new ranch and who in no time became quite friendly

PIONEER RANGER

When Stevenson-Hamilton was made a Justice of the Peace, the authorities saw fit to give him a public prosecutor and a clerk of the court so that the duties of his court could be carried out properly and with dignity. Unfortunately, they decided that one man should do both jobs.

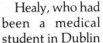

G. R. Healy.

After several misfits were sent him, including one who drowned in the Crocodile river after a week of high life, G. R. Healy turned up.

Healy, who had been a medical student in Dublin and had come to South Africa as a second lieutenant with the Irish militia in the Anglo-Boer War, decided he liked the country, and stayed on afterwards as a South African Constabulary corporal.

"He was a tall, loose-limbed, angular youth, with a head small for his body, fair hair and large innocent blue eyes which coupled with a simple expression, were apt to give strangers a mistaken idea of his character, and thus tempt the rash to try to get the better of him, to their subsequent regret," his chief said.

In 1908, four years after being assigned to Stevenson-Hamilton as a court policeman, Healy left the police force to take up the post of ranger in charge of the 6 500 km² of country between the Sabie and Olifants rivers.

He got the job only because Stevenson-Hamilton, who particularly wanted him above other possibles, cut his own salary and made other reductions in the Reserve's budget to be able to afford to pay Healy's wages.

Healy was later set up in a wood-and-iron home, one of the loneliest in South Africa, at what is today Satara camp.

In 1914, the year World War I began, Healy went off to receive a commission in the King's African Rifles and was killed in 1916.

The Selati railway line from Komatipoort to Tzaneen, via Sabie Bridge (Skukuza) was officially opened on 7 November 1912. The line was re-routed along the Park boundary in 1968-69 and the last train passed through Skukuza siding in 1973.

with James Stevenson-Hamilton.

Having ensconced their cattle in the Reserve, the company executives began to put on the pressure. Claiming that they had every right to shoot game there, they ordered Mr Crosby to start doing just that. Mr Crosby, by now a supporter of the Sabie Reserve, told this to Stevenson-Hamilton, who said he had better get on with it.

"Thereupon, with great deliberation, he proceeded to dispatch a lone bull wildebeest, and then duly reported his

action at my office."

The ensuing court case was a milestone in the history of the Park. The land company's argument was that the wildebeest was eating grass and that grass was a plant within the meaning of the law which allowed landowners to kill wild animals caught in the act of destroying plants or crops.

The magistrate thought otherwise and ruled against the company, which then appealed to the High Court. There the judge upheld the magistrate's verdict,

stating that if he ruled that grass was indeed a plant in that sense, it would make mockery of all the game laws.

The second event which helped swing the tide was the visit to the Sabie Game Reserve of a remarkable South African figure, Colonel Denys Reitz.

Reitz had been a protégé and friend before and during the Anglo-Boer War of the renowned Boer leader and fighter General Jan Christiaan Smuts, and his books on his experiences during the war and later are classic insights into that tragic affair.

So disillusioned was he by the terms of the Treaty of Vereeniging which ended the war – as he saw it, in Britain's favour with virtual surrender by the Boers – that he emigrated to Madagascar.

There he worked as a transport rider until General Smuts's wife, Issie, persuaded him to come home. Back in the new Union of South Africa he soon became Minister of Lands in General Louis Botha's Cabinet.

Reitz had known and admired the Lowveld before the war in President Kruger's day and was a strong advocate of conservation. In 1923, at the invitation of Stevenson-Hamilton, he visited the Sabie Game Reserve with his official entourage. The trip around the Sabie in mule carts through the bush convinced him that the Reserve must not die and he swung the full weight of his authority into the fight to save it.

This Reitz achieved after much wheeling and dealing by excising some of the privately owned land and adding, instead, some government land. In the process the Sabie and Shingwedzi section became united and the Reserve gained permanency, but at the cost of a chunk of choice territory for certain species of animals in the western region, south of the Olifants river.

This move was to be ratified by the government in a law to make the whole Reserve a national park. But before the bill was put before Parliament, the government was defeated in a general election.

Fortunately, the Reserve had powerful supporters in the new government led by General J. B. M. Hertzog, so there was no change of direction, only delays in getting the conservation measures entrenched in law. Meanwhile other things happened which gave further momentum to Stevenson-Hamilton's plans.

In 1923 the South African Railways began a winter tourist service to show off the Transvaal. Called the "Round in Nine" tour because it took nine days, it was a cheap, efficient and comfortable train ride which took in the magnificent views of the Great Escarpment, the citrus plantations in the warm Lowveld around Nelspruit and other scenic splendours, culminating in the short run from the border to Lourenco Marques, famous for its beaches and night life.

Stevenson-Hamilton discovered that these trains were due to travel on the Selati line through the Sabie Game Reserve by night without stopping. He suggested to the railways manager that this section of the journey be made by day, or that the train stop while travelling through, to give the tourists a chance to see the wildlife.

The railway executive was astonished at the idea that anyone might want to look at game and said it would be a better idea to let the tourists shoot a

H. S. Trollope, ranger from 1925 to 1928.

buck or two. Stevenson-Hamilton, astonished in his turn, eventually squeezed a deal by which the train would stop for the night at a siding on the other side of the river at Sabie Bridge and leave one hour after sunrise. There the rangers would arrange a campfire for the travellers.

To everyone's surprise, this stop became the most popular event in the entire "Round in Nine" tour. The camp-fires were an immediate hit, with passengers from the train sitting around while the eyes of wild animals glinted from the dark in the flicker of the flames, singing songs and thoroughly enjoying the bush experience. To add to the fun one of the train staff would wrap himself in a lion skin and crawl into the firelight while the Sabie Bridge police sergeant realistically imitated a lion's roar through a tube.

This was the first solid evidence Stevenson-Hamilton received that tourism could be his most powerful ally in winning public support. One of the first big backers it won over was the South African Railways, whose leadership thereafter gave the Reserve its full support.

In 1925 Prince Edward, Prince of Wales, came to the Sabie during his visit to South Africa. He did not stay long – merely passing through it by train with a one-hour stop for dinner at Sabie Bridge – but his presence brought a huge and welcome charge of publicity which helped put the Reserve firmly on the map.

This was enhanced by growing exposure in the South African media and abroad, including backing by conservationist bodies in America and Europe. The only fly in the wildlife ointment in these years of progress was the lions . . .

In 1925 farmers along the south bank of the Crocodile river, the Reserve's southern border, began to raise Cain about the damage being caused by wild animals crossing the river. Lions, they said angrily and with much justification, were killing their livestock, and water-buck and the wily kudu were causing havoc in their orchards and other crops. Farmers began to demand the right to come into the Reserve to shoot lions at will.

Stevenson-Hamilton had to act fast before the tide of complaint became a flood the authorities could not ignore. So he hired a ranger, one Harold Trollope, who had much experience as a hunter, and ordered him to thin out the lions. From Stevenson-Hamilton's description, Trollope was a remarkable man: "He shot all his lions single-handed, preferring to gallop after them on horseback where the country permitted, and shoot them at bay, or in the act of charging. He was a dead shot, possessed

of the coolest nerve, and so successful was he, and so entirely confident in himself, that latterly he used to go out of his way deliberately to invite charges, finding the sport not sufficiently exciting otherwise."

Yet, Stevenson-Hamilton remarked, Trollope was lucky. One misfire in a cartridge and he would have been dead.

The leopard is one of the most dangerous antagonists in Africa – a killing machine of awesome power . . .

That taking the slightest chance in the bush can be fatal was demonstrated in an unfortunate episode, when Trollope took along on a leopard hunt his elderly father-in-law, Mr Glen Leary, a retired magistrate.

The leopard is one of the most dangerous antagonists in Africa, a killing machine of awesome power equipped with fangs, claws and dazzling speed. Tangling with one has been described as akin to boxing against a buzz-saw.

Trollope's dogs had flushed a leopard. He told his father-in-law to remain in a position where he could clearly see and shoot the leopard if necessary. Trollope himself followed the dogs.

Mr Leary became bored or impatient and walked forward to try for a better view. This took him into long grass and before he knew it, the leopard was upon him. He fired one shot, missed and fell backwards into a dry stream bed with the animal on top of him. Trollope arrived seconds later and shot the leopard dead, but by then it had inflicted terrible wounds and Mr Leary died in hospital.

Efforts were continuing to elevate the Reserve's status to that of national park and yet another figure came on stage: Mr Piet Grobler, who replaced Colonel Reitz as Minister of Lands when the Hertzog government was elected. Grobler was a grand-nephew of the late President Kruger and an equally ardent conservationist. When Reitz put the original National Parks Bill before Parliament, it was Grobler who proposed that not only Sabie and Shingwedzi but all future national parks should fall under the control of a board of honorary trustees elected by the four provinces and other involved groups such as the then Game Protection Society.

Grobler was also tough. He told the landowners whose land inside the Sabie it was proposed to buy out, in terms of Reitz's original plan, that they would be compensated in either cash or in other land outside the Reserve. But, he added firmly, if they did not accept, the government would simply expropriate their property. They agreed to sell or swop.

As the National Parks Bill approached Parliament again, those opposed to it raised the controversy afresh, arguing once more that the Reserve was a repository of disease.

On 31 May 1926, Grobler moved the second reading of the National Parks Act in Parliament and called for a board of control because "politics must be kept out of it". He was supported by General Smuts, then the Leader of the Opposition, and the measure became law with unanimous support.

* * *

Thus was born the Kruger National Park – a name chosen in honour of the old President by the Cape Town newspaper *Die Burger* with the support of Stevenson-Hamilton and Harry Stratford Caldecott, a writer/artist who with his tireless campaigning perhaps did more than anyone else to swing the limelight onto the Sabie and Shingwedzi Reserves.

This step set South Africa on the road to becoming a world leader in the field of conservation. As Dr "Tol" Pienaar, present warden of the Park, wrote: "Nature conservation in South Africa had at last a firm foundation on which to build and three men will always tower in the background: Paul Kruger, James Stevenson-Hamilton and Piet Grobler. They represent three phases in the history of nature conservation in this country: the Word, the Deed and the Law."

That was the watershed for conservation in South Africa. It came the year after the first African national park was proclaimed – the Albert National Park in the then Belgian Congo, now Zaïre – and a few years later Minister Piet Grobler proclaimed three more in this country: the Kalahari Gemsbok National Park, the Bontebok National Park in the southern Cape and the Addo Elephant National Park in the eastern Cape.

Men of vision

The struggle for the elevation of the Sabie Reserve to National Park status benefited enormously from the solid support of two successive Ministers of Land. They were Colonel Denys Reitz **(left),** a former Boer fighter who was persuaded to return from self-exile in Madagascar and joined the government of General Louis Botha, and Mr Piet Grobler **(above)** a grand-nephew of President Paul Kruger who succeeded Reitz when the Hertzog government came to power in 1924.

The Park – early years

For James Stevenson-Hamilton, the proclamation of the Kruger National Park heralded a new and tougher era. Gone were the days when he and his minuscule band of six white and 60 black rangers had to concentrate only on protecting wild animals while fending off poachers and the criticism of soured farmers and land-hungry entrepreneurs. Now they had to turn this long tract of bush into a tangible public asset – tangible to all South Africans and not the conservationists alone.

In its stilted officialese the National Parks Act states: "The object of the constitution of a park is the establishment, preservation and study therein of wild animals, marine and plant life and objects of geological, archaeological, historical, ethnological, oceanographic, educational and other scientific interest and objects relating to said life or the first-mentioned objects or to events in or the history of the park, *in such a manner that the area which constitutes the park shall, as far as may be and for the benefit and enjoyment of visitors, be retained in its natural state*" (my italics).

When the Act was promulgated the Park staff regarded it with a somewhat jaundiced eye. That the Park should be "retained in its natural state" made good sense, but hardly so the reference to the "enjoyment of visitors". Not surprisingly, since at that stage the Park had such a small staff of rangers and assistants and endless wilderness – no roads and not a single rest camp. Mechanised travel was almost non-existent; when rangers travelled they did so on horseback or by oxwagon or donkey cart. The only cash inflow into the Park was a meagre state allowance for management and maintenance and equally meagre staff salaries and wages.

But Stevenson-Hamilton realised that without public support and revenue from visitors, the Park's future was severely limited, if not doomed.

That meant enticing tourists to look at and live among animals. Taking stock in 1925, Stevenson-Hamilton had estimated that the Park contained perhaps 100 000 head of game (a small part of its population today). There were two breeding herds of elephants in the Letaba river area, rarely seen nyala at Pafuri, a few impala between Skukuza and Satara

Members of the first National Parks Board of Trustees gathered for a photograph with Minister Piet Grobler. They are from the left: R. A. Hockley, Senator W. J. C. Brebner, A. E. Charter, W. A. Campbell, Minister Grobler, G. S. Preller, C. R. De Laporte (acting warden), Colonel Denys Reitz and Dr A. K. Haagner. Absent were Sir Abe Bailey and Mr Oswald Pirow.

and none in the Pretoriuskop area, and here and there a giraffe.

A small herd of buffalo which had survived the rinderpest was increasing satisfactorily, there were sable and a few roan antelope and, occasionally, eland could be seen. Only blue wildebeest, zebra and waterbuck were common. And, of course, there were plenty of lions in the Pretoriuskop area, where grass had been burned to attract buck to the new growth.

His dilemma was that the Park could not have visitors without roads but the money to build roads could not be raised without visitors. The only road in the Park at the time was one built by ranger C.R. de' Laporte, who had joined the Sabie Reserve in its early days, between Crocodile Bridge and Lower Sabie for his own use, for he had acquired a car. It was also used by Stevenson-Hamilton in his own "tin lizzie". But it was a rough, rutted track and not the kind of road the public could use.

Being a man of ingenuity, Stevenson-Hamilton found ways and means of cutting red tape and hiding costs in his management books in order to divert money into road building, and in 1927 he began the project in earnest. By their own blood, sweat and tears he and his men cut a reasonably decent road from Lower Sabie to the Olifants river, a

second from Sabie Bridge (now Skukuza) to Pretoriuskop and a third from Sabie Bridge to Crocodile Bridge – a considerable feat in that heat.

Now at least parts of the Park were accessible to visitors. Staying in the Park overnight was another problem. There was no accommodation. People had to make short day trips or camp – and camping was decidedly a risk for those unaccustomed to the wild Lowveld.

One thing the Park did have in abundance was thornbushes – so the rangers built thornbush enclosures inside which visitors could erect their tents and camp around their cooking fires. It was primitive, but perhaps the best way to bring the public close to the nature the authorities wished to preserve – and it was an immediate success.

In 1927 the first motorists entered the gates for a fee of £1 per car and in that year the Park cleared the munificent sum of £3. One of the first cars to arrive had as a passenger seven-year-old Douglas Jackaman, who liked the Park so much he later joined its staff as a tourist officer.

Soon they were coming in their scores, then hundreds and the Park began building accommodation for them in the form of the traditional, round, earth-floored, thatched huts known as *rondavels*. In 1930 Stevenson-Hamilton esti-

Every meal was a wonderful family picnic

A flooding river could not stop our morning drive.

Stevenson-Hamilton after a lion shoot.

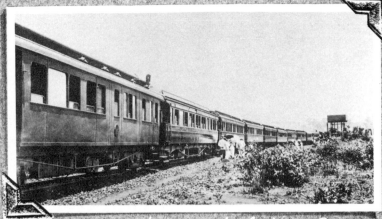

A tourist train at Sabie Bridge.

Men of Steinaecker's Horse at Komatipoort.

Stevenson-Hamilton with "Ballboy".

Harry Wolhuter and his dog.

Snapping game is really quite easy.

mated that the Park had about 100 of these. "Unfortunately," he wrote, "the builder suffered from a severe lion complex.

"He pictured the lions as ravening monsters, whose most ardent desire in life was to rend in pieces defenceless tourists. So instead of having windows, by which of course the lions might easily enter, I arranged that all the walls should not meet the roof, and that there should be an air space of about a foot between, which it was considered the lions might haply not observe, while seeking entrance elsewhere.

"To serve the double purpose of lighting the interior and to allow the inmates to gaze fearfully out upon the raging but baffled beasts outside, a round hole was cut in the middle of each door.

"In practice, I regret to say, that these holes were used rather by inquisitive persons desiring to see what was going on inside them than by the inmates for the purpose intended, and were generally found stuffed up with old socks, shirts or paper."

There are still some of these old huts in the Park, at Pretoriuskop and Balule, now minus the holes in the doors and equipped with rather better facilities.

Even with huts, Park life was far from luxurious. The beds in them were simple wooden frames carrying "springs" of waterbuck or wildebeest thongs and, initially, no mattresses. All cooking was done over open fires and water had to be fetched from wells or rivers – not a chore for the nervous at night.

Pretoriuskop, close to the Park's western border, soon became the most popular area for tourists and a permanent camp was built there which by 1931 had become so large that a full-time camp manager, Captain M. Rowland-Jones, was appointed. At about the same time it was decided that Pretoriuskop should remain open throughout the year – because it was free of mosquitoes and thus the risk of malaria, and its road network was good – while the rest of the Park opened only in winter.

Viewing his "Cinderella" (as he liked to call it) in the late 1930s, Stevenson-Hamilton had some belated second thoughts about ensuring its survival by attracting tourists. Tourism had come

Left: Portrait of a conservationist extraordinary – Colonel James Stevenson-Hamilton.

to stay, but had brought with it some of the ugliness of cities and civilisation.

By this time the Park had furnished huts, catering, stores, bathrooms with hot and cold water, service stations, some electric lighting and a telephone office.

"During the height of the tourist season," Stevenson-Hamilton wrote, "processions of cars cloaked the roads in clouds of dust; on Bank holidays merry trippers, packed in crowded lorries, shouted and discarded bottles and orange peel as they careered along; at night the rest camps resounded with the strains of broadcast music, of gramophones, and of community singing."

Initially nobody knew quite how the wildlife would react to the noisy, smelly vehicles of those days, and visitors were allowed to bring in one gun per car for their own protection. The guns were sealed at the entry gates and if a seal was broken, the owner had some explaining to do.

As it happened, the animals virtually ignored the cars, presumably regarding them as just another strange and noisy species, but when people broke the rules and stepped out of their vehicles, the animals instantly became alert and wary.

In 1927 the grand total of visitors was three carloads. The next year it rose to 180 and in 1929 to 850. In the first two

A fine photograph of Satara camp in the early 1940s. Note the tent area to the right of the picture.

years they could visit only the Pretoriuskop area and in the third the Olifants river as well.

Frequently they descended upon the rangers for accommodation, not a very happy situation because the rangers usually had to give up their beds to the visitors. One remarked to Stevenson-Hamilton: "I didn't mind so much their using my soap, towels, plates, knives and forks, but I do wish they hadn't used my toothbrush."

The Kruger National Park was then a raw wilderness without any of the amenities like supermarkets, restaurants, filling stations and luxury accommodation which it offers today. Roads were so rough that there was no problem with speeding drivers. Night travel had to be stopped, however, because so many animals were dazzled by headlights and killed.

The Park's staff began to build simple accommodation and by the end of 1929 there were 18 rather primitive huts at various camping sites. Nearly 200 km of roads were completed in 1928, 617 km by the end of the next year and by 1936 the network had grown steadily to more than 1 400 km. Tourist traffic rose apace to over 6 000 cars and trucks in 1935 carrying about 26 000 people. By 1938 the traffic had grown to about 10 000 vehicles carrying more than 38 000 visitors.

That year was the Park's best before World War II and saw many people visiting from overseas as well as from South Africa and its neighbouring territories. By then the amenities had so improved as to include hot and cold running water in comfortable bedrooms, clean sheets, electricity and catering.

The visitors contained, inevitably, some who wanted luxury and scorned the facilities, writing letters to the press criticising the "sordid conditions", the "hard life unsuitable to civilised human beings" and the "utter absence of all comfort".

But they were a small minority and the Park's popularity grew. After World War II the tourist numbers soared, placing much strain on the Park's few camps with limited facilities and its mere 1 600 km of gravel roads. Despite these shortcomings it hosted 38 376 visitors in 1946 and in the mid-1950s the number passed the 100 000 mark for the first time.

The authorities were overtaken by their own success. They suddenly found themselves not only looking after wildlife but increasingly involved in the business of catering for people – most of them amiable but many impatient, intolerant, sometimes arrogant and often totally ignorant of basic rules of behaviour in the bush. The park staff were, in short, up to their necks in the hotel business.

New and bigger camps had to be built as air, road and rail transport to the Lowveld improved and brought yet more visitors. Camp facilities had to be expanded to meet all needs and tastes, from those of the campers to the seekers of five-star comfort. Power and water had to be brought in, offices, shops, restaurants and bathrooms built, and many more roads carved through the bush. Within the space of a decade the Park operation had expanded far beyond the wildest dreams, or nightmares, of Stevenson-Hamilton.

The tourist flow rose to 101 058 in 1955, 216 680 in 1964, 304 346 in 1968 and in 1980 to 428 840 people who came in 108 157 vehicles. It peaked to 463 853 in 1981, then dipped, but rose again to 451 780 in 1983 and in 1984 topped the half million mark with 509 173.

The Park had a couple of setbacks during its early period of growth, the worst possible being bad publicity caused worldwide by a single incident.

In March of 1929 a special train arrived at Crocodile Bridge carrying a party of American tourists from one of the luxury cruise ships then visiting South African ports. Half of these visitors were loaded aboard the two trucks brought on their train to make the 60 km trip to Lower Sabie and back in the morning, with the second half due to make the same journey in the afternoon.

The day was hot and still and the local ranger warned that the weather could break, but the railway staff accompanying the tourists decided to send off the first batch anyway. A few kilometres later a tropical storm burst over them, turning the road into mud. They tried to

THE REMARKABLE MRS LLOYD

If life in the remote bundu is difficult for the ranger's wife of today, it is luxury compared to that of 60 years ago.

In the 1920s the pioneer head ranger, Colonel James Stevenson-Hamilton, hired a new ranger, W. W. Lloyd, and stationed him at Satara camp, which was then entirely isolated and accessible only on foot or horseback or by wagon across trackless veld. He took with him his very young wife and their three small children.

Lloyd appears to have been something of a novice to Lowveld conditions. He was so energetic, Stevenson-Hamilton wrote, "that he quite forgot that he was 56 and not 26 years old."

Being an energetic man with a good horse, Lloyd often patrolled great distances in a single day and sometimes slept out in his wet clothes, which consisted of lion skin pants, a thin shirt and a hat.

"It was hardly to be expected that, acting thus, poor Lloyd would last long in the Lowveld climate. It appeared that having arrived at home one hot day, after a long ride, soaked with perspiration, he sat on the veranda till he cooled, and not surprisingly contracted pneumonia."

Lloyd's wife, who sadly remains unnamed for she was an amazingly brave woman, did what she could for her husband but, in that remoteness and with no medicines, this was nominal and Lloyd stood little chance of survival.

Stevenson-Hamilton was awakened at 3 a.m. one morning by a weary messenger who had run and walked nearly 90 km in 24 hours through lion-infested country bringing him a brief note from Mrs Lloyd. It said simply that Lloyd had died the previous night and that she was alone at Satara with her children and a black policeman.

Leaving his Sabie Bridge headquarters before dawn, Stevenson-Hamilton reached Satara at the end of that day on horseback to find that Mrs Lloyd, with the help of the policeman, had already buried her husband under a tree near the house. (The grave is still there, off a path between the camp's tourist area and the section ranger's house. Park legend has it that Lloyd's headstone, placed there some years after his death, was put at the wrong end – at his feet.)

"Anything more pathetic than the situation of this young woman, absolutely isolated as she was in the savage wilderness with her three children, and with her husband lying dead in the house, it is difficult to conceive," Stevenson-Hamilton wrote in admiration, "but she rose to the occasion and throughout maintained an attitude of level-headedness and composure."

His wagon arrived two days later and he packed on it and on Lloyd's donkey cart all the family possessions, with the poultry hanging on top, and set off for Sabie Bridge with horses, donkeys and dogs.

The head ranger noted that in rain, shine or when it was over 40 °C (104 °F) in the shade, the three small Lloyd boys wore only a single cotton shift.

"Clearly, in their contempt for external conditions they were the true sons of their father."

What became of the remarkable Mrs Lloyd and her children is not known.

Mr J. H. Orpen and his wife Eileen, the two best-known benefactors of the Park. They donated seven farms comprising 24 500 hectares between 1935 and 1944.

turn back but found a dry creek they had crossed was now a roaring torrent. One truck tried to make the crossing but the force of the water tipped it on its side, blocking the road totally.

By now the group was thoroughly sodden and as the day dragged on they heard lions roaring. The tourists decided on discretion and climbed whatever trees they could find. Most of these, unfortunately, were full of thorns, which did little for the already bedraggled summer dresses of the womenfolk.

There, shortly before dark, the Crocodile Bridge ranger found them. He ushered them into a couple of huts at Gomondwane, lit fires inside and left the women in one hut, men in another, to dry out. They stayed the night and left the next day, no worse for wear and excited by their adventure.

Some, however, contracted malaria, which led to lurid accounts in American newspapers of the Park as a death trap. Sensibly, the National Parks Board decided to close the area to the public thereafter during the wet season, except for Pretoriuskop which was fairly free of mosquitoes.

Mosquitoes, in fact, were by far the main hazard for visitors. Despite the breaking of safety rules by people leaving their cars to try to take pictures of lions from close up (which usually caused the lions to move off smartly), despite their

Tourist cars leaving Skukuza camp across the Sabie river by ferry. The water level in all Park rivers is generally much lower today.

teasing the seemingly "playful" baboons and despite all the other idiocies of people contemptuous of regulations yet ignorant of wildlife, there was only one serious incident between animal and man in the first eight years after the Park opened to the public.

The casualty, Stevenson-Hamilton recorded, was a man who saw a sable antelope bull standing behind a tree and left his car to get closer for a photograph. A sable is recognised by experts as probably the most dangerous of all antelope, feared even by lions for the deadly accuracy and speed of its huge, curved horns driven by powerful shoulders. The visitor apparently did not know

that, nor that this particular sable had been injured in a fight with another – so badly that it died a few days after this incident.

The man was about six paces away when it lowered its head and charged. It drove one horn through his thigh and certainly would have killed him had not his wife and a servant come running to his rescue and driven it off by hurling sticks and stones.

Initially tourism was confined to the former Sabie Game Reserve region of the new Park, but in 1931 Stevenson-Hamilton decided to open up the Shingwedzi area because of its different character, terrain and animal species which were rare in the south, like elephants, roan antelopes, tsessebe, eland and nyala.

A new road had been carved northwards and soon visitors had access to a broad spectrum of countryside and wildlife, spiced by such adventures as crossing rivers like the Olifants, Sabie and Crocodile on ferries – fairly primitive platforms supported by pontoons on which they had to gingerly drive their vehicles to be hauled across to the other bank. The last of these was replaced by a bridge only in 1945.

Eventually the road reached the northernmost ranger station, Punda Maria. The name is curious. It was given by the first man stationed there, Captain J. J. Coetser, in 1919 and the common belief was that it was the Swahili name for a zebra, *punda milia*. In fact, it was a

Peter Pan of the Game Reserve,
In a house with a tiny door;
Never to grow any older;
For always to nearly be four:
To have her own wee white garden;
To play as the wild things play;
To care for baby lion cubs
While their mother has strolled away:
Where gnomes, and elves, and fairy folk
Are a matter of every day.
So she joins with ours her wishes,
For her little day is done.
Dawn of opal and gassamere
Vanishing into the Sun.

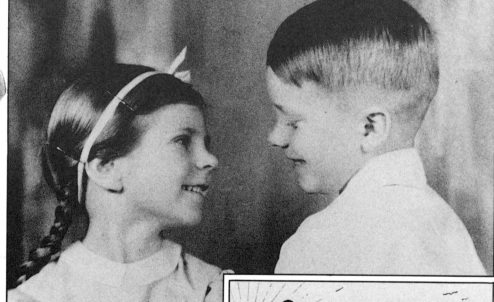

THE FIRST WARDEN'S WIFE

Hilda Stevenson-Hamilton, born Cholmondeley (pronounced Chumley) in Australia in 1901, was a tall, majestic woman who towered over her short husband. Through all the years she was always his most vigorous supporter in his drive to put the Kruger National Park on the world map.

She studied fine art in London and some of her paintings now hang in the Stevenson-Hamilton Memorial Library at Skukuza. To mark family occasions she created personal family greeting cards, usually depicting her children surrounded by friendly wild animals. Some cards were poignant such as one sent to friends on the death of her small daughter Margaret **(top)**, another on the retirement of her husband as Park Warden **(right)**.

Mrs Stevenson-Hamilton was also an accomplished photographer and in the days of 16 mm movie cameras, captured a famous sequence of lions catching a leaping impala. The photograph of her children Jamie and Anne **(above)** is said to be hers.

She met Stevenson-Hamilton in London in 1928 when he was 61 years old and recuperating from a serious bout of malaria. They married two years later and had three children, one of whom died of meningitis.

combination of this and Coetser's wife's name, Maria.

After roads, the Park's main need was water – for both people and animals. Naturally a fairly arid area, away from the main rivers and pools it was practically waterless in the dry season. The result was that wildlife crowded to these watering points and for kilometres around them grazed the veld bare. However, where there was no drinking water there was still plenty of grazing. The answer, obviously, was for the Park administration to provide more watering points, but this was not easy.

The problem became serious in the long drought periods of the 1920s and 1930s and the first to respond to the crisis was a Mr J. H. Cloete of Clocolan in the Orange Free State, who offered money for a borehole and windmill. A regular visitor to the park, Mr Bertram Jeary of Cape Town, then took up the cudgels and launched a nation-wide campaign with the help of newspapers to raise money for more boreholes.

The response was overwhelming. Money poured in from many sources, ranging from individuals to municipalities, including Mr J. H. Orpen, a member of the National Parks Board, and his wife Eileen who also, between the years 1935 and 1944, bought and gave to the Park seven adjacent farms totalling 245 km² (95 square miles) – the biggest gift to the Park yet.

By the end of 1935 two drills had sunk about 20 boreholes from which windmills were pumping between 1 400 and 9 000 litres of water an hour (300 to 2 000 gallons). So effective has the public "water-for-game" campaign remained since then that today the Park has more than 390 borehole sites, some with two or more windmills and many with concrete storage dams, plus 12 large dams and about 50 smaller catchment dams in seasonal watercourses.

So pleased was Stevenson-Hamilton by the public's rally to the Park's need for water that he said: "The immediate response made by the public was the most encouraging thing that had yet occurred in the history of the Park. It showed that people were genuinely interested in it ... that they regarded it as a national asset."

Oddly, the reaction of the wild animals was not always so euphoric. The lions loved the new concrete troughs beside

Colonel Stevenson-Hamilton with is wife Hilda after his retirement as Park Warden.

the windmills and would lie around them all day, lolling in the sun, occasionally drinking and, of course, deterring many other creatures from coming too close.

In any case troughs were not quite the thing for those animals which liked to wallow and paddle as well as drink, so Stevenson-Hamilton had the borehole water pumped instead into natural pools where the animals could feel free – and lions did not like the muddy edges.

At first other animals would not use the new waterholes in the dry seasons. Genetic memory ingrained by the habit of generations would make them trek every dry season to the familiar old natural waterholes, there to compete with all the other animals for the grazing and to die when it ran out, although the veld around the new artificial watering

points was lush with food. Today, however, they have become accustomed to the new system and use it.

During 1938 foot-and-mouth disease swept in from Mozambique and to combat it thousands of head of cattle were destroyed in the Lowveld. Wild animals were spared by the veterinarians but because they might be carriers, all cars travelling within the Park had to drive through troughs of disinfectant and visitors had to walk across sacking doused with it.

World War II began in September 1939 and there was immediate panic in South Africa generated by rumours, later proved to be groundless, of German agents slipping through neutral Mozambique to cross the long, unguarded border into South Africa through the Kruger National Park.

The war years were good for the Park. Poaching virtually ceased because ammunition was unavailable, being needed elsewhere to shoot people. More visitors than ever came in that time and many were foreign soldiers stationed in or passing through South Africa, so the Park's fame spread wide.

The war's end heralded further major changes, the biggest of these being the retirement at the age of 79 of the redoubtable James Stevenson-Hamilton after 44 years of work in which he had steered the Kruger National Park to becoming one of the world's greatest attractions.

Just before stepping down in 1946, he expressed deep concern that the Park might become a "glorified zoo and botanical garden, dotted with scientific experimental stations of every kind, hotels and public recreation grounds, which are all preliminaries to the liquidation of the last vestige of wildlife.

"The first object should be to educate the public in the rudiments of natural history; to show people what the wild animals of their country look like and how they act in their natural state, free from the terror of Man; to cultivate a spirit of sympathy with them; to let it be revealed that they are more admirable alive and in their natural setting . . . than converted into the rags and bones of hunters' trophies or confined, listless prisoners, behind bars."

Of course, the Park has become so renowned that it has to cater for many more people than in 1946. But in doing so it has spread the knowledge and appreciation of wildlife, as Stevenson-Hamilton urged, to millions of people the world over. Outside the luxury of the modern camps the wilderness is just as wild, the lions kill as they always have and the elephants remind you to keep your distance.

Stevenson-Hamilton died 11 years later at the age of 90, after a full and rewarding life, on 10 December 1957, leaving as his permanent memorial the Park's main camp, Skukuza. His ashes were scattered on his favourite place, Shirimantanga hill south of Skukuza, as were those of his wife Hilda, who died in 1979.

After World War II

Stevenson-Hamilton was succeeded by a South African Air Force officer from Cape Town, Colonel J. A. B. Sandenbergh, whose experience came from hunting and from owning a large, well-stocked game farm in the Lowveld. He held the post for seven years.

Sandenbergh too faced challenges, though none as formidable as his predecessor's. The Park's popularity was growing apace but it had only 13 rest camps and less than 1 600 km of roads, all gravel-surfaced. The Park was completely unfenced, which meant that game wandered in and out at will, all too often to be shot outside. Water came from just eight dams which held water for any length of time, plus 13 boreholes and eight wells equipped with windmills. Except for Pretoriuskop, the entire Park was closed during the summer months. Catering was poor and accommodation was limited and not always of the best.

At the time Sandenbergh took over, the Lowveld was gripped by drought so severe that the Letaba river stopped flowing for the first time in the Park's history. He launched another "water-for-game" campaign, which tapped a further flood of money that paid for another 46 boreholes. These brought some relief but, ironically, the drought persisted until Sandenbergh resigned in 1953.

One of Sandenbergh's major contributions to the Park was administrative:

. . . today these links extend worldwide, for conservation knows very few political boundaries.

he urged that specialised jobs be done by specialists, from road building to research. Before his tenure rangers performed virtually all duties.

The kind of *ad hoc* Park management in which decisions of necessity had to be made on experience – then the only yardstick available – was eventually ended by a major step taken by the National Parks Board during Sandenbergh's tenure.

In October 1950, the Board set up a research section in the Park staffed by

scientists of various disciplines, whose task it was to investigate the natural processes behind the ecology of the Park, and to interpret these so the knowledge could be used in running the Park.

Biologists, veterinarians, ecologists, entomologists, ornithologists, botanists and others were brought together over the years with a strong team of technicians and modern technical facilities. So began the vast and complex research programme which the Park continues to operate and which has changed the whole character of its management.

Today decisions on when and where to burn off old grass are based on topographical, climatic, animal migratory and all sorts of other detailed information meticulously garnered on scientific levels and all assimilated by computer. The same goes for animal translocation, culling, road building and the siting of new camps.

The selection process for the Park's managers has also altered in this process. Gone are the good, simple old days when a man needed to be only a hunter or to have lived long enough in the bush to qualify as a game ranger. Such experience is still valuable, in fact essential, but now it must have an overlay of skills which can be gained only through higher education and training.

The research unit soon established links with many universities and other institutions to collaborate on a large number of projects and today these links extend worldwide, for conservation knows very few political boundaries.

This approach necessitated a revision of the Park's administrative structure and in 1952 the Board appointed a team of accountants headed by Professor P. W. Hoek to do that. The Hoek Commission recommended the appointment of a full-time Director of National Parks.

The man chosen out of more than 80 applicants from all over the world was Mr Rocco Knobel, son of a Bechuanaland (now Botswana) missionary. He had a sound administrative background, a deep love of nature and pronounced leadership qualities and directed the country's national parks for the next 26 years.

This step left the warden of the Kruger National Park purely in charge of conservation and, to further relieve him for this task, a tourist manager was appointed.

In 1953 a senior ranger, Mr L. B.

Col. J. A. B. Sandenbergh.

Mr L. B. Steyn.

Mr Rocco Knobel.

Mr A. M. Brynard.

Prof. Fritz Eloff.

Steyn, succeeded Sandenbergh and the Park increasingly adopted the Sandenbergh-initiated new style of scientific management. Veld-burning experiments began, camps and roads were improved, the trading stores and restaurants, until then run privately, were taken over by the Park, new research projects were started and a campaign launched to wipe out all non-indigenous plants in the Park, such as syringa, lantana, prickly pear and sisal.

The mammoth job of fencing the entire Park began in 1959 – a highly expensive business because fences had to be game-resistant, and that meant deterring even such giants as elephants, which can push over large trees with casual ease. The southern boundary was fenced first, followed by the western and northern boundaries and finally the eastern boundary – 1 800 km in all and the longest game park fence anywhere in the world. On the eastern or Mozambique boundary the elephants have been effectively barred passage by an extraordinary fence whose posts, over 2 m tall, are lengths of railway track deeply embedded in concrete. Strung between them as "wires" are 2 cm thick steel cables of the kind used on the gold mines to hoist skips up and down their vertical shafts.

Continuous control by shooting off the lion and other carnivora stopped in 1958, to be replaced by a policy that such control would only be used in special circumstances. In 1959 an aircraft was used for the first time to count wild animals – and, much to the surprise of the Park staff, they found that the elephant population numbered almost 1 000.

Warden Louis Steyn retired in 1961

to be succeeded by the chief biologist, Mr A. M. Brynard, who was made nature conservator, with the rangers and researchers united under his control into one Nature Conservation Department. At the same time Henry Wolhuter, son of the famed Harry, became senior ranger.

Dr U. de V. Pienaar became chief biologist and in 1970 became nature conservator when Brynard was made assistant director of National Parks. In 1978 Dr Pienaar was promoted to chief warden of the Kruger National Park and a year later, on Knobel's retirement, Brynard succeeded him as chief director of the National Parks Board. Knobel was subsequently made a Commander of the Order of the Golden Ark by Prince Bernhard of the Netherlands for his work.

The Chairman of the National Parks Board, Fritz Eloff, recently retired from the post of professor of zoology at the University of Pretoria.

His scientific work includes a 26-year study of the Kalahari desert lions. He is also vice-president of the South African Rugby Board.

Today a team of research scientists, backed by nearly 300 rangers and field staff, continue their delicate and absorbing task of maintaining and improving the stability of the Park's ecosystem so that within it nature can follow its own course as normally as possible.

Guarding the future

At all times the Park – still powerfully backed by the local Wildlife Society, the World Wildlife Fund, the International Union for the Conservation of Nature and similar bodies – is sensitively alert to threats from outside.

Some it has had to submit to, such as the twin row of power line pylons that march across the far north of the Park from Mozambique's Cahora Bassa dam to South Africa's internal electricity grid. They represent visual pollution but they do not materially affect the life of the Park.

Other such threats the Park has successfully staved off – so far. The biggest of these was a government-backed scheme for South Africa's Iron and Steel Corporation (Iscor) to mine coking coal in the north of the Park close to the Luvuvhu river.

"No," said the Park's authorities and South Africa's conservationists, with strong public opinion supporting them. The original National Parks Act included a specific clause that "no prospecting or mining of any nature shall be undertaken on any land included in a Park."

"Yes," said Iscor and the government, because coking coal is of strategic value to South Africa.

To date the Park has won the argument and Iscor has, in the face of public opinion, turned its search for coking coal elsewhere. But the threat remains.

So too do the dangers from closer by. The Lowveld outside the Park is a fast-developing region of agriculture and industry. All the rivers traversing the Park run through this region first, with the result that the river water in the Park is becoming increasingly polluted with factory effluents. The water, in fact, is steadily diminishing as industry and farming inland use more from the rivers.

The struggle for the Park's survival is therefore extending well beyond its borders. But that is a new challenge for coming generations using new tactics.

Maintaining an ecosystem

"Those placed in charge of the great Kruger National Park should take care that development, improvement and scientific research do not form cloaks for the sinister word exploitation," wrote the mentor of the Park, Colonel James Stevenson-Hamilton, shortly before he retired as its first warden.

His philosophy was that of a true wilderness lover and dweller, suspicious of the modern industrial age with its hunger for profit and development. It was ingrained in him after 44 years of an ultimately successful struggle to ward off those who would have abolished the Park and its tremendous wealth of nature simply for profit.

But no national park anywhere – especially one as large as the Kruger National Park and which is not a completely self-contained ecological unit – can survive without expert knowledge. And that demands scientific research plus scientific skills to employ the knowledge it yields.

"In spite of Colonel Stevenson-Hamilton's mistrust of a certain type of 'scientist'," wrote the present warden, Dr "Tol" Pienaar, "the National Parks Board became aware of the fact that a huge wilderness such as the Kruger Park cannot be managed properly without scientific knowledge and guidance of the correct kind. With this in mind the Board created its own research station in the Park (based at Skukuza) in 1950 and staffed it with research personnel with the required ecological training.

"It is certainly true to say that some of Stevenson-Hamilton's successors, if not all, erred in some of their management policies, but their errors of judgement were made in good faith and were certainly not irrevocable.

"Improvements which have been made and development projects undertaken have rarely been aimed at nature itself

Left: A close-up view of one of the Park's more truculent characters, the buffalo.

unless they were steps to perpetuate the quality of the habitat and safeguard it from destructive influences . . ."

Today the Park has a resident research staff of some 16 and about 25 assistants representing a wide range of scientific disciplines who maintain continuous investigation into almost every aspect of the Park's life, from the internal parasites of zebras to animal migrations, to carnivora control, to plants and the insects that feed upon them.

The range is enormous, and to support their research the scientists have the collaboration of many leading universities and other investigative institutions.

The main job of the Park's research unit is to define and explain the natural processes governing the functioning of the Park's ecosystem, states Dr Pienaar, so that the Park staff can accurately plan the Park's management.

The question has often been asked: Why manage the Kruger National Park at all? If nature was able to look after it successfully for so many thousands of years, why not trust nature to look after itself now?

The answer is simple: the Park is no longer part of the far larger wilderness that existed before the advent of modern man, where nature was able to sustain and balance itself because the sheer size of the area made it able to withstand localised calamities.

Then, animals could migrate freely over great distances if there was drought or a population explosion. Now they cannot because the Park, despite its huge size, is still a very small part of that former vastness and is sealed off by the creations of man: agriculture, industry and national boundaries.

Warden Pienaar wrote: "The maintenance of the diverse habitats comprising the ecosystems of the Park in a sound and productive state and the preservation of the various unique qualities which led to the original proclamation of the conservation area, represent major objectives of the Park's administration.

"To achieve this the conservation staff

Dr U. de V. "Tol" Pienaar, present Chief Warden of the Kruger National Park.

entrusted with the task of administering this national heritage have opted for a management strategy from the earliest days, and not for the option of *laissez-faire*."

For example, he states, a "leave it to nature" policy of providing water for the wildlife would have been tantamount to reckless irresponsibility because in long periods of severe drought, such as that of 1961 to 1970, many rare and irreplaceable speices might have been lost.

"It is considered the most responsible and the wisest choice, in the management of such conservation areas, to intervene in such a manner that both animal population peaks of an explosive nature and population declines of 'crash' proportions are guarded against. This ensures a healthy and productive environment which can accommodate long-term natural changes, is conducive to relative stability and ensures regulating strategies of manageable proportions.

"The primary objective of the management plan adopted in the Park is therefore conservation measures aimed at maintenance of the quality and diversity of the habitats which make up its ecosystems, but this is achieved with minimum interference in natural processes and regulating mechanisms."

The result of a management policy based heavily on scientific research is a healthy and productive environment still able to cope with long-term natural changes.

When the Park's research department was begun in 1950 the ecosystem was virtually unknown, so initial research concentrated on compiling inventories of animals and plants to provide a baseline or starting point. This led to the founding at Skukuza of the internationally famous Stevenson-Hamilton memorial reference library, museum and herbarium situated at the research complex, which houses all the data on the different forms of life found in the Park so far. It is by no means a complete list, as new life forms are still being discovered.

When they had identified and documented the most important ingredients of the ecosystem, the scientists then switched their research to studying the lifestyle of the ecosystem itself. This was done in three phases: collecting information and storing it for long-term monitoring, finding out what made the ecosystem tick, and closely observing the management based on these studies to see how efficient it was.

Management rests on the control or manipulation of four factors – water, fire, populations and disease – but in broader terms conservation rests on seven pillars: boundary control, soil conservation, water conservation, grazing control, disease control, population control and protection against man, as one Park expert put it.

The new generation of scientific managers made fencing the Park a priority, not only to avoid the possibility of disease – primarily foot-and-mouth – spreading to adjacent farming areas, but also to prevent crop and other damage by wild animals, to facilitate anti-poaching operations and generally to clearly define the Park and improve control.

"In the beginning the fence was

An aerial census team of rangers and scientists is briefed before take-off by chief research officer Dr Salomon Joubert (left), who is also the pilot.

breached in many places," says Warden Pienaar, "but gradually the animal populations in the Park adjusted to the artificial barrier and adapted their seasonal movements within the closed system of the Park.

"The complete fencing of the Park brought with it a new dimension in the management of this conservation area, as the man-made obstruction effectively prevented animal populations from executing their traditional migrations or seasonal movements and also prevented them from escaping the effects of natural regulating mechanisms such as disease epidemics, local pressure from carnivora, water depletion during times of drought and lack of grazing following extensive dry spells or fires."

The result was a steep increase among those animal populations with few natural enemies – like elephant, buffalo and hippo – and consequent overgrazing. But within the new management policy the Park staff were able to counteract and control this effect.

"The fence obviously imposed a finite limit to the number of animals which can be supported by the enclosed natural resources of habitat, food supplies and water and accentuated the need for sensible regulation of animal numbers," Warden Pienaar wrote.

After much deliberation the Park management decided they could not leave nature alone to control populations

by starvation, thirst, disease and other means. Experience elsewhere in Africa had shown that unnaturally high populations of elephant could devastate the environment, as happened in Kenya's Tsavo, Zambia's Luangwa Valley and Botswana's Chobe National Park.

Taking stock from the air

Stocktaking is as much a part of business in a park as it is in a supermarket. But how to count heads of even such large animals as elephant, hippo, buffalo or lion in a place as big, bushy and rough as the Kruger National Park? Early counts were largely guesswork – patrolling by foot or vehicle to count animals within sight and then extrapolating the tally to a larger area, in which the particular animals were known to move, to obtain a vague idea of their numbers.

In 1959 the Park made its first aerial count from a light aircraft. The survey was incomplete, but it recorded 986 elephants – many more than expected.

In 1964 the first count of elephant and buffalo was made using a helicopter – the ideal aircraft for the job – and Park staff discovered they had 2 374 elephants and 10 514 buffaloes.

Today the Park operates a fleet of four aircraft, two Cessna fixed-wing aircraft and two Bell Jetranger helicopters.

Counting from the air is a sophisticated operation demanding more than a quick eye and rapid arithmetic. Some rangers who are susceptible to air sickness hate it, especially on hot days when flights are bumpy.

Two types of aerial census are done annually. One is the census of elephant, buffalo and hippo for which a helicopter is used because of its tremendous manoeuvrability, which makes a high degree of accuracy possible.

The helicopters fly less than 200 m above the ground; this has been found to be the best height for counting these animals. There is little margin of safety at this height, however: if a Jetranger suddenly loses power at below 200 m it becomes difficult if not impossible for the pilot to put the machine into auto-rotation, in which the rotors are kept spinning by the downward pressure of the air, and drops the helicopter reasonably gently – like the falling seed pod of a propellor tree.

Should the buffalo herds be particularly large (they often number well over a thousand), photographs are taken and the count done later from blown-up prints.

The count of elephant and buffalo is broken down into breeding herds with young calves and solitary bulls.

All the other species are counted – and a wide spectrum of environmental conditions recorded – from a six-seater Cessna aircraft. In this survey the Park is criss-crossed in parallel strips about 800 m wide and ideally not longer than about 10 km, flying at heights of 60 to 80 m above ground. Taking part in this census are a research officer, recording the vegetation conditions, water distribution and the position of animals on a 1:100 000 scale map, and four scientists and rangers who do the actual counting. They pride themselves on their accuracy and seldom differ by more than one or two, even with a big herd.

The observers aim for a total count of each species but realise this is impossible because of habitats, different colourations and other limiting factors. However, they make a considerable effort to achieve consistency from year to year, ensuring that the same proportion of each species is recorded.

Each transect takes about three to five minutes to cover. Counting becomes tricky when groups of more than 30 to 40 animals are found because the pilot must circle to allow the rangers to com-

plete their tally. This is when sensitive stomachs get queasy.

All the major herbivores – sable, roan, tsessebe, eland, waterbuck, kudu, giraffe, impala, wildebeest, zebra, warthog, white and black rhino, as well as ostrich and the turkey-sized ground hornbill – are counted and their positions plotted.

Other features recorded include grass height, plant cover, how green or dry the veld is, how much dead vegetation is about, the intensity of fire damage, the distribution of water, and elephant damage to vegetation.

All information is fed into the Park's computer, which prints out a map on which varying degrees of shading give the density of animals and other relevant features. With this and a host of other data the Park executives then decide whether there is overpopulation by any kind of animal, and whether the excess game should be culled.

Killing to save

The deliberate killing of animals to save their own and other species is, and always will be, a highly contentious issue, since it appears to clash violently with the principle of conservation: to preserve life.

It is even more contentious when the culling is so effectively organised that a park derives revenue from it. But culling

is unavoidable if a park is to survive and it is, in fact, no different from a cattle rancher sending his excess stock to the abattoir to limit his herds to the optimum carrying capacity of his land.

The routine practice of thinning out lions and other carnivora in the Kruger National Park by shooting stopped in 1958 amid torrid public debate. The authorities ruled that thereafter lions would be shot only if there were so many that they were threatening the survival of other species and then only on a temporary and localised basis.

A tragic event in the dry year of 1961 demonstrated the necessity for humane culling in extraordinary circumstances. The Letaba river virtually ceased to flow, forcing the hippos to crowd into a few remaining pools, and face a lingering death by starvation because their concentrations had so denuded the surrounding grazing. The Park staff had no alternative but to destroy some to let others survive. They shot 104 – the first animals to be killed in the Park because of overcrowding.

This was done on an experimental basis to determine population structure and dynamics on which to base a strategy for culling if and when required.

On 30 November 1965, the National Parks Board held a conference in Pretoria to discuss overpopulation in nature. The conservation experts present adopted a

This is a small section of an aerial census map after a morning's count. Each letter represents an animal species, with the numbers indicated next to it. Most are identified in Afrikaans such as: V – Vlakvark (warthog), K – Kwagga (zebra), O – Olifant (elephant), KP – Kameelperd (giraffe), KO – Koedoe (kudu), W – Wildebeest, T – Tsessebe, SR – Swartrenoster (black rhinoceros). Bracketed numbers denote calves and underlined prefixes indicate bachelor bulls.

resolution that the Board should order the cropping of those herbivorous animals whose populations had risen to the point where they could no longer be supported by the Park's natural resources.

The dominant species which would have to be cut back, the Park authorities decided, were elephant, buffalo, hippo, wildebeest, zebra and impala – although the latter three would be affected only in times of drought. The scientists worked out the maximum permissible numbers of each and drew up a programme for them to be thinned out between 1966 and 1968.

It was a difficult decision but there was no alternative. Nor was it a precedent; other parks in Africa and elsewhere in the world had been faced with similar population explosions and had been forced to take similar steps.

In Uganda, East Africa, the cropping of hippo in the Queen Elizabeth National Park and of elephant in the Murchison Falls Park was already under way. Kenya was thinning out elephants in the Tsavo National Park and in South Africa's Natal province impala, zebra, wildebeest and warthog were being shot. In the world's oldest national park – Yellowstone Park in the USA – protection had allowed the elk herds to grow so much that they were being culled. Conservationists recently calculated that 25 000 elephants would have to be cropped in Zambia's Luangwa Valley out of a population of about 100 000 if widespread environmental damage was to be avoided.

The number of elephants in the Kruger National Park has been growing steadily since it was the Sabie Game Reserve, thanks to protection. In 1905 Stevenson-Hamilton found about 10 elephants sheltering in the then wild and inaccessible area between the Letaba and Olifants rivers, where they had escaped the attention of the ivory poachers who had all but eliminated the Lowveld elephant.

By 1912 the number of elephants had risen to about 25, by 1917 to about 135, and by 1946 to 450, according to Stevenson-Hamilton's last tally before he retired. In 1947 they had increased by about 100 to 550 and in 1954 the count was 740.

In 1959 the research unit estimated there were 1 000 elephants in the Park

Right: An artist's impression of an elephant culling operation in progress. A maximum of 15 animals are killed at a time.

and the 1964 aerial count showed 2 374. In 1967 there were 6 585 elephants, of which 11 per cent were under a year old. The next year the population reached 7 701 – an indication of the elephants' capacity for survival after almost being wiped out and also of how swiftly their numbers multiply the more of them there are. They can reproduce remarkably fast considering their gestation period of 22 months. Elephants reach sexual maturity early, have a low infant and adult mortality rate and can live to 60.

In the severe drought of the 1960s, buffalo numbers increased sharply and they and the elephants dominated the waterholes on the Lebombo flats to the detriment of the rarer grassland animals. Not that they might have wanted to share: after a herd of buffalo has trampled through it, a waterhole is distinctly unattractive to other animals.

Surveys at this time revealed that the drought was so severe that in the far northern part of the Park, in the Punda Maria and Shingwedzi regions, not one sable calf of the previous breeding season had survived.

Also in the 1960s, the number of elephants began to grow south of the Olifants and Sabie rivers at the same time that the number of tourists increased. There was initially some fear that elephants would clash with tourists because the south was the main visiting area. Nothing happened, however, and tourists and elephants continue to get along well together – provided the tourists do not take unnecessary risks.

In 1967 culling began, to keep numbers down to the scientifically determined maximum that the Park's resources can carry without suffering damage: about 7 500 elephant and 22 000 buffalo.

Public reaction was instant and angry: "It's a monstrous step . . . The beginning of the end . . . The death of the Kruger Park." Such were the angry letters to the Press condemning the Park's action and predicting that the word "sanctuary" would become meaningless.

The public was equally incensed by the National Parks Board decision that all meat and animal products stemming from culling operations had to be used

to best advantage, and for this purpose a by-products plant had been built on the Park's boundary near Skukuza. There, canned meat, biltong (dried meat), carcass meal, skin, ivory and trophies were to be produced from the culled animals under strict veterinary supervision.

This caused widespread public criticism because the idea of an animal products factory in a game reserve seemed so contradictory. But what were the Park authorities to do with the animals they killed? They could not simply leave them to rot.

Having committed themselves to a

policy of cropping excess animals, the Park authorities then had to find the quickest and most merciful way to do so. They tested two drugs which were commonly used in capturing animals, but these left a residue which was unacceptable under South Africa's strict Pharmacy Act, as the meat was to be marketed for human consumption. Also, the drugs were slow-acting and animals would scatter widely before succumbing.

Next they tried shooting the animals, the way culling is done in most other African countries. The rangers are by and large good marksmen and they used high-powered rifles to good effect; inevitably, however, some animals were merely wounded and escaped into the bush – particularly buffalo, which are not the easiest of targets. A wounded buffalo lurking in the bush intent on revenge is one of Africa's most fearsome dangers.

The researchers then came up with another, more practical, less traumatic drug: succinyldicholine chloride, or scoline for short. This triggered yet another public furore.

In experiments on humans scoline was found to paralyse the subject completely while leaving him mentally alert and fully aware of pain. There were frightening accounts in newspapers by people who had been injected with the drug and of their ensuing mental terror at being unable to move or talk. Medical and veterinary experts argued that scoline caused muscular paralysis and slow death by asphyxiation and was therefore unacceptable.

But the Park officials insisted that it was still the most efficient and humane method of immobilising and killing and, moreover, scoline did not contaminate the meat. The tactic adopted by the Park was to dart elephants from the air and as soon as they fell, have rangers on the ground quickly kill them with a bullet – if their hearts had not been stopped already by paralysis. Buffaloes presented no problem: they died from scoline overdose in little more than a minute after the dart struck home.

To allay public fear and criticism the National Parks Board invited the Wildlife Society of Southern Africa, the S.A. Veterinary Association and the International Society for the Prevention of Cruelty to Animals to send observers on a culling operation. They did, and the observers came away convinced that the darting-with-scoline technique was the best available to modern science.

Culling has, incidentally, greatly helped research on the biology of elephants and provided much information to scientists of many disciplines which will be used as an additional means of ensuring their survival. Since 1975 the Park's researchers have fed statistics on elephant population, movement and other details into a computer, with which they are now able to predict future populations accurately and thus plan control measures carefully.

In 1980 the Park darted 16 elephant bulls and an equal number of cows and attached radio collars to them so their movements could be monitored. They found that the cows each ranged over more than 400 km² where the habitat was good, but where it was poor they extended their domains to 700 km². The bulls' territories averaged 600 km², but one covered more than 1 000 km².

Such data plus information on water resources, plant life and other ecological factors enabled the Park to focus its elephant cropping very specifically on clans of elephants. Culling begins only after the scientists have determined how many must be killed and where, and only after both the warden and the National Parks Board have given their approval.

Culling is done discreetly in remoter areas of the Park well away from tourist routes so that visitors are not aware of it. In the late afternoon a ground crew of drivers, a supervisor, butchers, loaders and a veterinary department inspector will set out for the chosen site with a mechanical horse and trailer, a tractor with a power winch, a heavy duty tip trailer and other equipment.

They wait while a helicopter reconnoitres the area to pinpoint the elephants and then lands near the ground crew to pick up a marksman and an observer. The marksman sits behind the pilot, while the observer sits as a counterbalance on the port side of the helicopter. The right door is removed, and the shooter secures himself with a special "monkey chain" to avoid being thrown out during the aerial gymnastics the helicopter has to perform to stay close to the elephants. His weapon is a modified 20-gauge double-barrelled shotgun that fires aluminium darts.

No more than 10 to 15 elephants are culled in one operation, the maximum number that can be handled by the ground crew and animal products plant.

If the herd is larger, the pilot flies in low over them causing them to scatter, and typically they do so into small families which can survive independently. The team in the helicopter then chooses a group at random and the pilot herds it away from the others towards an area selected earlier for the darting, usually a clearing in the bush.

Ranging back and forth like an airborne border collie, keeping higher and behind the group when it is moving in the right direction, swooping in low to turn them when they veer off, the pilot steadily drives the elephants to the previously determined culling site.

When they reach it, the helicopter closes in low, flying over bushes and trees of wildly varying heights through which it has to dip and dodge. The marksman braces himself and fires the first dart from a distance of between 30 and 10 m and the second from as close as 10 to 3 m. While he reloads, the pilot circles the group to keep it close together and then comes in low again. The process is repeated until the whole group is darted, at which point the ground crew is called up by radio.

The operation takes no more than about 10 minutes and by then the deeply drugged animals lie within an area rarely more than 100 m in diameter. The supervisor moves quickly to shoot each of the elephants in the brain at close range just in case the drug has not already killed them, and after him comes the veterinary inspector who carefully checks the animals for any signs of disease.

The butchering staff take over and within 45 minutes the carcasses are on their way to the animal products plant.

Elephant bulls with good tusks are usually left alone during culling because they are tourist attractions. A survey in 1980 indicated that some three per cent of the Park's adult bulls have tusks of about 45 kg (100 lb) each.

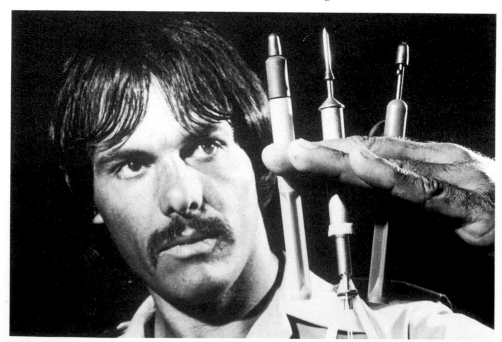

Three types of dart are used in the Park. They are, from left, a disposable mass-immunisation dart, a short-range large volume capture or culling dart, and a long-range capture dart fired from either air or ground. The latter two are usually recovered for re-use. Holding them is assistant instrument maker, Johan Jacobs.

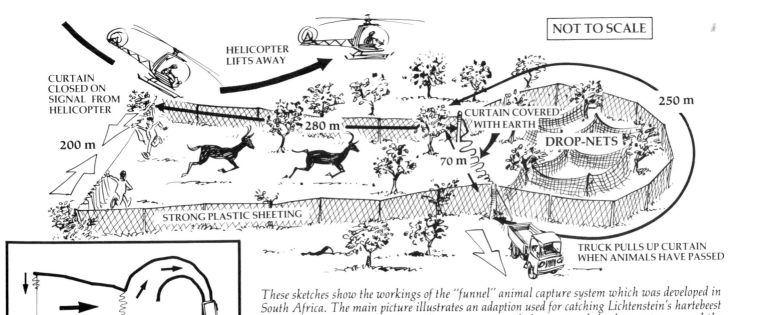

HELICOPTER LIFTS AWAY

CURTAIN CLOSED ON SIGNAL FROM HELICOPTER

CURTAIN COVERED WITH EARTH

250 m

DROP-NETS

200 m

280 m

70 m

STRONG PLASTIC SHEETING

TRUCK PULLS UP CURTAIN WHEN ANIMALS HAVE PASSED

These sketches show the workings of the "funnel" animal capture system which was developed in South Africa. The main picture illustrates an adaption used for catching Lichtenstein's hartebeest in Malawi in July 1985. Once the running animals have entered the drop-net zone, and the curtain is pulled up behind them, rangers hidden around the perimeter suddenly show themselves, causing the milling animals to rush into the drop-nets and become entangled. Inset is the more commonly used system where the animals are shepherded down a curving funnel straight onto the backs of trucks or trailers, thus minimising the handling of the game.

Similar tactics are used for culling buffalo, but then groups of up to 35 are killed during each flight. With both elephant and buffalo the practice is to herd those to be darted well away from the others, causing the least possible disruption, and to kill the entire group – males, females and the immature – so that it is completely eliminated. No unattached animals are left to roam around. The operation has been so streamlined that helicopter flying time from reconnaissance to the end of the darting averages 15 minutes for each elephant and only three minutes for a buffalo.

Hippo have to be culled for the same reasons as elephant and buffalo: they too make a major impact on the environment. They spend most of their days in the water but at night they emerge to graze. With their voracious appetites and bulky bodies, they can each consume 100 kg of grass and other vegetation a night. The researchers have concluded that the Park can sustain a maximum of 2 200 hippos, with each river being given a maximum allocation.

The hippo are not darted – high-powered rifles are used to kill them outright with brain shots. The operation is difficult and is done at first light. A team of four rangers approaches a pool. One goes downstream and the other upstream to shoot any wounded animals that might try to flee; the other two go straight to the pool.

When the bullet strikes home the dead hippo sinks immediately and the carcass can be retrieved only when it

floats to the surface between 30 minutes and an hour later.

The reason for culling in the early morning is that the hippo then have full stomachs and the fermenting gases bring the dead animals to the surface. Some, however, never reappear and are presumed to be eaten by crocodiles.

The animal products factory about 10 km from Skukuza, well off the tourist roads, has grown from a modest start to a sophisticated establishment that contains separate abattoirs for elephant and buffalo, a carcass meal plant, cold rooms, deboning rooms, a plant which cans both corned meat and braised steak, drying rooms for making biltong and also sections which process hides and ivory.

Nothing is wasted. Most of the meat is canned to the specifications of the South African Bureau of Standards and is either issued to Park staff as rations or sold to tourists in the Park's shops; the balance is dried into biltong, a favourite South African delicacy always much in demand.

Skins, feet, tails, tail hairs, ivory and horns are sold by tender to tanneries and curio manufacturers, but the best ivory of trophy calibre – tusks weighing over 30 kg – is sold by public auction. South African firms take the bulk of it for the jewellery market. Animal fat is sold to soap makers.

Revenue from the products of culled animals is about R900 000 a year and all of this is paid straight back into the Park's conservation funds.

Bringing them back alive

The translocation of wild animals is no less important in the Park's management than culling. It involves bringing into the Park from other parts of Southern Africa animals which either used to be there but were shot out or which disappeared many years ago.

It is easy enough to determine from old records, Bushman paintings and the stories of early explorers which animals once inhabited the Lowveld, and according to these the variety was matchless. But by the time the Kruger National Park was born the depredations by hunters, poachers, farmers and other factors had taken a heavy toll, not only of the individual animals but of entire species.

The Park management made it a priority to reintroduce these locally extinct creatures and they began with a species which is today one of the main attractions: the white rhino. Reintroducing these was suggested as early as the first National Parks Board session in September 1926.

This animal, distinguished from the prehensile-lipped and irritable black rhino by its wide, square lip and more benign manner, although they are actually the same colour (the "white" is thought to be a corruption of the Dutch word *weyde* meaning "wide"), vanished in the Lowveld under the guns of hunters as far back as 1896.

Negotiations began between the National Parks Board and the Natal Parks

RECAPTURING THE PAST

The Lichtenstein's hartebeest, named after a German explorer in the early 1800s, differs from the fairly common red hartebeest of South Africa's Cape and Orange Free State provinces in that it is somewhat larger and heavier, rufous-golden in colour and has horns which rise conically direct from broad bases on the skull and then curve sharply to point backwards.

Scientists differ somewhat about its classification and it is believed by some to be more closely related to the wildebeest than the hartebeest family. The Lichtenstein's hartebeest is quite common on the Central African plateau but was not known to exist in South Africa until recent research suggested that it was once found in the north-eastern Transvaal – specifically north of the Soutpansberg and in what is now the northernmost area of the Kruger National Park.

Further research by Park staff uncovered several references from the early part of this century to the presence of Lichtenstein's hartebeest in the far northern Transvaal, and a collection of trophies in Pretoria of animals shot in the Transvaal includes a pair of Lichtenstein's hartebeest horns.

In their quest to restore the Park as closely as possible to its original pristine state, the Park authorities decided that its wildlife should include these animals too. The decision led to a long and arduous international operation to capture and bring back Lichtenstein's hartebeest.

Politics, war and distance precluded all sources except Malawi, where these hartebeest still abound. Dr Anthony Hall-Martin, for the National Parks Board, and Mr Moses Kumpumula, for the Malawi Department of National Parks and Wildlife, entered negotiations and finally agreed that in exchange for these hartebeest, the Park staff would capture and translocate a variety of animals within Malawi.

The operation, under the overall command of the Park's chief ranger, Bruce Bryden, involved getting men, trucks, camping and catching equipment to Malawi overland through Botswana and Zambia. One of the Park's Bell Jetranger helicopters was sent up after fuel depots had been established for it *en route*.

The animals to be captured inhabited the Kasungu National Park in the west-central area of Malawi. When the Parks staff found a herd near a convenient site, a 300 m keyhole-shaped catching pen with nylon net walls was erected there and a road hacked through to it.

The helicopter – flown by Piet Otto – herded small groups of the hartebeest through the funnel, into the round end of the pen and headlong into rope nets strung crosswise. Once entangled in the nets, the catchers quickly moved in to subdue them – an exercise which produced many bruises and cuts from flashing horns and, for one man, a gored leg. The Lichtenstein's hartebeest was rather more vigorous and powerful than they had expected.

Once firmly down, the animals were injected with tranquillisers and then carried bodily by four to six men at a time, using stretchers, to a truck which ferried them to a carefully screened and isolated holding pen.

During this period of about 15 minutes between entanglement in the nets and loading on the truck, the hartebeest were treated against stress, parasites and other possible ailments.

While the animals were in temporary quarantine in Malawi, blood samples were sent to be tested for communicable diseases in Britain and South Africa.

Six weeks after the start of the project, a big Hercules transport aircraft took off for South Africa with ten Lichtenstein's hartebeest aboard.

Four of them died from stress while in transit, or shortly after release in the Park. On landing at Punda Maria, the survivors were taken by road to the Nwashitsumbe quarantine camp north of Shingwedzi where they are thriving.

There are now seven in the Kruger National Park (one produced a foal shortly after arrival at her new home).

After about three-quarters of a century an animal that became extinct in South Africa under the guns of hunters is now back, alive and well.

Photographs taken on the capture operations: **Left:** *Dr Anthony Hall-Martin desperately holds a struggling Lichtenstein's hartebeest until other rangers and veterinarians can help subdue it.* **Above:** *The Jetranger helicopter flown by Piet Otto comes in low.* **Top:** *Dr Hall-Martin and chief ranger Bruce Bryden help carry a tranquillised, blindfolded hartebeest on a stretcher to waiting transport.* **Top right:** *A close-up of one of the captured animals.* **Right:** *Veterinary technician Ben de Klerk calls for phials while taking blood samples.*

Board, which had done wonders in protecting the white rhino in Natal and substantially increasing its numbers. In 1961, thanks to modern drug-immobilising technology, four of them were darted in the Umfolozi Reserve, loaded in their very dazed state on big road transporters and driven through the night to the Lowveld.

On 13 October 1961, 65 years after they had vanished from the Park, white rhino were back again. Two bulls and two cows, still groggy from the drug M99, staggered down ramps from transporters into a specially built corral just south of Pretoriuskop – nearly 900 km from their previous home in Natal.

The cows were named Umfolozana and Kwangolatilo and the bulls Mfohlozi and Charlie. Umfolozana distinguished herself in the Park by producing six calves – four bulls and two cows – but the last calf died in 1981 because the ageing mother, then between 30 and 35 years old, could not provide milk for it. Umfolozana herself died shortly afterwards.

So successful was this translocation that the Park obtained more white rhino in increasingly large batches from Natal: first another 38, then 100 and then 200, so that today imports and births in the Park give it a total of more than 900 white rhino.

Black rhino had last been seen in the Park area in 1936, so replenishments of

M99 – DRUG TO THE RESCUE

One of the greatest single technological advances in wild animal conservation has been the development of drugs which immobilise the animals and keep them calm so they can be easily captured and transported.

Before the current range of drugs commonly used was produced, capture was mainly physical. Game catchers had to use nets, lassoos and a variety of other tools to literally entrap the animal and hold it until it was either exhausted or could be wrapped in ropes and sacking until it was powerless.

The procedures were dangerous for both animals and men. In East Africa, for example, black rhino were caught by roping them around the neck in pursuit from the back of a heavy truck – and then standing on the back of the truck and clinging to handholds while the enraged animal charged and battered it repeatedly until it was so tired it could be pinned down with more ropes around its neck and legs.

The effect of stress on the animals led to high death rates and many broken limbs, and often catchers were gored or had bones broken and were even sometimes killed.

When drugs were first attempted in the 1950s and early 1960s they were difficult to use because those types had to be given in large doses, which were also difficult to determine for different species. The catcher had to get very close to an animal to be able to deliver a heavy syringe with sufficient dosage.

Then in 1963 scientists synthesised a powerful new compound with the interminable name of 6, 14-endoetheno-7α(1-(R)-hydroxy-1-methylbutyle-tetrahydro-oripavine hydrochloride, known for short as etorphine hydrochloride. It quickly became widely known in the conservation world by its trade name: M99 (Reckitt), or simply M99.

About 200 times more powerful than morphine, of which it is a derivative, it could be used in small syringes, so veterinarians working together with technicians soon devised weapons from which to fire standardised sizes of syringes – adapted shotguns or rifles powered by gas or blank cartridges. They were also adapted to crossbows, which give a range of up to 120 m in still conditions.

M99 can be used on all ungulates and many other larger animals. Experience has taught catchers that the dose has to be varied not only for size but also for species. The gemsbok or desert oryx in Namibia, for example, has a metabolism so finely attuned to survival in those arid conditions that it can take relatively huge doses of the drug with little immediate effect.

But when properly administered, M99 knocks out an animal within one to five minutes. One major advantage of this is that the injected animal does not wander far before the M99 acts and brings it to a standstill. One level of dosage has such a remarkable tranquillising effect that a creature as aggressive as a black rhino becomes quite tractable. Further dosage produces a state of virtual catatonia without apparent loss of consciousness.

M99 has to be handled with great care because as little as half a milligram in a man's bloodstream can kill him. Game-catchers treat it with the utmost respect and its sale, transport and safekeeping are covered by severe government restrictions.

The Kruger National Park uses M99 and also two later drug developments, Fentanyl and Super-Fentanyl. These are not quite as strong as M99 and are derived from slightly different bases. They are used chiefly when catchers want to immobilise an animal but keep it on its feet – to restrain it rather than knock it flat.

None of these powerful narcotics is administered to animals in its pure state, as this can have damaging side-effects such as depressing breathing rate and causing excitement in certain species.

The syringe that strikes into an animal's rump or shoulder from the dart gun actually contains a kind of chemical cocktail: the narcotic plus various tranquillisers which react with it, each enhancing the effect of the other. In this way even sensitive animals have been kept for days without harm in a state of such placidity that one can pat them on the head.

When catchers have completed their task with a drugged animal – examining it, dosing it, doing surgery or simply transporting it – they inject an antidote to the immobilising drug, which acts so fast that catchers have to make sure they get away extremely quickly. After the antidote is injected a downed elephant or rhino can rise to its feet within seconds and is usually in a very bad mood.

Thanks to such drugs and the ingenious devices invented to deliver them, many thousands of animals which might otherwise have become extinct or vanished from their normal environments have been saved and restored.

these too were brought up from the rhino cornucopia in Natal. This was a trickier task because the black rhino, though smaller than the white, is an evil-tempered character who does not take kindly to capture bids, but in 1971 two were translocated and released along the Nwaswitshaka river, west of Skukuza. The following year another 12 were brought from the Zambezi Valley in what was Rhodesia to supplement 18 more from Natal, which later supplied another 30. Today the population of at least 100, all in the Park's southern and central districts, is stable and growing.

Subsequently all sorts of animals were reintroduced into the Park to roam free as they had decades before, or brought to augment existing populations. They included cheetah, red duiker, oribi, the rare and shy Livingstone's suni, mountain reedbuck, grey rhebuck, nyala, eland, roan and sable antelope, tsessebe and, most recently, Lichtenstein's hartebeest, until finally the Park re-attained the whole range of large animals which lived there when the first Voortrekkers arrived.

In 1982 the Park began translocating samango monkeys into the riverine forests along the banks of the Luvuvhu river at Pafuri, from State pine forests west of the Park near Louis Trichardt. These monkeys had vanished from the Pafuri area 23 years earlier, and in the State forests they were doing much damage to the young pine trees.

The Park is now in a position to give or sell some kinds of animals, such as young elephants which are caught at the age of about two years during culling operations, to private and public game reserves elsewhere. More than 300 young elephants have been "exported" so far, some back to the generous Natal parks and others to places like the new Pilanesberg National Park in Bophuthatswana.

During culling operations the young elephants are darted with M99, dewormed, then revived with an antidote. They are caught usually in winter in the late afternoons to prevent any excessive suffering from heat stress during the often long and slow drive to the holding pens at Skukuza – where they become quite tame. They have to be quarantined for 30 days and during this time are fed on a healthy diet of lucerne and other feedstuffs before being sold by tender to

Peter Retief, quantitative biologist for the Park, is responsible for programming the wealth of data stored in the computers used by the research staff.

interested buyers for up to R2 000 each.

Other animals are sometimes darted but more usually herded by helicopter between ever-narrowing "funnels" of cleverly concealed hessian walls into drop-nets or, as in the case of zebra, straight onto the back of a waiting truck. Animals are sometimes transferred to holding pens where they are left for a time to calm down and are then loaded onto trucks for their move. So skilled have Park staff become at this technique that the mortality rate among animals from shock and other causes has dropped to below one per cent.

The menace of disease

Disease is obviously endemic and in-eradicable in a wilderness the size of the Kruger National Park, with its thousands of wild creatures of many different species, especially as the Park is bordered to the east and north by territories where organised disease control is meagre.

The Park's policy is to try to eradicate really dangerous diseases which could reach epidemic proportions and endanger agricultural activity outside, but the ideal is unattainable. Thus the objective is to maintain an early warning system so that any outbreak can be swiftly countered.

Many sicknesses and parasites are in any case a natural part of the cycle of wilderness life and have their role in the ecosystem and it would be futile to attempt to create a sterile environment within the Park.

The Park has its own veterinary department and also the services of a State veterinarian and a number of assistants. Their work is multi-faceted.

They have to monitor wild animals constantly because of the possibility of disease spreading from across international borders. Buffalo, for instance, can be carriers of several notifiable diseases such as foot-and-mouth and theileriosis, a tick-borne disease which is usually fatal in cattle. Warthogs can transmit African swine fever, usually fatal in domestic pigs.

With the help of rangers and research staff they must check animals that die in the Park, especially those not killed by predators. Blood samples are taken from carcasses of animals and brought to the veterinarians for study.

Other work includes research into wildlife diseases which might pose a threat to domestic livestock; treating animals caught in snares, injured by vehicles or wounded by poachers; treating sickness in endangered species (but not in others, where nature is allowed to

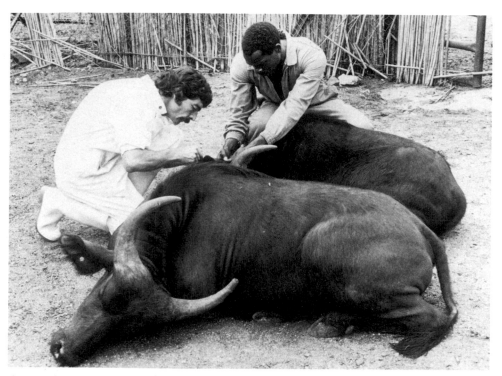

State veterinarian Dr Roy Bengis, assisted by Elvis Mdhluli, injects M99 antidote into a sedated buffalo. It had been deliberately infected with foot-and-mouth disease for research into its transmission from wild animals to cattle.

take its course); helping with animal translocation; and giving aid and advice to private game reserves bordering the Park.

Foot-and-mouth disease is one of the more persistent pestilences in Africa. It is not dangerous to wildlife and not usually fatal among cattle, but its impact is economic. So feared is it in Europe that countries there will not import beef from an area where foot-and-mouth has broken out.

The disease spread through domestic livestock in Mozambique and the Lowveld in 1938, when State veterinarians ordered the large-scale extermination of all infected cloven-hoofed domestic animals in the Lowveld. Many which were not infected, or had recovered, were also destroyed.

This seemingly pointless slaughter of thousands of animals for a passing sickness was incomprehensible to the public (and unforgivable to the blacks whose lives revolved around their stock) and caused a major outcry.

Fortunately, there was no move to butcher wild animals although they were suspected of being carriers, and the only precaution taken in the Park was that all visitors had to drive their cars through shallow troughs of disinfectant and wipe their shoes on sacking soaked with it.

When foot-and-mouth disease broke out again in 1944, the veterinarians did not order mass killing; they simply set up a cordon which isolated the affected areas.

In 1960 anthrax caused havoc in the Park. This ancient disease seems to be ineradicable and lies dormant for years until suddenly conditions occur which are just right for its recurrence.

The difficulty in combating it is that it spreads so quickly and easily. Vultures, hyenas and jackals feeding off a dead animal carry the infection far and wide and seed it at the place where all animals gather at some time during the day or night – the waterhole.

The 1960 outbreak spread through the Park like wildfire, attacking every kind of animal, drastically reducing the numbers of rare species such as roan antelope and killing large numbers of kudu in particular and even elephant, hippo and lion.

For nearly six months the Park staff carried out one of the most expensive and intensive anti-disease campaigns in the Park's history. They tracked down and burned all carcasses, deliberately burned off large tracts of grazing in infected areas to force animals to move elsewhere, drained waterholes and seared them with fire and disinfectant,

or simply covered them, and with the aid of South African troops set up a radio network to help trace outbreaks.

They won that round, and are ready for the next if it comes. Part of their readiness for any future anthrax outbreak can be seen in the form of purple blotches on the rumps of many roan antelope. This is a marker dye left by a dart which the rangers fire into the haunch of the animals to inoculate them against the disease. The dart injects the shot of serum and then falls free. This roan antelope anthrax protection programme costs the Park about R20 000 a year.

Non-epidemic sickness is commonplace in the Park and visitors might see a lion with mange or a giraffe with unsightly warts on the neck and shoulders. While the veterinarians will take note of such cases, they will not necessarily do anything about them, because their task is to control disease in populations, not individuals.

Diseases and parasites play an important part in population control and are as natural to an ecosystem as fire and water, taking toll of the weakest, as do the lions and hyenas. In a sick or wounded animal the number of parasites will increase dramatically and, if the predators do not get it, the parasites will. It is a grim scenario, but one in which only the strongest and fittest survive, thus ensuring the continuance of the species.

Research into parasites is extensive and sometimes necessitates the killing of as many as 50 animals from one species so that they can be minutely studied in laboratories. Parasite levels have been monitored in eight of the Park's species and the zebra alone have been found to carry an average of 25 million internal parasites each.

Field staff routinely check for any signs of foreign parasitic invasions as well as disease and have a stockpile of 25 000 disposable darts, made by Park technicians for R5 each, to "go to war" if necessary.

Sometimes the veterinary side of the Park's operations brings spin-offs which benefit others. Research into *besnoitia* cysts in wildebeest, impala and kudu – which are caused by larvae from eggs deposited under the skin by insects – led to the development of a very efficient vaccine now used to prevent the same thing in cattle.

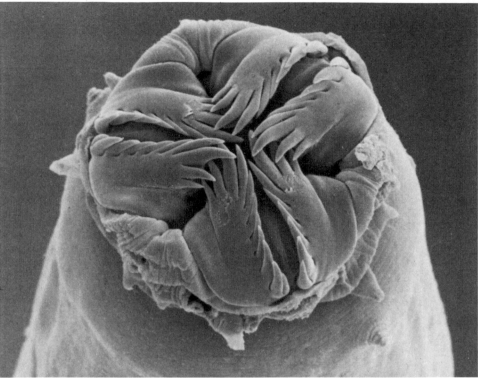

Scientists have discovered that zebras are hosts to millions of parasites. More than 100 million were counted in just one carcass from Namibia. Over 100 species of these worms have been recorded and the heads of two are shown here photographed through a scanning electron microscope. **Right:** Cylico auriculatus *with mouth parts everted magnified* 360 ×. **Above:** *The* atracid Crossocephalus *magnified* 550 × *(photographs by H. J. Els).*

Water – life-blood of the Park

The Park's major rivers include the Crocodile river as the southern boundary and the Limpopo river as the northern boundary. Between them are the Sabie, Olifants, Letaba and Luvuvhu rivers. These are all perennial, rising to the west along the Drakensberg escarpment and beyond. The Shisha, Mphongolo, Shingwedzi, Tsende, Timbavati, Nwaswitsontso, Nwanedzi and Mbyamiti only flow during the rainy season, but normally hold water in a series of permanent or semi-permanent pools. All drain into the Indian Ocean.

The rivers are the Park's arteries. Interfere with them and everything in the Park is jeopardised. Already agriculture and industry drain off valuable water supplies which, as the demand for water increases in the future, could mean the Park will need to resort to artificial means of maintaining its river ecosystems. Even a large perennial river such as the Letaba has regularly stopped flowing since 1964. Unfortunately the Park is not getting the share of water it did at the turn of the century because so much is absorbed upstream by industry, agriculture and settlement.

The 1920s and 1930s were drought years in the Park. One by one the waterholes dried up; desperate fish flapped in the black soup and death came slowly as the mud hardened. Many of the rivers became isolated pools.

There was another problem: animal migrations. In those early years the Park was unfenced. The animals did not appreciate the establishment of the sanctuary and in a drought they trekked in search of water.

Then came the donation from Mr A. J. H. Cloete and especially Mr Bertram Jeary's "water-for-game" fund, which started a flood of money from the public. Since the sinking of the first borehole funded by the campaign in 1933, it has paid for most of the nearly 400 windmills now groaning and clanking all over the Park as they draw life-giving water from the earth.

In 1946 the Park was once again in the grip of a serious drought, and once again the Park warden launched a campaign which raised money for 46 new boreholes.

From the first "water-for-game" campaign it was obvious that the Park needed stable water supplies. The original idea saw an adequate network of reliable waterholes for the thirsty herds during drought. It was eventually taken further to ensure that many natural pools and springs would only dry up in extreme droughts. Low weirs and sluices were built in the major rivers so that there was some flow of water in the dry season. Their importance cannot be over-estimated, as they prevent the extinction of many species dependent on running, well-aerated water.

Attention had to be given to large pools in the seasonal rivers which not only provide water for large herds of buffalo and elephant but sanctuary for fish, frogs, waterbirds, hippos, crocodiles and other forms of water life. To ensure that they would withstand several years of drought, a series of dams was built in the rivers.

Researchers had to take into account the wet and dry cycles. They found that after good rain years, animal numbers built up to a point at which, unless they were buffered against the bad times, their numbers could plummet so fast that they might never recover.

Although the "water-for-game" programme cannot eliminate drought, it can and does soften its impact sufficiently to avoid total disaster.

Without the artificial watering holes, many rare and irreplaceable species probably would have been lost during the desperate drought of the 1960s. Even 1970 saw great mortality of game where natural waters dried up. Nature might possibly have put paid to such rare antelope as roan, tsessebe and eland had man not intervened.

Severe drought, cold spells, floods or epidemic diseases are all cataclysmic but natural ways of keeping the wildlife populations in check. However, if they were allowed to wreak their full destructive force in a conserved area, recovery would take centuries.

River pollution in the Kruger National Park is like blood poisoning in man: unless it is countered quickly it can kill. The many rivers and streams in the Park are its life-blood; without them it could not carry anything like its present populations of wildlife and would be entirely without the rich variety of creatures which can exist only in the rivers and the profusion of growth along their banks.

Unfortunately, none of the bigger rivers has its source within the Park. All of them bisect it to link with others or flow directly into the Indian Ocean through neighbouring Mozambique. All rise well inland and course, with great seasonal fluctuations in volume, through farmland, past towns and villages, beside factories, over dams and sluices and between the crowded fields and huts of tribal lands – picking up pollution on the way.

The vulnerability of the rivers underlines the fact that the management of the Kruger National Park should extend beyond its borders, firstly because it is not a unitary, self-sustaining ecological area and also because the specific needs of the Park often clash with those of the growth and development taking place outside it. Two recent events illustrate this.

In January 1983, the Park's worst fish disaster in 63 years occurred in the big Olifants river when hundreds of tons of sludge-like silt clogged the river from upstream. It came from the Phalaborwa dam west of the Park when the sluice gates there were opened.

Nobody was to blame for this. The dam's sluice gates have to be opened from time to time, especially when the river is in flood, and unfortunately the Olifants brought down large volumes of silt caused by heavy rains along the escarpment it crosses and, ironically, by severe drought in Sekukuniland, through which it flows.

But, by partially choking the river, the silt drastically reduced the numbers of almost all of the 20 species of fish found in the affected 80 km of the Olifants inside the Kruger National Park.

"Not in all the years I've been in the Park have I seen anything like it," said

RAINFALL MAP

Warden Pienaar, who described the event as catastrophic.

"Tens of thousands of fish rotting on a hot afternoon produced a smell so bad that the stink could be noticed at the Satara camp 75 km away. Eighty years of conservation and preservation were lost within hours and every form of river life has been affected. Eels 20 to 30 years old perished and even the hippo have been deprived of their pools by sludge."

River pollution, Warden Pienaar said recently, had obliged the Park to build silt precipitation dams and other structures which did not properly belong in a conservation area.

The problem had been severely aggravated by the increasing withdrawal of water from the rivers' upper reaches outside the Park. The only answer was to build storage dams on the upper reaches like the one on the Olifants – which had kept it flowing when formerly perennial rivers like the Sabie, Crocodile and Letaba had at times stopped flowing entirely.

Compounding the impact of the silt was simultaneous pollution by acid effluent and concentrations of minerals such as aluminium, iron and nickel, all from an unknown source.

"We are sitting at the end of the chain. Everything people pump in upstream comes down to us."

Fisheries officers said that despite the extent of damage, fish life would return to the Olifants – provided the silt and effluent problem could be stopped at source.

The other example involves pollution of another kind: the "green cancer" of water hyacinth, Kariba weed and water lettuce which menaces dams and rivers all over Southern Africa, including the giant Lake Kariba between Zimbabwe and Zambia.

In recent years these lovely but noxious weeds began spreading fast in the waters of the Sabie (water lettuce) and Crocodile (water hyacinth) rivers, which flow in the Park's southern sector. Kariba weed established itself in a tributary of the Sabie but was recently eradicated.

Water hyacinth is a bright green plant with attractive flowers once much favoured for enhancing garden fishponds. It has the ability of being independent of land and can float free, its leaves acting as sails so the wind can drive it over large surfaces.

Its danger is that it is extremely prolific, spreading so fast that the floating islands of plants soon blanket the entire surface of river pools or dams. It causes hydrological changes, reducing oxygen content and otherwise altering the chemical balance in the water so that certain kinds of fish cannot survive, thus depriving crocodiles of part of their main diet. At the same time it deters

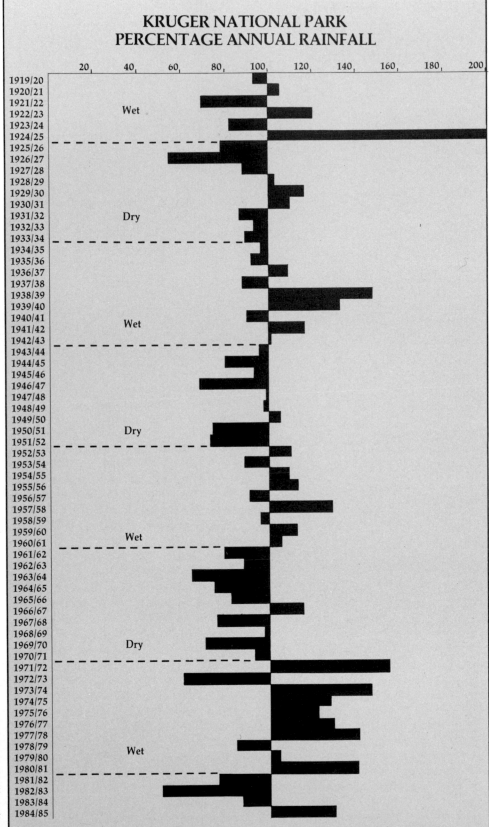

KRUGER NATIONAL PARK
PERCENTAGE ANNUAL RAINFALL

	20	40	60	80	100	120	140	160	180	200
1919/20										
1920/21										
1921/22										
1922/23	Wet									
1923/24										
1924/25										
1925/26										
1926/27										
1927/28										
1928/29										
1929/30										
1930/31										
1931/32	Dry									
1932/33										
1933/34										
1934/35										
1935/36										
1936/37										
1937/38										
1938/39										
1939/40										
1940/41										
1941/42	Wet									
1942/43										
1943/44										
1944/45										
1945/46										
1946/47										
1947/48										
1948/49										
1949/50										
1950/51	Dry									
1951/52										
1952/53										
1953/54										
1954/55										
1955/56										
1956/57										
1957/58										
1958/59										
1959/60										
1960/61	Wet									
1961/62										
1962/63										
1963/64										
1964/65										
1965/66										
1966/67										
1967/68										
1968/69										
1969/70	Dry									
1970/71										
1971/72										
1972/73										
1973/74										
1974/75										
1975/76										
1976/77										
1977/78										
1978/79	Wet									
1979/80										
1980/81										
1981/82										
1982/83										
1983/84										
1984/85										

This graph shows how the annual rainfall in the Kruger National Park has, over the years, followed a definite pattern of 10-year wet and dry cycles. Armed with this knowledge, Park scientists are able to roughly predict future rainfalls. This helps them in their long-term management planning such as deliberate veld burning which is done more frequently in wetter years than in dry. A variance in the numbers of buffalo to be culled and the control of waterpoints are also possible considerations.

water birds and limits their range of diet.

Mining, pulp milling, food canning and a host of other real or potential industries west of the Park, upstream along its rivers, will increase the danger of pollution in the years ahead. The only solution is joint, long-range planning of counter-measures for the mutual benefit of all, man and wildlife.

The rains

The rainy season in the Transvaal Lowveld begins in early October and lasts until April. The wet months are followed by exceedingly dry ones. About 80 per cent of the rainfall comes from heavy, erratic thunderstorms which rend the skies. These are particularly hazardous as the accompanying lightning can set the parched veld alight during the spring months.

The relatively high regions around Pretoriuskop record the greatest mean annual rainfall, about 760 mm, while about 500 to 540 mm falls in the central areas each year.

At Punda Maria rainfall rises to about 640 mm, but at Pafuri – less than 50 km to the north-east – the mean rainfall is only 210 mm, the lowest in the entire Park.

Rainfall records for the Skukuza area stretch back to 1908 and since 1964 it has been a First Order meteorological station recording soil temperature, windspeed and direction, humidity, evaporation and rainfall. These details are telephoned to the Weather Bureau in Pretoria twice daily.

Satara, Shingwedzi, Letaba, Lower Sabie and Punda Maria are Second Order meteorological stations which measure daily temperature gradients, humidity, rainfall and windspeed. Each ranger's post records rainfall only.

Meteorological research in South Africa has revealed good and bad rainfall cycles of roughly 10 years. The climate of the Park falls into a similar pattern, although the cycles appear to be shorter. Records show a wet period up to 1925 when the rainfall was 119 per cent of the average, followed by eight years in which the figure dropped to 90 per cent of the average, and then five successive periods of 8 to 10 years of wet and dry.

The 1970s saw the highest rainfall for 60 years but in 1979 rainfall dropped and the Park entered a blistering, freak-

Soaking summer rain, lifeblood of the Park, washes over a lone impala ram. Although it can be very hot, the Park is at its most beautiful at the height of the wet season – in January.

But not the wildebeest and zebra. Again, as in the 1934-42 period, their populations declined. At first the public and many experts blamed the lions, the familiar old scapegoat. The lions were indeed exploiting the weakened condition of these animals, but a little careful research disclosed that something else was taking its toll of them: paradoxically, the good rains. Both these species thrive in drier periods because their favourite food is short grass. The abundant rain made the grass grow too tall for them to cope with so they began to weaken from lack of sustenance.

Fire as nature's aid

Colonel Stevenson-Hamilton found early in his career that fire could be a useful tool. He noted in 1903 that the Sabie and Shingwedzi Reserves "were almost annually harassed by devastating fires passing the borders from outside where they were started by natives, hunters and stock farmers."

Natural fires were nothing new. Early Portuguese explorers nicknamed Southern Africa *terra dos fumos* (land of smoke) because of the frequency of bush fires. Before the advent of man these must have been started chiefly by lightning, a commonplace occurrence in the Lowveld.

The first people to exploit fire in the veld were probably the nomadic Bushmen using it to flush game and thus help them in their hunting. Then came the black tribes who settled in the area and used it for the same reason. From 1838 the Voortrekkers moved in and used fire to clear the countryside to make grazing easier for their cattle.

The pattern of deliberate veld burning in the Park was haphazard in the early days as wardens tried to work out the best system. Stevenson-Hamilton burned firebreaks in 1934 but, as he ruefully admitted, these did little or nothing to stop runaway bush fires. So instead he changed to a pattern of burning the veld every second autumn.

His successor, Sandenbergh, saw no advantage in burning. "Deliberate burning in an area which should be kept in its natural state must upset the balance of nature," he wrote, and was right in part.

In 1949, the National Parks Board

ish period of drought which killed many animals, especially impala, kudu and warthog, and lasted until 1984, when it was broken by cyclone weather. Ironically, this last drought was a boon to tourists because a great variety of animals came to graze on green grass shoots growing at roadsides where moisture accumulates on and runs off the asphalt. Even usually nocturnal hares could be seen by day.

The average rainfall difference between wet and dry periods is 13 per cent. This does not appear to be a significant variance, but it causes game populations to see-saw. In the wet spell of 1934-42 the numbers of wildebeest and zebra dropped so much that there

was widespread public concern. However, their numbers quickly rose after the wet period as nature again asserted its own checks and balances.

The year 1970 was the driest in living memory and a disastrous one for the Park. Animal mortality was heavy as all the major rivers except the Sabie and Olifants stopped flowing, hippo crowded into the dwindling pools and starvation took its toll. That summer the drought broke with exceptionally heavy rains and for the next eight to nine years the vegetation thrived in good rains and with it most animals such as waterbuck, kudu, nyala, giraffe, sable antelope and warthog, whose numbers soared, in many cases to record levels.

ruled that no area should be burned more than once in five years. They also ruled that it should be burned only after the first spring rains, which accorded more closely with natural lightning fires.

In 1954 the Board changed this ruling and laid down that "until it is proved to be wrong" (an indication of how conservationists were still feeling their way), the whole of the Kruger National Park would be divided into sections and that all long and rank grasses must be burned every three years.

The Park staff then began a programme that lasted into the late 1970s, of cutting a network of more than 4 500 km of firebreak roads which divided the Park into some 400 blocks for fire control and other veld management purposes.

It has long been recognised that fire plays almost as significant a role in an ecological system as does rain. From the mid-1970s veld-burning was tailored to fit the needs of the Park's vegetation – on which all Park life is dependent – as well as its game, and almost 80 per cent of the Park was burned during each span of three to four years by burning blocks of land in fixed rotation.

In 1979 fire was "tamed" by a computer and the authorities were able to adopt the far more flexible burning policy that operates today. A data bank is maintained within the computer in which all fires, natural or otherwise, are recorded, as well as their cause and effect. Other information such as local rainfall, veld conditions, soil types and even the underground rock strata is also stored, enabling Park scientists to accurately evaluate the most suitable blocks or areas for a burning programme – and to also give special attention to sensitive parts containing, for example, an endangered plant species.

Chief warden, Dr "Tol" Pienaar, explains it thus:

"Fire is regarded as a natural ecological factor, the effects of which can be used to advantage in wildlife management systems, particularly in maintaining or manipulating habitat conditions to suit the requirements of particular species or animal communities as a whole.

"A controlled burning regime, based on long-term research, has therefore been instituted in the Park which is aimed at maintaining optimum habitat

A controlled fire lit by rangers lights up the evening sky at Shingwedzi.

conditions for all species, simulates the natural incidence of fires as far as possible and ensures the achievement of particular management objectives in certain areas – such as the protection of unique botanical areas and stimulating seasonal grazing patterns."

In their deliberate burning of the veld the rangers emulate another paradox of nature: that there are more natural fires during the cycles of good rainfall than in dry cycles. These occur, of course, in the dry months of the year and the explanation for the phenomenon is that in the wet cycles the grass grows much faster than during dry times and in the process there is a faster build-up of dead, combustible material.

The rangers use butane gas burners to ignite their fires, a nerve-racking job for a conservationist, who fears nothing more than a runaway fire. Hence even controlled burning is banned when the humidity drops below 50 per cent and the noon temperature tops 30 °C.

Rangers – the guardians

The popular image of a game ranger assiduously cultivated by film and fiction is of a strong, silent man, deeply tanned beneath his broad bush hat with its band of leopard skin, roaming the African wilderness with his faithful black bearers, trusty rifle slung from one shoulder, meditatively sucking at a battered briar as he ponders whether to

FIXED-POINT PHOTOGRAPHY

A valuable visual record of veld conditions is achieved with fixed-point photography. In 1977, 74 sites were selected throughout the Park and since then a series of six to eight pictures are taken periodically through 360° from various set vantage points. The photographs are compared to reveal aspects of plant dynamics such as the effects of fire or impact of elephants on the veld.

This photograph was taken near Satara on 11 April 1977 and shows veld that was burned annually over a long period.

The same view taken after burning was stopped for three years. Encroachment by low shrub is clearly evident. Note the ring-barking by elephants of the marula tree on the right.

This was taken exactly one year after burning was resumed. Most of the shrub has been burned back but the veld condition is improved. The damaged marula is dead.

rescue the lady in distress or shoot the rogue elephant first.

Reality, regrettably, is not quite like that. Today's ranger is indeed physically fit, bronzed by the sun if he happens to be a white man, clad in practical khaki and carrying a rifle. But there the Hollywood-style glamour ends.

He will have a minimum qualification in nature conservation plus the skills of a rancher or farmer and a great deal more. His work is demanding, strenuous and requires stamina, initiative and an intimate knowledge of his wilderness environment. It involves long hours at odd times of the day and night, and on top of this he must have the patience and tact to deal with people. The pay is not high, compared to many other jobs demanding considerably less ability.

More often than not a ranger is a person of retiring, rather modest personality. By nature he is the kind of person who enjoys the remoteness and solitude of life in the bush, away from other people except his own family and assistants. His motivation is an intense affinity with the wilderness and he tends to be dedicated and sensitive.

The life of today's ranger is somewhat easier than that of his early predecessors like Fraser, Duke, De Laporte, Healy, Gray and Wolhuter, but a ranger and his family still have to make sacrifices of certain amenities, especially if he lives at an outpost, which comprises his house and a compound for his assistants, often up to 40 km from the nearest neighbour.

And yet the post of game ranger in the Park is still one of the most prized in South Africa. When a single vacancy occurred a few years ago more than a thousand fully qualified people applied for it, some with academic qualifications well above the minimum required.

And what are the criteria for the job? A matriculation certificate plus a university science degree, or a diploma in nature conservation obtained by taking a three-year course at a technikon in Pretoria, Johannesburg or Cape Town, which includes 24 months of study and 12 months of practical experience.

Each year the Park takes in a wildlife diploma student doing his 12 months' practical training, but without any guarantee of work should he pass the course; many take "lesser" positions in the Park just to get a foot in the door in the hope of future vacancies.

These are not common because rangers tend to stay in their careers for many years, in a way of life which absorbs and satisfies more than most in spite of its hardships.

There are only 22 section rangers in the Park, so that each is responsible for an area of between 40 000 and 60 000 hectares (400 to 600 km²) – more territory than most other parks and reserves in South Africa.

Each is obliged to have an exceptionally close knowledge of his piece of the Park, which he gains by covering it constantly on foot, by truck or motorcycle and, when the opportunity arises, from a light plane or helicopter.

Helping him to keep in touch with his "parish" are between seven and 14 locally recruited black rangers, carefully selected for their ability and then trained in the Park. These men are his eyes and ears when making regular patrols, always in pairs and armed, on foot or bicycle along outlying firebreaks and fences, and they help him with routine chores like repairing equipment.

The ranger's duties are manifold but his priority is always conservation, which means closely studying the daily life in his slice of wilderness until he reads its variations like a newspaper. He must watch the movements and behaviour of animals, noting what they eat and when, keeping alert for signs of sickness. He must become equally familiar with the plant, bird, insect and reptile life. Together with the Park's other experts he must make aerial tallies of the animals in his area, help in the culling of surplus stock, help in the capture and translocation of animals and assist the scientists and veterinarians in their research.

He must maintain the law, which means everything from curbing speeding motorists to combating poachers. The latter can be a risky job because today's poachers no longer rely exclusively on home-made weapons and standard hunting rifles, but also on modern automatic weapons, with which they sometimes defend themselves.

In the south-western part of the Park there is still an occasional shoot-out between poachers and rangers. In 1977 a ranger from Pretoriuskop was shot dead, and in 1981 another was wounded in the stomach by a 0,22-calibre bullet. A group of selected rangers has now been given para-military training to

enable them to tackle poachers.

Assistant rangers are stationed at pickets along the Park's borders to patrol daily in search of poachers or their spoor and to guard against illegal entry for any other reasons.

White rangers patrol on foot or by vehicle. Their philosophy is that in order to know an area, one has to get close to it, and to do that one must walk it. But these areas are so large that they must make frequent use of vehicles.

The technique is for a ranger to drive out to a selected area of his section, and then set up a base camp from where he can patrol on foot or by vehicle. This can

Rangers at work. **Top:** *Tom Yssel left, and Johan van Graan guard a drugged sable during capture operations.* **Above:** *Louis Oliver and Ted Whitfield select a capture site.*

keep him away from home for up to 10 days.

On such patrols he checks that life is carrying on as usual and, if not, he investigates. He is, in a sense, not unlike a cattle rancher, the difference being that a rancher can move his livestock about at will from one grazing area to another, which the ranger cannot do with wild animals. Like the rancher, he must continually assess the water, grazing and other local conditions and act fast to counter outbreaks of disease.

"Within seconds we were surrounded by flames and suffocating smoke."

But much more than any farmer, a ranger has to be as familiar with his environment as a stockbroker is with Wall Street. He must be a competent "bushcraftsman" and yet always willing to learn more.

His many other chores include planning and clearing firebreaks in conjunction with the Park's research department, burning off old veld when the researchers say so, coping with accidental fires caused by lightning or perhaps by a cigarette butt thrown by an unthinking tourist, and maintaining the windmills that pump water for the wildlife.

The ranger is also expected to fulfil many other roles: an ambassador for the National Parks Board; a bush mechanic able to repair his own radios and vehicles; a communicator who can educate people about wildlife and spread the message of conservation; and a permanent student furthering his own expertise.

There are risks too – from injured or simply bloody-minded animals, as well as poachers. Every turn in the bush hides a potential hazard, although a ranger might go a lifetime without experiencing one. A renowned researcher on lions, Ian Whyte, has been chased up a tree only once, and that was by a hippo in his own garden.

To meet risks, however, a ranger must be a skilled shot and hardy enough to withstand the extremes of climate and exposure to the Lowveld's blazing summer sun and chilly winter nights.

Beneath the exalted rank of chief ranger there are three other grades: junior, senior and regional rangers. They are distinguished from all other employees in the Park by a special breast badge,

and also one other privilege – only rangers are allowed to have dogs as companions, for security at home and to help them in their field work.

Apart from that of ranger, there are many other categories of employee, from the scientific staff to the engineers, caterers, camp managers, mechanics, bricklayers, cashiers, cooks and even traffic officers, all of whom are jointly necessary for the welfare of the Park.

One of the ranger's greatest fears is the bushfire, whether caused by lightning, poachers or tourists. In a matter of a few hours, or even minutes, fire can turn beautiful veld into a blackened desert.

On one typically wild night of thunder and lightning, ranger Tom Yssel of Nwanedzi in the central part of the Park had been working with teams of other rangers to douse fires which flared in several places over a huge area when he learned that two fires had started just north of his own post. This is his story:

"It should have been easy to control the fire with the help of three teams, but then the unforeseen happened.

"The wind changed from a moderate north wind to a strong south-easter and drove the fire in our direction.

"It was an incredible sight when the flames shot up in the air to scorch the tree tops. Within seconds we were surrounded by flames and suffocating smoke. I remember seeing a number of panic-stricken impala fleeing from the fire.

"At that stage it was a case of running for one's life. Arrie Schreiber (another ranger) and his truck almost burned out when he tried in vain to burn the last bit of the block before the fire reached it.

"With singed hair and moustaches we started anew. This time we succeeded in checking the fire."

The result: 35 impala burned to death, a kudu and a giraffe so badly burned they had to be destroyed, uncountable smaller animals from birds to tortoises killed, and huge tracts of grazing and other plant life lost.

For the man at an outpost in particular, the life of a ranger is an extraordinarily enriching one which places

great demands on both him and his family. A ranger's wife is all-important to his success and contentment. If she cannot cope with the remoteness and loneliness when her husband is away, she will not cope with an outpost and, therefore, nor will her husband. Conditions are far better for families today than they were 20 years ago – a radio call will bring help by road or helicopter if necessary – but they are still demanding.

Children, for instance, often have to go to hostels from the beginning of their school careers. The National Parks Board, however, provides transport which makes it possible for even high school children to be with their parents every weekend.

The war against poaching

Hunting is one of mankind's oldest and deepest instincts, translated in modern society into the quest for wealth, power and status. But that it still lies close to the surface of civilisation is reflected in the lucrative world-wide safari industry catering for people who spend small fortunes to hang another head or skin trophy on their living-room walls . . . an industry which, ironically, is a strong inducement for conservation because most of its clients are affluent wildlife enthusiasts who spend large sums of money on hunting.

The lust for hunting is manifested in two other forms: hunting for sport or profit, and hunting for food.

The most devastating are the greedy hunters – those who kill out of wantonness or for profit and have ravaged so much of Africa, especially in its last two or three decades of turbulence. Such people are those who in the last century in single hunts killed thousands of animals in the name of "sport" by shooting randomly into the now vanished herds which migrated across South Africa's great plains. They are also those who, armed with modern, automatic weapons of war, have wiped out entire local populations of animals in orgies of killing allegedly for meat – and then left most of the carcasses to rot.

Such, too, are the hunters who in many parts of Africa annihilate entire herds with their guns, traps and snares, purely for the profit – killing elephants for their ivory, rhinos for their horns, crocodiles for their skins, birds for their

feathers and even wildebeest and giraffe for their tails. Equally a part of this destruction are those far removed from the killing but who encourage it: the traders selling the prized ivory and horn to the Far East, the leopard skins to furriers in Europe and North America, the lion claws and hippo tusks to the manufacturers of tourist gewgaws.

The international trade is enormous and only its tip shows above the surface in the illicit dealings brought before the courts, or in the tusks and skins loaded aboard Arab dhows in ports like Mombasa and Dar-es-Salaam.

But its effects on once renowned game parks and reserves in Africa are obvious. During Idi Amin's bloody reign of lawlessness in Uganda, the beautiful Queen Elizabeth Park on the shores of Lake Edward was virtually destroyed by the wholesale slaughter of hippo and buffalo, and untold thousands of animals were butchered in the Murchison Falls Park. A wholly illegal but tacitly tolerated

Shawu (above), the elephant that had the longest tusk recorded in the Park (3,17 m), died from an infection believed to have been caused by poachers' bullets. The top picture shows chief ranger Bruce Bryden with the tusks.

ivory trade struck Kenya like a disease and the once great Tsavo National Park is today a shadow of its former self. Until Zambia banned hunting in 1980 a rhino was killed every day, and their numbers dropped from 8 000 to 12 000. A World Wildlife Fund investigation there, at a cost of R3 million, revealed that commercial poaching involved at least 26 mercenaries using Kalashnikov automatic rifles and similar modern weapons. Between them they had the firepower to kill 600 elephants and 240 rhino monthly.

Tanzania has lost 80 per cent of its rhino to the horn trade, which has also wiped out some 90 per cent of Central Africa's rhino population. In six years Kenya lost about 10 000 rhino, leaving less than 1 500 there today.

Much of this occurred while Africa's new black governments were struggling with taking over administrations left behind by colonial powers, trying to apply new ideologies and resolving innumerable human conflicts and problems, leaving them little time, money or manpower to look after their wildlife.

Today there are signs that the tide is turning. The Queen Elizabeth Park in Uganda has been sheltered and its waters and bountiful bush are being rediscovered by hippo, elephant and buffalo. The Mozambique government – horrified to discover that the more than 45 white rhino in its Maputo game reserve in the far south were reduced by poachers to about six – is trying to clamp down heavily on hunting there, and now regards poaching as a national disgrace. Zaïre is giving new attention to its old and once famous parks.

In Africa this was always the rule. People grew crops but hunted for meat, for skins from which to make clothing, for hair and sinew with which to sew them, for bone to make needles and other implements, and to satisfy spiritual needs – like the Masai *moran* who spears a lion not merely to protect his cattle but also to prove his manhood. The hunter killed no more than enough to meet his needs, as does the little Kalahari Bushman today, who uses every vestige of his prey, and the limitations of spear, arrow, club and trap made wholesale slaughter extremely rare.

The advent of guns has dramatically changed all that at a pace which has increased drastically in the last two decades, during which political instability has rocked Africa and Eastern bloc countries have poured millions of guns into the continent without control or care about whose hands they fall into.

Like all parks, the Kruger National Park has always had to contend with poaching by the "food-for-the-pot" hunter whose impact has been minor and has never really threatened the Park's survival. There has long been understanding of the motives of the unsophisticated tribesmen living adjacent to the Park.

It has never been necessary in the past for them to grasp the concept of conservation. Wild animals were always there, available for hunting, part of a way of life which goes back little changed to the very origins of man. It is too much to expect the hungry man with hungry children to appreciate suddenly the aesthetics or scientific logic of preserving a roan antelope on the other

further hold as more and more blacks, chiefly from the urban areas, make use of the Park's tourist facilities.

But at the same time the profit and food motives, coupled with political change, unrest and administrative decline in neighbouring states, are threatening to bring to the Kruger National Park that kind of wanton slaughter which has struck large parts of the rest of Africa.

The hungry man on the Mozambique side of the border no longer has to go to the labour of butchering his prey and utilising all of it, now that guns are so commonly available. So sharply has the price of ivory risen in recent years that one kilogram can keep him and his family fed for a year and one average tusk can keep his whole village content for as long. All he need do is shoot an elephant, cut off some meat if he wishes, leave the rest to rot, and disappear with the tusks to be sold on the underground market.

These factors lie behind the killing of nearly 200 elephants in the Park since 1981 – and also the death of six poachers, shot by rangers and police.

The start of the killing appeared to coincide with a southward surge of civil unrest in Mozambique to the region outside the Park's eastern borders: wild and undeveloped territory neglected even during the pre-independence era of Portuguese rule. The National Parks Board reported that poachers launched a period of "bloody and unscrupulous onslaught" on the Park's elephants for their ivory and threatened the survival of the "Magnificent Seven" – huge old elephant bulls famed worldwide for their great size and weight of tusks.

One such giant was named Shawu, whose tusks weighed 52,6 and 50,8 kg (116 and 112 lb), one being 3,17 m (10 ft 4 in) long. Dr Anthony Hall-Martin, a senior research officer, had been monitoring Shawu's movements by means of a radio transmitter collar attached to him. When the transmitter revealed that the elephant had stopped moving, Dr Hall-Martin went in search of him and found his carcass. An examination showed that he had died of an infection probably caused by bullet wounds.

Another member of this famous group, a bull named Dzombo, was gunned down in October 1983 at Dzombyane windmill near Shingwedzi. The

... Eastern bloc countries have poured millions of guns into the continent without control or care ...

Kenya has attempted to curb the wholesale poaching of its wildlife by banning the sale of all curios made from hide, hair, horn, tusk and other animal products.

The other major form of poaching is illicit legally but not morally, and will remain ineradicable until society is able to spread its benefits sufficiently to make it unnecessary. It is the hunting by man simply to provide food for his family and himself.

side of the game park fence. To him it is meat on the hoof and a park is, at best, a larder and, at worst, a playground for the wealthy.

Conservation is able to live with this kind of hunter. In any case, he is slowly fading away in those parts of South Africa adjoining the Park as the tribespeople gradually turn more to farming for profit instead of purely for subsistence and enter the cash economy. The principles of conservation are taking poachers were disturbed and ran off without the tusks.

Most of the poachers' victims were bulls and according to the Park's chief ranger, Mr Bruce Bryden, they took only the tusks and abandoned the rest.

When the authorities began pursuing them the poachers changed their tactics, and using automatic rifles like the Kalashnikov, began firing indiscriminately

Elephant expert, Dr Anthony Hall-Martin.

into breeding herds. In one incident, just north of the Letaba river, two cows and a calf were found dead and the carcasses of another three cows were found 3 km away.

"The ground and vegetation were spattered with blood. In both groups there must have been a total of some 12 animals. Those elephants that were not killed outright, scattered and fled headlong, their bodies riddled with bullets.

"One young bull was later found dead in the Engelhard dam. Its carcass was torn with bullet wounds and it was clear that it had died a prolonged and painful death. The other wounded animals are either dead or are still suffering intense pain somewhere in the bush," it was reported.

Similar incidents occurred in several parts of the Park within striking distance of the Mozambique border, but the Park staff were able to save one wounded prize bull named João, whose tusks were then estimated to weigh about 50 kg (110 lb) each (see p. 122).

In 1982 Dr Hall-Martin immobilised João with a drug dart in order to replace a defective radio transmitter. Examining

the drugged elephant, he found that at least four bullets had penetrated his trunk, tusk, palate and ear – someone had shot at him with an automatic rifle. João's wounds were treated and he was given injections of long-term antibiotics. He recovered.

Calculating that until March 1982, 102 elephant bulls were killed by poachers, the following year 52, 27 during 1984 and in 1985 between 10 and 15, each of them carrying at least 25 kg of ivory, it is estimated that well over R500 000 of ivory has been removed from the Park since the beginning of 1981.

The rangers admit that the poachers are highly skilled in bushcraft. All are rural inhabitants and some undoubtedly had the experience of years of guerrilla warfare against Portuguese rule. They are courageous and will dare to track down a wounded elephant at night . . . a task dangerous enough by day.

On 10 March 1981, two rangers were doing routine observation work in the Makhadzi area to the east of Letaba camp when, in the dusk just after 5 p.m., they saw three men approaching their camp. All three were carrying what, in the fading light, seemed to be guns.

When the newcomers saw the rangers, one immediately raised a semi-automatic rifle and began firing wildly at them. The rangers returned the fire, killing all three. They were found to have an old SKS Russian-designed rifle, a panga and a hatchet and one wore a Mozambique army camouflage jacket. They were, undoubtedly, poachers.

In later incidents, two more poachers were shot dead and three more caught, two of them wounded. Another wounded man escaped across the border but died later. Fortunately there were no casualties amongst the rangers.

After that, said warden Pienaar, the knowledge that they could be captured "or even worse" acted as a strong deterrent to poachers, whose raids then became sporadic.

They were, he said, a "completely lawless element" because of the civil unrest on the other side of the border and because the Mozambique government, preoccupied with other matters, could not afford to operate anti-poaching patrols in all its parks and reserves, most of which had been overrun by poachers.

Almost all the killing took place in

remote areas of the Park well away from the beaten track because the poachers wished to avoid detection and make their strikes quickly. It had little or no effect on the normal life of the Park, which is now sheltered by strong patrols along the affected borders.

The hope now is that poaching from Mozambique will diminish sharply as a result of the Nkomati Accord signed between Mozambique and South Africa on 16 March 1984, which took the heat out of the political confrontation between the two states.

Mozambique has launched moves to bring order to its remoter areas and to stop infiltrations into South Africa by tighter patrolling, which means that poachers too will be curbed.

Hunting down the modern poacher adds a new and highly dangerous dimension to the life of a ranger, whose training is to conserve life in all its forms – not to contend with guerrillas who do not hesitate to fight and who are likely to be armed with extremely efficient assault rifles capable of spewing bullets at the rate of 400 rounds a minute.

The following is a graphic firsthand account of a pursuit of this kind of poacher, recorded by chief ranger Bruce Bryden in *Custos* (Vol. 12, no. 3), the official journal of the National Parks Board.

"It was morning of the fourth day out of the base camp. Our concentration was beginning to lapse as we thought of the luxuries that awaited us when we got back there, like a bath in a bucket, a stretcher to sleep on, not the hard-baked ground of the Lebombos, a change of clothes and chance to get at the ticks, some of whom had been travelling companions for days and were already gorged on our blood. The flies that constantly sought to find moisture at the corners of our eyes or mouths would still be with us, and so would the heat.

"Our food was running low and we were looking forward to a change in diet. Each member of the patrol had his own carefully selected special tins back at camp. Tinned pears and peaches seem to be a general favourite. Not ice cold, as you might eat the fruit at home, but at least cooled, by hanging it in a wet sock in the breeze for an hour or more.

"Then it happened. To my left my partner stopped, cupped his hand and held his fingers downward. He had found a spoor. Immediately several questions came to mind. How many? How old? What direction?

"Close examination showed the spoor of three people, one barefoot and the other two wearing *mapashans*, which are home-made tyre-soled sandals. The spoor was about 14 hours old and heading east, towards the Mozambique border. They obviously hadn't shot anything yet, or we would have heard them.

"We started tracking; one of us on the spoor and the other keeping a look-out slightly behind and to one side of the tracker. We relieved each other about every hour, giving eyes and concentration a rest.

"From the spoor we deduced that they were only after elephants, as every elephant spoor they crossed, they examined minutely. We could picture the three poachers arguing about the age of the spoor and the size of the tusks that the elephants were carrying. These men were experts in bushcraft, of this we had no doubt. They would decide on a spoor, then stick to it till they killed or severely wounded the elephant. Our determination to catch up with these *skelms* (rascals), as we call them, grew as we progressed on the spoor.

"The sun was merciless. There are few trees on the northern Lebombo flats, just scrub mopane – seldom more than two metres high. Even if there was shade, we could not afford to rest. The poachers had a big lead on us and we had to cut it down as fast as possible if we wanted to catch up with them before they encountered any elephant. The spoor, however, went straight to a windmill and water reservoir, and mercifully gave us a chance of filling our waterbottles. We were now carrying seven litres of water each, which should have seen us through the day.

"Not 500 m from the reservoir, we found their camp – if you could call it that. It consisted only of a few branches placed in the fork of a tree and the ashes of a very small fire. We deduced from the imprints left by the gun butts where they had leaned the rifles up against a tree that two of the poachers were armed. They were hungry too, as they were now deviating more and more from their original direction. Obviously they were looking for something small to shoot for meat. Marula trees are scarce in this area, but every one within sight had been inspected for fruits. All these were good signs. We were making up for lost time and closing in on our quarry.

"As the sun climbed higher, dehydration became our biggest problem. Would we have enough water not to have to leave the spoor to top up our bottles? The heat and the angle of the sun – almost right above us – now made tracking more difficult, as the tell-tale shadows in the spoor were harder to see. There was only one consolation; the poachers were suffering under the same conditions, including the swarming flies that sucked the moisture from our sweating faces. To them the flies were a nuisance, but to us they were becoming unbearable.

"The water we drank out of the bottles was now more suitable for making tea than for cooling a parched throat. Even if we might have talked, silence being of the utmost importance, we probably couldn't, because our mouths were too dry. Food was unimportant. We had eaten the night before and were now probably running on high octane adrenalin, as neither of us felt hungry. Just the heat, the flies and thirst. Fortunately our packs were almost empty, except for a litre or two of water, which we would make last as long as possible.

"At about 6 p.m. the spoor suddenly changed direction. Apparently something must have startled the poachers as they were now heading straight for the border. The obvious question was whether they would reach it before we did, in which case we would have to turn back. They were also moving faster than before, but we still had to stick to the spoor, because if they suddenly changed direction, we would lose them.

"We pushed on as hard as possible. Weeks of hardship depended on our catching up with them in time.

"This group, we were certain, had been responsible for killing at least 11 elephants, among which were a pregnant cow and a calf not much bigger than a Great Dane. How many others they had wounded with their Russian-made guns, we didn't know.

Right: This illustration depicts rangers looking for tell-tale "sign" while on anti-poaching patrol along the Mozambique border.

"Then, at this crucial stage, we lost the spoor in the rocky Lebombo terrain. We spent some time carefully looking for any sign of them, but when we at last picked up their track again, they must have been about 10 minutes ahead of us. It was clear that they had moved with great speed towards the border.

"With a last desperate burst of energy we moved towards the border fence. We couldn't throw all caution to the wind as the skelms were still armed – and close to hospitable country. The tracks led – as we now were certain they would –

Although he was bleeding heavily and in great pain, Aron walked nearly 2 km to an outpost . . .

straight to the fence. Another failed mission.

"We turned our backs on the border with the sick taste of disappointment in our mouths. Tomorrow we would try again, in this seemingly unending battle against the poachers. For the moment we could only think of the long walk back to base camp."

On another occasion ranger Rafael Chiburre of the Stolznek area (near Pretoriuskop) was cycling to an outpost to deliver a message when he saw a small herd of fleeing buffalo. He stopped and saw two men stalking the animals, one with a rifle and the other with a spear.

Rafael quietly laid down his bicycle and crept nearer until he reached an open patch with little cover. He stood up, ordered the men to stop and fired a warning shot above them. The men promptly fled in different directions. A second warning shot simply made them increase speed.

The ranger chased after the fugitive armed with the spear and caught him after a dash of 300 m.

Two other rangers in the vicinity heard the shots and came running, suspecting poachers. As they approached they saw the man with the rifle making for the Park's boundary. He spotted the rangers and it became a race for the fence. One of the rangers, Aron Nkuna,

caught up with the fleeing man, challenged him to stop and fired two shots in the ground near his feet.

The poacher immediately dived behind a bush and shot the ranger from about 13 m, hitting him in the stomach. The other ranger returned the fire but the man managed to escape across the border.

Although he was bleeding heavily internally, and in great pain, Aron walked nearly 2 km to an outpost where his colleagues applied first aid. By good fortune, two doctors happened to be guests at Pretoriuskop camp and rushed to Stolznek. They gave Aron emergency treatment and he was taken to hospital in Nelspruit, where he recovered completely.

One indication of the levels of poaching in the Park comes from the Shangoni area between Mphongolo in the north, Nalatsi in the south and a distance of 15 to 20 km from the western border of the Park. Shangoni is an outpost closed to visitors so few people know of it, but ranger Paul Zway recorded that in a 13-month period between 1983 and 1984, 92 poachers were arrested there plus 11 other people for damaging trees, plants and the like. He and his staff removed about 3 000 snares and three gin traps. From poachers they seized nine spears, 18 axes, about 30 knives, 15 fishing nets, seven fishing lines and five rifles. They had to destroy 31 of the poachers' hunting dogs.

Snaring, Zway recorded, was the commonest form of poaching and the fish poachers were usually women. Almost all the poachers caught said they were doing it for money – selling dried meat and fish. Some bring packs of up to 15 dogs into the Park.

Snares are a continuing problem around all the Park's borders and kill several hundred animals a year, according to warden Pienaar. But the problem is being contained.

To meet the poachers who come across the Mozambique border in the east armed with automatic rifles on equal terms, the Park now has a trained body of para-military rangers. They were put through a course of weaponry skills, combat drill, fieldcraft and basic discipline in mid-1984 with the help of the South African Defence Force. Their special assignment is to protect the Park's elephants – the poachers' favourite target.

Sometimes, of course, luck swings in the rangers' favour and they get help from unexpected quarters. In 1982 a poacher was visiting his line of snares to see what they had caught, and discovered that one had trapped something he did not particularly want: a very annoyed leopard. When it saw him it tore the snare free and attacked him with such energy that he ended up in hospital badly wounded.

Flying for science

Hugo van Niekerk has a great respect for elephants. As a dedicated conservationist he respects all animals, large and small, but more so elephants, perhaps because his close encounters with these kings of the Kruger National Park have convinced him that they have a sagacity beyond all others.

Since April 1974, Hugo has built up an intimate knowledge of the Park's rich range of wildlife from an extraordinary viewpoint: a pilot's seat, mostly in a helicopter but often in a fixed-wing aircraft. Today he is chief pilot for the National Parks Board and, with the only other full-time pilot, Piet Otto, is called upon to operate the board's two Bell Jetrangers and two Cessna six-seaters in any of the 11 national parks scattered all over the country. Four other Park staff members are also qualified to fly fixed-wing aircraft and give valuable assistance to the team.

"There are few people in our country who can tell us anything about conservation flying or game catching . . ."

But Hugo mainly flies in the Kruger National Park from the headquarters camp, Skukuza, where he lives with his wife Deirdre. Piet Otto also lives there with his family, as does their engineer, Mike Rochat, whose meticulous job it is to keep their aircraft flying safely in unusually tough conditions.

Flying in a game reserve might sound like fun and in many ways it is. Primarily, though, it is an arduous, taxing, often dangerous and always exciting full-time job for the three men which none of them would easily give up for the softer work of commercial aviation, although they are fully qualified for it and would certainly earn more money there.

"Both Piet and I are conservationists," says Hugo. "We help in the culling of animals but if we did not believe it had to be done, we would not do it. I could not kill an animal myself. We have both had fantastic offers from outside but we still prefer to work here.

"We do this job because we want to be here. We love nature and we also love flying. There are very few people in our country who can tell us anything about conservation flying or game catching."

Hugo came to the Park after 10 years of service in the South African Air Force, where he came second in his helicopter training course and ultimately flew the big Super Frelons, followed by a brief spell of running his own crop-spraying company in the Tzaneen area of the Lowveld. Piet Otto also came from the SAAF, where he flew Alouette and Puma helicopters.

Both have a wilderness background. Hugo's father's sport was hunting big game in various Southern African countries and on camp with him the son found an affection for wildlife burgeoning in him which became a lifelong obsession. He always intended to eventually become a conservationist and the crop-spraying company (which earned its next owner a great deal of money) was a stop-gap while he waited for a vacancy in the Park.

When the Park staff first began using aircraft for game counting or catching they hired charter planes and pilots. It worked, but not as well as they would have liked because game catching and culling, especially, are exercises demanding uniquely specialised skills.

So finally the Park bought its own helicopter and decided to take on a full-time pilot; when the post became vacant Hugo, who had repeatedly applied, eventually got it. He has never looked back and as the use of planes expanded, Piet and Mike were taken on to maintain the growing little fleet.

Hugo has now accumulated some 8 000 hours of flying time, of which more than half is on helicopters, and Piet about 6 000 hours, also more than half on helicopters. This is a considerable record when it is considered that the law permits helicopter pilots a maximum of eight hours a day, 30 hours a week, 100 hours a month and 1 000 hours a year.

They average about 500 hours a year but when taking the annual animal census they fly about seven hours a day for four days in a row, with the next four days off doing their desk work and squeezing in golf or other pastimes.

"Game flying", as these pilots call it, demands even more skill than combat flying in a helicopter, according to Hugo. He attributes the skills he and Piet have developed, however, largely to their air force training: "It wouldn't be wise to put a civilian-trained pilot into game flying."

one gave the rhino the antidote to bring it back to normal.

"The helicopter was ready to take off and the rhino was just lying there, as if dead," he said. "Its nose was flat on the ground and it lay there flicking one ear and then the other. But from that position the rhino got up within one second and charged the helicopter. I just managed to pull up and it passed underneath me."

Now he never parks closer than 40 m to a rhino or other large drugged animal, leaving it to the rangers in attendance to administer the antidote and then set speed records to the helicopter before

"Mike, naturally, took off at high speed for the nearest cover, as did everyone else . . ."

They have done many hundreds of Park operations without an accident under the most strenuous conditions (saying this, Hugo quickly touches wood) although they have had a few close shaves. Two of their predecessors were not so lucky: both were flying Jetrangers when big bateleur eagles came crashing through their windscreens, sending perspex, feathers, blood and bone flying into the cabins. Both survived without crashing but were bruised, both mentally and physically. One was Dickie Kaiser, whose wife figures elsewhere in this book in a dramatic battle between three men and a crocodile (see p. 101).

Since then Hugo and Piet have observed that when an eagle or other kind of bird takes fright while in flight, it invariably drops, presumably to gather speed and escape. So their firm rule is never to fly below eagles, vultures and the like but to go around or over them.

Piet once had a narrow shave in a Jetranger. He was flying low when the craft suddenly lost power and plunged straight down into the middle of a herd of buffalo. Piet hauled at the "collective lever" for more power. Centimetres from the ground, the helicopter recovered sufficiently for him to gain height and return to base.

Hugo himself came close to disaster while his helicopter was on the ground, parked, with engine idling. It was during an operation to dart-drug a black rhino for examination. He landed close to the sleeping, immobilised animal while the scientists did their work and watched as

the irate animals can get back on their feet.

Even engineer Mike Rochat has had an uncomfortable encounter. It was during the drugging of the huge bull elephant Mafunyane, a member of the Park's Magnificent Seven (see p. 126) for examination.

Mike was one of the people on the ground when the bull, with the help of a push from a front-end loader, struggled to its feet and began flapping its sail-like ears and looking for a target for its awesome tusks. Mike, naturally, took off at high speed for the nearest cover, as did everybody else in the immediate vicinity.

Mafunyane began lumbering away from the scene and as fate would have it, headed in the same direction as Mike and another man – probably not deliberately, but simply wanting to escape the area.

Hugo was already airborne in a Jetranger and brought it down low between the bull and the running men, hovering close to the ground. The racket and the dust and debris the plane churned up turned Mafunyane away.

Hugo, however, decided that the two men needed exercise and flew close to them, pointing with a finger down into the bush behind them. Both took this to mean that Mafunyane was still in hot pursuit and turned on the speed. When Hugo picked them up in a clearing about 400 m further both were gasping for breath – and not at all amused to learn they were not being chased.

Chief pilot of the Kruger Park, Hugo van Niekerk, right, with engineer Mike Rochat.

If there is a flying operation in the Park which might be called tedious (although pilots unaccustomed to it would not think so) it is game counting – taking the annual census of those animals that can be counted from the air such as elephant, rhino, buffalo, giraffe, roan and sable antelope, wildebeest, zebra and, if necessary, animals as small as impala. However, it can also be hazardous.

The Cessnas fly at 250 feet, the optimum game-spotting height for their type and speed. At this height the pilot has almost no glide distance to find a suitable clearing should his engine cut out.

The Park's chief research officer, Dr Salomon Joubert, and his five passengers flying on a counting operation were lucky to escape when the crankshaft of their Cessna suddenly sheared. Hugo had his helicopter parked about 2 km away when he heard Dr Joubert call over the radio that the plane was losing power and heard the changed note of its engine. He urgently told Dr Joubert to find a road and try to follow it back to base, so he could land on it in emergency.

But Dr Joubert had no time. Power dropped further. Hugo told him to find any clearing and commit himself to a landing there. Dr Joubert did just that, with just enough power coming from the failed motor to slightly boost his glide. As he went down Hugo jumped

into his helicopter and took off so fast for the crash landing site that when he got there the dust clouds were still settling.

All six people who were in the Cessna were standing around it, shaken but unhurt. "All I saw was clouds of smoke," Hugo said. "They had one pipe between them and they were all smoking it . . . puff, puff . . . passing it around between them, non-smokers and all!"

Dr Johann Botha, then editor of the National Parks Board magazine *Custos*, flew with a game counting team and brought back this graphic account (*Custos*, Vol. 11, no. 6):

"Before taking off, the pilot steered the Cessna towards the farthest end of the runway so that the aircraft would clear a number of grazing blue wildebeest and zebra.

"The fat zebras trotted away somewhat indignantly, but the wildebeest had to make a scene as usual, galloping away, snorting and with streaming manes and tails. Coming to a standstill they stood shaking their heads as though commenting on man's crazy conduct.

"It was still early on a winter's day – 07h20 – the right time to take off if you have four to five hours of monotonous back and forth flying over the bush ahead of you.

"Usually two shifts are flown. The first shift lasts until about 11h00, when the pilot takes the aircraft down to allow the crew to stretch their legs, have a smoke break and quench their thirsts. The second shift ends at about 13h00, for by then everyone has had his fill of sitting in the cramped space.

"By that time perceptibility is deteriorating since the animals tend to seek shelter from the sun in the shade of the trees and shrubs. The flight becomes more bumpy and the cabin of the Cessna 206 sultry. And it is then that the symptoms of nausea appear. Clearly time to land.

"But the early morning is virginal. Africa in its primeval purity slips past below the aircraft and one is overwhelmed by a profound affinity for this harsh landscape.

"On our way to where the day's grid flight plan had to start, we sat back, relaxed and observed the area as a whole from a height of about 400 m. At this height no particular detail of shrub or animal stands out, but neither does the earth acquire the clinical features of a map.

"The primary contours of the area are prominent, the folds and undulations, catchment areas and run-offs, the criss-cross of roads and game paths. From that height, the extensive savannahs resemble patches of illuminated white against the brown soil; massive lead-wood, marula and ebony trees shrink to flecks; a line of wildebeest becomes a mere column of ants on a kitchen floor.

"The aircraft descended in a wide circle to a height of 60 metres and the first leg of the grid flight commenced.

"Six pairs of eyes searched the area. It takes time to get accustomed to the new perspective in which the various species are seen, but at 60 metres above the ground the proportions are more normal.

"In these early hours of the day the animals seek out the warmth and they stand in clearings on the sunny side of trees and thickets. The early light reveals them in perfection.

"The observations are called out.

'Zebra, six. Giraffe, five; calf one.'

'Impala, about 45. Male kudu, two.'

'Steenbok, one. Wildebeest, 17 . . . 18. Correction, 21.'

"These ecological aerial surveys occupy about three months a year.

"The whole Park is systematically criss-crossed. The habitat is continually assessed according to a system of 13 criteria – the length of grass, grass cover, stage of growth, accumulation of dead vegetation, utilisation, trees and shrubs and their seasonal stages, the effect of veldfires and other factors.

"The larger herbivores are counted and data on distribution, herd sizes and herd composition are recorded for feeding into the computer every afternoon. Particular attention is paid to the rarer species, such as roan and sable antelope, and where practicable, as for instance when counting giraffe, offspring under the age of one year are noted separately from the adults.

"The flight pattern which is followed has the advantage that it can be repeated. Probably only a few of the smaller species such as steenbok, duiker, nyala, bushbuck and reedbok are observed, but the numbers remain consistent, and the information forms part of the broad spectrum of knowledge according to which the total ecosystem can be moni-

tored and sensible management decisions can be made.

"Against the background of the vegetation surveys – the charting of soils and water sources, and a special census taken by helicopter of elephant, buffalo and hippo – a fund of knowledge is built up which will be of immeasurable value for many generations to come.

"In the meantime the legs of the grid flight are completed one by one. It is done with meticulous care but not without some light-heartedness: there is time for jesting and good-natured derision.

'Female kudu, six, no seven . . . Blast! Correction, they were waterbuck.'

'It looks like the A team has had it. Throw them out, Sollie, they're useless!'

"That is why six pairs of eyes are needed – to supplement and correct, to generate a spirit of companionable competition and introduce a bit of variety into a task which could, conceivably, become an exhausting and boring routine.

"To these men it never does. They appreciate that it is a privilege to be involved – no matter in what capacity – in the furthering of an ideal which materialises daily in the concrete form of a black rhino calf, the nest of a white-headed vulture, a bateleur in flight, a roan among sable antelope – the ideal that it is an eminently worthwhile job to help keep this world of ours intact."

While counting animals takes up a large slice of the flying year, it is by no means the only task for the pilots, nor the biggest. They routinely have to do a variety of jobs in other national parks and even abroad (such as a recent operation to catch Lichtenstein's hartebeest in Malawi), take Park staff on inspections of fences and rivers and dams, help them chase poachers, fly senior staff from headquarters to and from various parks, and make mercy flights to take heart attack victims, road accident casualties and the like to hospitals.

Most of their work comes under the generic heading of "chasing animals". They fly to herd animals from the air into fences of netting so they can be caught for any one of a variety of reasons, from sending them to other parks, to translocating them within this Park, to enabling scientists to examine them.

Helicopter pilot Piet Otto.

They fly so that sharpshooting rangers can fire darts into them, to immunise them against disease, immobilise them for inspection, or to knock them out for slaughter in culling operations. Nobody particularly likes this last task but it has to be done: the Park has a finite carrying capacity for all of the species it shelters (see p. 71).

After all their years at it Hugo and Piet are probably the most specialised practitioners of animal chasing in helicopters in Southern Africa. It is, in pilots' language, the most "hairy" work. It is rough on machines as well as pilots: engines are well protected from dust by particle separators, but main and tail rotors get severely eroded by the sandblast effect of dust kicked up as they whirl and swirl close to the ground, and the strain on gearboxes is enormous – so far above the normal that they make inspections and replacements of certain parts like freewheel units far within the time limits set by helicopter manufacturers.

The planning, siting and execution of culling operations revolves very closely around the helicopter and its pilot. In close collaboration with the ground staff, he chooses a suitable site to which the animals to be culled will be driven across terrain suitable for what is virtually stunt flying, and the right kind of weather conditions. Safety and efficiency are the priorities.

During culling operations, the helicopter is often only a metre or two above the ground, dashing erratically from one side to another, up and down and even backwards. The pilot rarely has time to check his speed or fuel gauges, although his eyes constantly flash across those giving vital information like temperature and oil pressure. His eyes rove from instrument panel to animal to bush to ground and from side to side to watch for a tree or slightly higher bush that might clip his skids or his tail rotor. Hugo and Piet have found that of all the physical strain such flying imposes, their eyes suffer most.

There is one Park operation, in Hugo's view, which takes the prize for toughness: darting roan antelope. These rather rare creatures – their population numbers about 340 – are susceptible to anthrax and have to be regularly immunised, which is done by giving them injections from the air.

A roan antelope at top speed can reach about 60 km/h and they are notorious for their ability to change direction at almost right angles in mid-stride as they swerve and weave through the bush.

"There is nothing as difficult as chasing a herd of these, say 17 animals, over constantly changing terrain for several kilometres," says Hugo. As they gallop, the hooves of the roan throw up clods of earth which splatter on the windscreen of the helicopter, which is following so low its skids are at the height of the animal's withers. This demands precision flying of a very high order and takes much out of both pilot and shooter.

After all his years of flying of this kind, Hugo thought he had learned as much as one can about animals from the driving seat of an aircraft – especially about elephants, which he loves and respects. Recently he found he was wrong. Twice in quick succession two cow elephants taught him something new – and almost brought him out of the sky, as he tells in his book *Vlugkastele*.

He was on an elephant culling operation in the Nyando bush area of the Park, in the north between Pafuri and Shingwedzi, herding a group of 10 which included a large cow with a fairly young calf.

"On the way I noticed that one cow kept turning back towards the helicopter. The rest of the group would walk on for about 100 metres and wait for her.

"She did this quite a few times and I got annoyed because she was holding up the whole operation. She was walking from tree to tree and then she went

behind one tree and I got the impression she was looking for a tall, thin one."

Hugo went up closer to try to "blow" her out – blast her into moving away with the powerful downdraft from the helicopter.

"She kept the tree between the helicopter and herself all the time. Every time I moved around the tree she moved around the other side.

"I was lucky because if she had waited another five or ten seconds . . . I was about to go lower when she suddenly jumped forward and pushed the tree over, straight at the helicopter! It slammed the ground with leaves and branches flying and came within two or three feet of the helicopter's skids. I actually went cold. I got goose pimples all over."

There is no doubt in Hugo's mind that the elephant cow tried to swat him out

of the sky. Two other Park rangers with him agreed. And that was not the only time.

A few weeks later it happened again about 10 km from the scene of the first swatting, again with a cow and her young calf.

"I noticed the same behaviour with this second cow, when she started dodging from the helicopter behind a big tree. I said to the rangers with me: 'Well, you didn't believe me the last time, so watch this!'

"And, by gosh, the same thing happened. She pushed the tree at me!"

This time Hugo was ready for it and kept out of trouble, and he had more witnesses to this extraordinary demonstration of elephant intelligence.

Now Hugo really respects elephants, and steers clear of them as well as of flying birds.

EAGLES IN THE SKY . . .

Martial eagle

Brown snake eagle

Crowned eagle

Black eagle

Wahlberg's eagle

Longcrested eagle

Bateleur

Tawny eagle

Fish eagle

Blackbreasted snake eagle

African hawk-eagle

Steppe eagle (migrant)

Lesser spotted eagle (migrant)

The Kruger National Park is richly endowed with eagles and 15 species patrol the skies, 11 of them local breeding residents while another four are seasonal migrants. As an aid to identification, 13 of the 15 wing profiles are shown here. The migrants not shown are the rarely seen Ayres' hawk eagle and the booted eagle.

(Printed by kind permission of the Transvaal Nature Conservation Division).

With up to half a million people visiting the Kruger National Park every year, it is inevitable that close encounters of an unnerving kind occur occasionally between them and the resident population of hundreds of thousands of large animals and innumerable small ones.

But in its long existence the Park has not lost a single tourist because of animal behaviour and, apart from road accidents, there have been exceptionally few clashes between man and animal.

People have died violently in the day to day business of the Park, although much fewer than in the average industry. However, the deaths have been among those whose livelihoods involve risk: the Park rangers and the poachers – the real people of the bush. Some have been the innocent victims of completely accidental confrontations with enraged or frightened animals but by far the majority owe their fate to their own carelessness, stupidity or foolhardiness.

Tourists have indeed had very close shaves. In the old days before camps were securely fenced, several people who slept on the ground outdoors, ignoring elementary rules, were mauled by scavenging hyenas – two of them very badly. Park rules giving elephants the right of way are to be taken seriously – as proven by the holed and battered cars of a few tourists who drove too close.

Baboons are extremely dangerous. Tourists too often refuse to believe that these clowns of the bush, however amiable they might appear, should not be fed snacks from cars.

As a result of this habit many baboons have lost their natural fear of people, especially near Skukuza, and expect to be fed. They will clamber all over a car and can do considerable damage to windscreen wipers and similar attachments. If teased or frustrated they can become viciously angry – and there are few things more frightening and dangerous than an irate baboon reaching into

Left: The King of Beasts has been the central character in most sagas of the Park.

a car window or even climbing inside to grab food. A full-grown male chacma baboon can weigh 45 kg; it has fangs bigger than a leopard's and its strength and speed are awesome.

Tragic encounters

Over the years encounters with lions, elephants, buffaloes and other large mammals have taken a small but steady toll of Park personnel's lives and some episodes have involved displays of incredible courage.

One such tragedy was the death of the wife of a ranger at Mahlangene, an outstation in the Park's central section. Like her husband, she was familiar with the ways of the wild and the dangers in the Park, but was trapped by unusual circumstances.

She and six other women were collecting firewood near the post and paused to drink from the Little Letaba river where it flows into the Great Letaba. Here a fully grown hippopotamus confronted her. She shouted and tried to escape but the big animal, which moves with surprising speed when alarmed, lurched forward and bit her once in the abdomen. She died before reaching hospital.

Park experts said the water level in the river was abnormally low at the time; this gives hippos a feeling of insecurity, which is probably why it attacked.

Carelessness caused by becoming too blasé caused the death of a worker near Punda Maria camp.

Louis Mathye and two colleagues had spent a weekend visiting friends outside the Park; when they arrived back at the entrance gate on the Sunday evening, they found they had missed the official transport to Punda Maria. Instead of staying overnight at the gate, they broke a cardinal rule by deciding to ride to the camp on bicycles after dark, with Louis on the carrier of the rear one.

It was nearly midnight when a lion charged the leading cyclist. It missed him but hit the following bicycle carrying

Louis. The man pedalling felt the shock as Louis was thrown off, screaming for help. His terrified friends pedalled furiously on for Punda Maria where they woke the ranger in charge.

Quickly arming himself with a rifle and rousing a driver and an assistant, the ranger drove back to the scene and found a black-maned lion tearing at Louis's body. He shot it and stayed on guard until daylight when police could be summoned, as they must be in all cases of unnatural death.

When morning came the ranger examined the dead lion and found it was missing its lower left hind leg, which had probably been caught in a snare, leaving a raw and septic wound.

It is this kind of injury which makes it impossible for a lion to tackle its usual prey, the fleet impala, zebra or wildebeest, and forces it to hunt the slowest and easiest: man.

Rescue from a crocodile

The ranger who shot the lion that killed Louis Mathye was Louis Olivier, who by a grim twist of irony, himself had had a brush with appalling death five years earlier.

The incident ranks as one of the most dramatic examples of courage, camaraderie, physical strength and fortitude in the Park's history and resulted in two men being awarded the Wolraad Woltemade Decoration for Conspicuous Bravery – South Africa's highest civilian honour for gallantry, equivalent to Britain's George Cross.

The Wolraad Woltemade Decoration for Conspicuous Bravery.

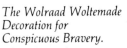

On the hot day of 21 November 1976, Louis Olivier and a fellow ranger Tom Yssel, inseparable friends, were sitting in a hut at Skukuza camp trying to prepare for an examination they had to write the next day. It was a Sunday when everybody else was relaxing and the heat was oppressive. Eventually Louis succumbed to Tom's plea to join friends who had gone out for a *braaivleis* (barbecue) in the open.

They drove 6 km from the camp to waterhole number six on the Sabie river, a quiet picnic place much favoured by the Skukuza staff, where lush trees shade the river bank beside a deep pool where they are allowed to fish.

Unaccountably both Louis and Tom broke the habit of years by leaving behind two things they normally never travelled without: Tom's Ruger 0,357 revolver and Louis's sheath knife. The oversight cost them dearly.

At the Sabie pool they found other Park staff already lazing in the shade sipping beers and preparing the meal. Among them were Hans Kolver and Dickie Kaiser, both helicopter pilots, Frans Laubscher, the Park's chief engineer, and their families.

Laubscher was already out fishing from a broad, sunbaked promontory of flat rock which extended almost across the Sabie from the far bank, hoping to land a mudfish or *tilapia* or perhaps even a fighting tiger fish from the deep pool. Louis and Tom decided to go out and watch him and splashed through the narrow shallows at the near side of the river onto the expanse of rock.

They idled there for some time – talking seldom, listening to the summer

Ranger Louis Olivier.

buzz of insects and the peaceful call of hoopoes and kingfishers – until a voice called to them from the river bank to come and eat. They waved acknowledgement and the trio started back through the thigh-deep shallows with Louis in front, Tom close to him and Frans coming up behind with his fishing tackle. Louis paused for a moment and turned around to say something to Frans, allowing Tom to take the lead. In that instant the water erupted behind him and he heard Tom scream. The

thought flashed through Louis's mind: "Damn Tom, he shouldn't play the fool like that, I'll give him hell later . . ."

Tom wasn't playing the fool. The crocodile slammed into him like a torpedo, clamped its powerful jaws on his right lower leg and immediately thrust across the shallows for the deep water of the pool – 5 m of killing machine strong enough to drown a buffalo swam effortlessly away with its prey.

Louis spun around and the expression on Tom's face burned itself into his memory: shock, pain and utter helplessness. He recalls that a remarkable, cold clarity took over his mind and everything seemed to happen in slow motion as he instinctively plunged into the boiling river after the crocodile and his friend.

Louis spun around and the expression on Tom's face burnt itself into his memory . . .

Louis flung himself full length straight at the beast. He sank under it and tried to grasp it around the body between the forelegs and hindlegs but its girth was too big and the rough, knobbled hide rasped his skin as he tried futilely to close his arms around it.

Bracing his feet against the river bed in the shallows, he desperately tried to slow the crocodile, to divert its attention, anything to make it relax its grip on Tom. But it swam forward like a powerboat, its long jaws firmly clamped on the

man, cleaving the water. Louis clung on grimly, most of the time with his head submerged. Once when he surfaced he saw the crocodile had shifted its grip from Tom's lower leg to his thigh.

Through the shock and pain Tom had the sense to gulp fresh air every time his head came out of the water and to cling tightly to the reptile's head and jaw to prevent its sharp, whiplashing movement from tearing his leg from his body.

Somehow Louis summoned unknown reserves of brute strength and forced the crocodile to turn about 15 degrees towards a sandbank, away from its course to deep water. Undoubtedly this action rescued Tom from the immediate danger of drowning. They had won round one of the grisly battle.

Louis suddenly remembered that he had a small penknife with a 5 cm blade in his back pocket. Holding on to the thrashing, plunging crocodile with one hand, he dug for the knife with the other, opened it with his teeth and tried to stab the beast in its stomach – "zip it open", as he said later. The blade buckled and Louis hurled it away.

Weeping with utter frustration and with the spittle of fury flecking his beard, he dragged himself on top of the crocodile and hammered at it with his bare fists until his knuckles were raw. His arms were still flailing when it pulled him and Tom underwater yet again.

When the crocodile seized Tom the picnickers on the bank heard his cry and ran to the water's edge. The women screamed, horrified but helpless. When Hans Kolver reached the scene he could not see Louis, who was at that moment beneath the surface and hidden by the turbulence.

What he did see, he said afterwards, was Tom's face with a desperate plea for help frozen on it. "I knew," Hans explained, "that Tom was finished, beyond help." But he did not hesitate for a second. He jumped straight in at the crocodile's head, thinking that Tom must not be allowed to die believing his friends had deserted him.

When Louis surfaced again he saw the astonishing sight of Hans clinging to the crocodile's back like a rodeo rider and bashing at its head with his bare fists.

Tom shouted at Hans with remarkable presence of mind for a man so close to death: "The eyes! Go for the eyes!" Tom

The victim of the crocodile attack, Tom Yssel, photographed recently with his wife Petro. He is now ranger at Pretoriuskop.

had tried himself, but so big was the reptile and so long were its jaws that his outstretched fingers could not reach the eyes set on top of its head.

Hans dug his fingernails into the eyes and the crocodile reacted with an explosion of energy. It whipped up its head and shook Tom as a dog shakes a rat, snapping his leg like a twig, then sank underwater with him still firmly in its grip. Louis saw Tom vanish. He surged forward drawing on yet further reserves of strength and lifted the head and Tom out of the water. Tom gulped air and gasped his gratitude.

At this stage Louis's rage was so great he actually believed he could defeat the crocodile; there was no question of him giving up. He tried again to lock his arms around it with the idea of hauling it from the river, but again he could not close his arms around the massive body. He seized its hindquarters to tow it to the bank but with one effortless flick of its tail the crocodile flung Louis metres away as if he were a feather.

Now winded and stunned, Louis realised that the crocodile was far too strong

for him. Staggering to his feet in the churning water he saw Hans still pounding the beast's eyes. With each stab of Hans's fingers the crocodile shook its head with a power that would have torn Tom apart had he not clung so tightly.

"That could be my daughter in its jaws," Hans thought and he struck with greater determination. He saw that at some point in the struggle the crocodile had again switched its grip and now held Tom around the waist, its teeth opening up a huge gash across his stomach.

A weapon, Louis thought. He splashed out of the river and ran to his truck, only to remember with numbing despair that they had left his knife and Tom's revolver behind. So frantic was Louis that he plunged straight back into the river without thought of finding something else, a knife or even a fork from the picnic.

In the water the scene had changed, but how nobody was able to remember afterwards. Tom was pulling on the fronds of a tree which drooped low over the water while Hans was trying to push the crocodile up against the bank,

hoping to hoist it far enough out to make it slacken its hold.

Tom seemed to despair. "I'll see you on the other side," he gasped to Hans. "Stop that talk!" Hans scolded. "You can't give up, Tom, you must help us beat this thing!"

Louis then tried a new tactic. Clambering back up the bank, he seized an overhanging branch, lowered his legs down the side and yelled at Tom to take hold. Dickie Kaiser, the other pilot, gripped Louis under his arms and pulled and a tug-of-war began between the men on the bank and the crocodile with Tom as a human rope. The beast's jagged teeth bit deeper into Tom.

Tom rose higher from the water and Louis and Dickie felt sudden hope. But it was not the crocodile giving up – it was Hans down in the water bodily lifting the beast with muscle-cracking effort. It could not last: their strength faded and the crocodile began to gain. Thorns cut into Louis's palms as he was dragged downwards and inches from his face he saw Dickie's heels gouge into the earth as he slid towards the edge of the bank.

At this point Dickie's wife, Corrie, rushed to their help. She pushed a small, short-handled spade into Louis's right hand. He dropped into the river beside the crocodile and swung the spade at the reptile's eyes – but Hans thought he was passing the spade across and reached out for it. The spade glanced off his hand and spun away into deep water.

The grim battle continued but it seemed nothing would make the crocodile release its grip. Hans and Louis tried

His blood was leaking into the water and shock and exhaustion had dulled his senses . . .

to force their fingers into the crocodile's tightly closed nostrils – "about as easy as pushing your finger through a leadwood tree," Hans recalled. The beast repeatedly freed Tom for a split second to snap at the other two men and grabbed him again before he could move a muscle to get away, each time puncturing his skin with another row of ugly perforations.

The struggle was taking a heavy toll of Tom. His blood was leaking into the water and shock and exhaustion had dulled his senses to beyond caring.

Realising that time was fast running out, Hans warned Tom that he was going to have "one more go" at the reptile's eyes and Tom braced his battered body for the inevitable bone-crunching shaking of its head.

Abruptly, the crocodile caught all three by surprise. With dazzling speed it dropped Tom and turned on Hans. All Hans could do in that split second was cross his arms in front of him to protect his chest. It grabbed his wrist, crushing it, then released it and bit into his upper arm as he turned a shoulder towards it. He said later, "it was better to lose a little bicep than a whole arm." Only the teeth at the very front of its jaw held Hans but the hold was like a vice.

Battered and broken, Tom stood chest deep in the water on his one good leg next to the trapped Hans.

"I was so relieved at being let go," he said later, "that I was suddenly full of adrenalin and because I felt no pain I wanted desperately to 'have a go' and help Hans."

He realised that there was nothing he could do when he looked down in the water "and saw with horror that one of my feet was facing backwards. With one hand I was holding my intestines in and the blood was snaking away downstream, and then I worried that other crocs would be there like a pack of wild dogs."

Now Hans was the helpless one trying to keep his head above the surface. When the crocodile caught him it immediately rolled in the water with the familiar corkscrewing motion they use to tear pieces off their prey. Only by keeping his head and gripping the eye cavity with his good hand, was Hans able to roll with it and prevent being torn apart.

With this new phase of the fight the resourceful and unflappable Corrie Kaiser again came to their aid, this time with a butcher's knife. Louis grabbed it and waded in pursuit of the crocodile. Realising that this was very probably the last chance, he felt with his free hand for the crocodile's eye with almost loving care.

He poised the knife over the eye and drove it in with such force that it sank deep into the socket and buckled the tip of the blade. The great beast shuddered, let go of Hans and swam slowly to deep water.

Pilot Hans Kolver – he also suffered injuries.

The battle was won. Now a new terrible fear overwhelmed the three men – that other crocodiles might have been attracted by the blood and commotion. The river bank where they stood chest deep in the water was so high that only the uninjured Louis could reach the overhanging foliage and pull himself out.

Hans summed up the situation and acted quickly. The slightly built pilot ignored his own injuries and bodily hoisted Tom first onto his shoulders and then, with a grip on the seat of his pants, literally hurled him up the bank

into the outstretched arms of Louis and the others.

Hans was now alone in the water. Turning, he pushed his way against the current for "the longest walk of my life" to a point 20 paces upstream where the bank was low enough for him to clamber out with the aid of the others.

Louis, still clutching the bent knife, was so swamped with fury that he went back into the river to find the crocodile and kill it, screaming for it to "come and get me". It had disappeared and the pool was still again. His strength and anger suddenly drained and, shaking almost uncontrollably, he waded out and helped the rest of the party load Tom into his light truck.

The entire drama lasted 15 to 20 minutes, Louis recalled later, "but it's hard to say – it felt like hours."

Louis drove the 6 km back to Skukuza "like a maniac" and was still so deep in shock when he arrived that the chief warden had to smack his face to bring him to his senses.

Tom's injuries were serious. His right femur was completely severed, the flesh of that leg was shockingly mutilated and his midriff was torn open. After receiving first aid he was flown by helicopter to the Nelspruit hospital and then a few days later to a private hospital in Pretoria for specialised treatment.

His life hovered in the balance for weeks. He was delirious from the deep infection of festering wounds in his leg, which should have been amputated but so weak was he that the surgeons dared not risk the operation. His life was saved by the dedication of the hospital staff and potent new drugs and he emerged six months later with terrible scars as mementoes of the epic fight. He is able to walk with a heavy limp although his thigh is still broken and he will need further surgery and grafts.

Hans's wrist healed completely and today he is a Boeing pilot with South African Airways. Louis and Tom are both back at their posts as rangers in the Park. Louis now has a young family and in 1982 Tom married Petro, whom he met at a party on the fifth anniversary of the crocodile attack. They now have two children.

Sitting in his neat home at Nwanedzi outpost, a stone's throw from the Mozambique border, puffing at his favourite pipe, Tom speaks of Louis with the affection of a close friend and of Hans with deep respect.

"When the crocodile took me I just knew that Louis wouldn't let me down," he says. "I knew he would be there because he was my best friend. We were always together in those days, wild bachelors of the daredevil kind, and Louis wouldn't have thought twice about diving straight in.

"But with Hans it was different. He took a brave, calculated risk that day. He had a wife, a family – they were sitting right there with him in the sun. From the moment he came into the water, until the croc let go of me, Hans was at my side in total control. He seemed to coolly make a plan all the time.

"Not once did he show panic, even when the croc finally took him. Several times when I was ready to give up, Hans would tell me not to be a fool, to hang on. He would also sense when the croc was about to take me under

After the award ceremony from left: Louis Oliver, Acting State President Dr Marais Viljoen, Tom Yssel and Hans Kolver.

and tell me to hold my breath, and when to hold tight for the terrible shaking. He is a hell of a brave man."

The crocodile was shot the day after the attack by another ranger – not from vengeance but because it was badly wounded, probably blinded, and helpless.

"We have nothing against crocs," Louis commented. "They fulfil a very important role in nature and we were just easy prey. Nowadays I go into water with butterflies in my stomach and an itch between my shoulder blades. For weeks I never went near water. We even sometimes go fishing – but oh boy, so carefully."

Nearly two years after that fateful day, on 8 September 1978, the entire staff at Skukuza, and many from elsewhere, gathered for a simple ceremony. Uncomfortable in formal suits, ranger Louis Olivier and pilot Hans Kolver were called to stand in turn before the then Acting State President of the Republic of South Africa, Dr Marais Viljoen, and were each presented with the Wolraad Woltemade Decoration for Conspicuous Bravery, for selflessly going to the rescue of their colleague and friend, Tom Yssel.

These three are more acutely aware than anyone of the pain and terror that another young Skukuza worker must have suffered on a fateful day in 1970.

Death in the river

Marius Meyer was a lean, bright young man who had been working for several weeks at the camp's post office as a clerk, a keen sportsman who was enjoying life in the wilds but did not have much experience of it yet, being behind a postal counter. Meyer also liked to go fishing in the Sabie in his off-duty hours.

On 18 January, a Sunday, Meyer and a couple of friends decided to fish a pool about 15 km from Skukuza, upstream on the Sabie. They were joined by Meyer's mother and stepfather who had come from the town of White River near the Park's western border to spend the weekend with him.

After more than an hour of fishing without success, Meyer told the rest of the party he was going to swim across the pool to fish from the other side.

His stepfather and the others remonstrated with him, saying the Sabie was renowned for its crocodiles, but Meyer insisted, stating that he had never seen any at this particular pool. He stripped off his shirt, waded in and swam across to clamber out through the reeds on the far side while his friends anxiously scanned the water.

But there was no sign of any crocodiles; not a ripple of danger.

Meyer fished for about another hour from the opposite bank without catching anything, then waded back into the river and swam to a rock in mid-stream, where he tried again. At about 11.30 his friends called him to come back because they wanted to go back to Skukuza for lunch.

Carrying his fishing rod, Meyer stepped down into the water, swam a few metres and then, when his feet touched bottom, rose and began walking up the river bed to the bank.

He was knee-deep when a crocodile rose from the water just behind him, plunged in again and closed its powerful jaws around Meyer's leg.

He shouted out in pain and the people on the bank looked up to see him being dragged down beneath the surface. A few seconds later his face and part of his body emerged fleetingly and then vanished again, leaving only a trail of bubbles which the slow-moving Sabie carried away.

There was absolutely nothing those on the bank could do. While Meyer's mother looked on in stunned shock, the three men ran up and down the bank, but they could see nothing. At the age of 24 Marius Meyer disappeared forever.

Police and Park staff searched from a helicopter and a boat, and also on foot upstream and down for kilometres but found nothing, not even a trace of clothing.

Post Office clerk, Marius Meyer, who died in the Sabie river.

Wolhuter and the lion

In 1904, when the Kruger National Park was still the Sabie Game Reserve, a ranger demonstrated courage so extraordinary that his story echoed around the world and brought the fledgling reserve international attention.

His name was Harry Wolhuter. In the Anglo-Boer War he served on the British side in the odd little private army known as Steinaecker's Horse (see p. 48).

When the war ended Wolhuter was offered a job as a ranger in the Sabie Game Reserve by Major James Stevenson-Hamilton. He took it, and so began a lifelong friendship between the two – the officer from Britain and the young man born and bred in the South African wilderness.

In August of that year, Wolhuter was travelling south after a routine patrol near the Olifants river in the middle part of the present Park with an entourage typical of those days: a couple of donkeys carrying all his worldly possessions, three dogs and four black policemen on foot. He was on horseback.

He intended to reach a certain waterhole on his second day of trekking, but when the party got there at about 4 p.m. they found it dry. Wolhuter decided to push on to the next waterhole some 19 km further, telling the policemen to follow while he rode on ahead.

As he wrote in his fascinating autobiography,

Ranger Harry Wolhuter.

Wolhuter was unperturbed about riding off alone even when darkness fell because he had never found lions in this area before. One of his dogs, Bull, stayed with him as he rode away.

This account of Wolhuter's remarkable experience is in his own words as it appeared soon after the incident in the

British *Journal of the Society for the Protection of the Fauna of the Empire*.

It was taken down verbatim from Wolhuter as he lay in hospital by the then resident magistrate of the Barberton district, who wrote that he could vouch for its truth.

"I know the man well, and sat with him at times when for weeks he was recovering in bed from his strange adventure; every circumstance was substantiated in the inquiries made, by an inspection of the spot, and in the post-mortem examination of the lion."

In it, Wolhuter describes what happened even more graphically than in his autobiography, written years later.

"... On 26th August, 1904, I had to do rather a long march on account of scarcity of water, consequently sundown found me riding along the bank some three miles short of my destination, Metzi Metz, accompanied by a large rough-haired dog (of no very special breed, but of tried courage), and carrying my 400 express rifle. My four natives and three donkeys were a few miles behind me.

"It was already pretty dark, twilight being a matter of minutes in these latitudes, and the path which I followed led along the banks of a small dry river bed. I had reached a place where a patch of long grass grew beside the path, when my dog Bull ran forward barking, and I caught sight of some indistinct forms which, from their general appearance, I took to be reedbucks; the very last thing I was thinking of was lions, having been fruitlessly tramping the country in hopes of securing one for some time. I therefore whistled to the dog, and the next moment was conscious of a lion close to me on the off side, and preparing to spring. I had no time to lift my rifle, but simply snatched my horse round to the near side, and drove the spurs in; he gave a bound which, no doubt, caused the lion partially to miss his spring, as his claws slipped on the horse's quarters, and though several ugly wounds were inflicted he lost hold. The concussion and the subsequent violent spring of the horse caused me to lose my seat, and simultaneously

I saw a second lion rushing up from the opposite direction. I absolutely fell into his jaws, and believe that he had me before I ever touched the ground. I imagine that these lions were after the horse in the first instance, there being no known man-eaters in the district, but finding me so easy a prey, this gentleman decided to accept what Providence offered to him.

"The next thing I recollect was being dragged along the path on my back, my right arm and shoulder in the lion's mouth, my body and legs underneath his belly, while his fore-paws kept trampling on me as he trotted along, lacerating the fronts of my thighs considerably and tearing my trousers to shreds.

"I had, of course, dropped my rifle, which I was accustomed to carry in a bucket in mounted infantry fashion. All the time the lion was dragging me along he kept up a sort of growling purr, sometimes like a hungry cat does when she catches a bird or a mouse, and is anticipating a welcome meal.

"My spurs kept dragging and catching in the ground till at last the leather broke. I cannot say that my feelings at this time were at all in accord with those of Dr Livingstone, who in his book, if I am not mistaken, expresses his feelings as those of dreamy repose, with no sense of pain. [The famous African explorer described in his book how he felt when caught by a lion.] I, on the other hand, suffered extremely in that respect, while I hope I may never have again to undergo such agony of mind as I then experienced; it seemed hard to die like that, and yet I could see no part of a chance, not the slightest loophole of escape.

"Suddenly, like a flash, I thought of my sheath knife; I always carried it in my belt behind my right hip, and on most other occasions when I had had a fall it had fallen out; was it still there? The lion holding me by the right shoulder, I was obliged to reach round and underneath me [with his left hand] in order to get at it. It took a long time, as it must be remembered that I was being dragged and trotted on by my captor all the time, but at last I managed it. How I held on to that knife! It was only an ordinary 3-inch blade of soft steel, such as one buys cheap at any up-country store, but it meant all the world to me then. I now no longer thought of death or anything else; all my mind and energy were concentrated on not letting go my one last road of escape. After dragging me nearly 200 yards, the lion stopped under a big forked tree with large roots [Stevenson-Hamilton later determined that the distance was actually 94 paces]; as he did so, I felt for where I judged his heart to be, and struck him behind the shoulder – one, two – with the energy of despair, using, of course, my left hand. He dropped me at the first stab, but still stood above me growling, and I then struck him a third time in the throat with all the force of which I was capable, severing some large vein or artery, as the blood deluged me. On receiving this last stab my adversary sprang away and stood facing me two or three yards off, still growling; I scrambled to my feet, and so we stood opposite to one another. I fully expected him to attack me again, and, recalling what I had often read about the effects of the human voice, I shouted at the pitch of my lungs all the most opprobrious epithets of which I was master. I fear much of what I said would be quite unprintable and quite unfit 'to point a moral or adorn a tale', but I don't think under the circumstances that even the most pronounced advocate of the *sauviter in modo* could have expected me to be polite.

"Perhaps the force and volume of my language helped what my good little knife had begun, but anyhow, after what seemed an age, and may have been only a few seconds, the lion turned and was lost to sight in the darkness. I could hear his growls turning to moans, which got fainter and finally ceased, and to my inexpressible relief I felt that I had probably killed him. Before this, however, I had lost no time in getting up the friendly tree as expeditiously as my lacerated right shoulder would permit me, and was hardly safely ensconced out of danger when the other lion, who had made a long and unsuccessful chase after my horse, with Bull sticking close and barking all the time, returned to the spot where it had parted from its companion, im-

Wolhuter's knife – 20 per cent larger than actual size.

mediately picked up my blood spoor, and came with a rush nearly to the foot of my tree. I now shouted to the dog to encourage him, and he went for the lion in great style, barking all round him, until the latter retreated and disappeared for a few minutes, at the end of which he returned and made an ugly charge at the dog, who cleverly avoided him, and nothing daunted, returned to the attack, encouraged by my shouts. Finding he could neither get rid of his diminutive antagonist nor yet get at me, Leo evidently thought he was giving himself a good deal of trouble for nothing, and so went off sulkily in the direction taken by his now dead companion.

"I was by this time feeling very faint and stiff, and fearing I would swoon and fall from the tree, I fastened myself to the branches as well as I could with neckcloth and handkerchief. Presently I heard voices which heralded the arrival of my boys. I promptly called to them, and with their assistance got down from the tree; it took an immense time. I was suffering from a raging thirst and in great pain, and we had 4 miles nearly to go to camp. Roughly bandaging my shoulder, we started off, carrying firebrands in case the lion should return. Never shall I forget that walk; often I fancied I heard stealthy footfalls in the darkness, and it seemed in my weakness and pain as if we should never arrive. I put the distance down at 14 miles, thinking I was estimating it very moderately, and even now it seems difficult to realize that it was barely four.

"However, all things have an end, and we got to the huts at last about midnight, I suppose. The boys ran off to get water, but, owing to the usual pools being dry, it took a long time finding any, and I lay enduring untold agonies of thirst. When at last the grateful liquid did come, I simply could not stop drinking, and don't know why I did not do myself some serious injury. High fever set in before morning. The

boys went out at sunrise and found my horse grazing quietly in the bush, and not much the worse, my rifle (a new one) uninjured, and the dead lion, which proved to be an old male, with a grey-flecked mane, his long canines worn quite flat at the points.

"His stomach was quite empty, and he must have been ravenously hungry. The other, I should say, must have been a much younger animal, from what I saw of him; I suppose they had had a run of bad luck hunting.

"After the boys had made a litter and I had rested a day, I was carried down to Komatipoort, and promptly forwarded to Barberton Hospital, which I reached six days after the accident, and here excellent attention and comfort awaited me."

It took them four days to carry Wolhuter to Komatipoort, now the border post between South Africa and Mozambique, by which time his arm and shoulder were hugely swollen and stinking with infection.

A doctor there patched him temporarily and a train took him to Barberton hospital where his life hung in the balance for weeks. Thanks to being supremely fit, he recovered. His only permanent disability was that he could not raise his right arm above shoulder height, but luckily still high enough to aim and fire a rifle in his long future career as a ranger.

The weapon Wolhuter used was a plain butcher's knife with a six-inch blade (although in the above excerpt he calls it a "three-inch blade") – the "Pipe Brand" made by T. Williams of Smithfield, London – which was popular in the Lowveld in those days. Wolhuter first saw it being used for cutting cheese in a friend's shop and, surprised that so good a blade should be wasted on cheese, he surreptitiously exchanged it for his own inferior one.

Today the knife and the skin of the lion he killed are mounted in the Stevenson-Hamilton Memorial Library's exhibition hall at Skukuza camp.

The trunk of the tree that Harry Wolhuter climbed after stabbing the lion still exists, although it is now preserved in concrete. It can be seen by tourists on the S35 Lindanda road east of Tshokwane picnic site.

Mdluli – a Park legend

Another brave and sagacious Park ranger was Nombolo Mdluli – who worked for 52 years in the bush until he retired in 1971 and became a legend in his own lifetime.

Because of his peregrinations on duty from one end of the Park to the other, from Malelane in the south to Punda Maria in the north, Nombolo knew it intimately and became an important source of information when the Park staff sought to reconstruct its early history.

He began working for the Park in 1919, seven years before it was officially proclaimed a national park, under ranger T. Duke at an outpost named Rolle about 55 km north-west of Skukuza. Rolle was abandoned many years ago and in recent years Nombolo was able to take senior Park staff exactly to the site of the old outpost, which they had sought in vain.

In 1921 Nombolo was posted to Malelane to join a new ranger, S. H. Trollope, who was later given the African nickname of *Mavukane*, "he who rose early".

An old record of that time, written by an unknown black member of the Park staff and recently unearthed by its researchers, states that because Trollope was a novice to the then game reserve he leaned heavily on Nombolo's store of knowledge and experience – as newcomers do even today with their established local African rangers, who are often far more experienced than they themselves.

Trollope had been at Malelane only a fortnight, the old document reveals, when some lions from the reserve crossed the Crocodile river and killed ranchers' cattle on the south side.

The ranchers naturally complained and Trollope decided to put a stop to the lions' depredations. He and Nombolo set out to hunt them along the bank of the Crocodile river.

It was Trollope's first hunt. They found the lions sleeping a few miles from their camp.

"Mr Trollope was a very ambitious man, he wanted to shoot the lions dead," the unknown historian wrote in Afrikaans. "They then came upon four lions, one male and three females, and all four were shot but one of the females was not killed and got up and ran away.

"Mr Trollope said to Nombolo, let's

Wolhuter was not the only ranger to kill a lion with a knife. Early in 1926 Stevenson-Hamilton sent a black ranger, Mafuta, to check on a place in the Park where lions were pestering railway workers. Mafuta, who was a good shot and had killed several lions before, took an assistant with him.

Inexplicably, he went off alone on his first day. He did not return and the next day a search party found his body. He had been savagely bitten in one thigh and a main artery was severed – clearly the cause of his death.

Stevenson-Hamilton hurried to the scene to investigate. A few hundred metres from Mafuta's body they found a dead lioness; under her lay Mafuta's sheath knife.

Reconstructing the tragedy from spoor and blood marks, Stevenson-Hamilton determined that Mafuta, instead of merely observing as he was told to do, had shot at and wounded a lioness. When Mafuta tracked her down she charged but he managed to fire only one shot before she was on him. They then fought, hands and knife versus claws and fangs, until he killed her. But his wounds were also fatal.

Nombolo Mdluli – a Park legend.

follow her; no, Nombolo answered, it's very dangerous when it's wounded like that.

"Mr Trollope was a courageous man, he again asked Nombolo to follow the lion which ran away. Nombolo again asked him to fetch the dogs before they followed it. They had walked only a few paces when the lion saw its enemies before they saw it."

The lioness pounced on Trollope, striking him in the thigh, but before it could harm him Nombolo raised his rifle and shot it dead.

So grateful was Trollope for this action "that he thanked Nombolo much and said, Nombolo, take my hand; so Nombolo took his hand and the ranger then said: Nombolo, you are certainly a man, a man I will never go without, a man I will never forget in my lifetime."

Trollope rewarded Nombolo with £5 – a considerable sum in those days – and every year thereafter, even after he had left the Park's service, wrote to Nombolo and sent him £2.

Nombolo then joined ranger Bert Tomlinson of Shingwedzi, with whom he worked for 17 years before his next posting.

Because Shingwedzi was a remote outpost not easily reached by supply wagons, Tomlinson had instructed Nombolo to plant a patch of maize. This Nombolo did, but he did not have the time to guard the crop, which was greatly coveted by marauding baboons and some browsers too.

So Nombolo employed a few *umfaans* (young boys) to guard the "mielies".

One hot afternoon when the boys were dozing off in the shade, yellow eyes were watching them from the bush. The eyes belonged to a lion, which slowly stalked nearer. Its attack was sure and swift. The *umfaan* stood no chance. In fear the others ran off to find their protector, Nombolo.

Alone – Tomlinson was away on patrol – and armed with a 0,303 rifle, the intrepid Nombolo crept towards the maize patch. He found the mutilated corpse of the boy, but no lion. Instinctively, though, he knew the animal was not far off. He waited.

Suddenly the beast charged. Nombolo stood his ground, hoisted the heavy rifle to his shoulder, took careful aim and squeezed the trigger. His aim was true, the lion skidded to his feet, dead.

For this second brave action Nombolo was awarded twelve shillings and sixpence, which then amounted to more than a month's wages for him. Colonel Stevenson-Hamilton was so impressed that he gave the man £5 and written permission to kill any other troublesome lions. But that was not the end of the rewards: Nombolo also received a Singer sewing machine which he greatly treasured and a hat made from the skin of the dead lion.

Ranger Bert Tomlinson who joined the Park in 1926. He was given the name Rivalo, *meaning "awe-inspiring".*

Forty-three years later Nombolo presented the rifle, the sewing machine and the hat to the Kruger National Park. They are now on display at Skukuza.

The old man became a fund of history for the Park authorities. One of the interesting things he was able to tell them was the African nicknames of many of the people who figured in the Park's life – some complimentary, some not, but all pertinently descriptive.

T. Duke, for instance, was known as *Makhoza* because he spoke the Xhosa language fluently. C. R. De Laporte was named *Makous* because he wore short puttees – from the Afrikaans word *kous* meaning a sock. W. W. Lloyd had two nicknames: *Kukuzele* or "looks as though he is stalking you", and *Nwashibeyilana* or "he who carries a little axe".

Tomlinson was named *Rivalo*, meaning "awe-inspiring", H. MacDonald was *Mahahalane*, meaning "scolds much", and W. J. D. Groenewald became *Majabula* because he always said, "You make me glad because you work so well."

Less flattering nicknames were given to Colonel J. A. B. Sandenbergh, former chief warden, who was called *Mashayimhala* or "rode over an impala" and also *Mehlomamba* or "mamba eyes". For ranger's wife Mrs De Laporte it was *Nwashihubutane*, meaning "she always wanted to storm and rage because of bad temper".

Battle for supremacy
Thrashing in the muddy water with their bodies at times almost completely submerged and their great scimitar horns scything viciously, two sable antelope bulls battle in the Manzimhlope waterhole west of Tshokwane picnic site. The two at first squared up in mock combat and then, suddenly, the challenge became deadly. Serious combat lasted for little over one minute and ended with one bellowing in pain from a deep thrust in the flank and submitting totally.

Animal oddities

Not all the funny, dramatic or strange episodes in Park life occur between men and animals. Most are between the animals themselves and over the years the Park staff have recorded hundreds of unusual incidents involving behaviour uncharacteristic of the creatures involved. The following are just a few of these:

In the early 1970s rangers caught several young elephants for the scientists to study which, it was thought, could be sold later to other game parks. In January 1971, they felt sorry for the last of these after its scientific duties and let it go free near the Transport dam in the Pretoriuskop area.

A short time later the young elephant was seen in the middle of a herd of about 400 buffalo, to which it had attached itself. Whether it had adopted the buffalo or vice versa is unknown – probably the former because the elephant began to emulate the behaviour of the buffalo.

Being inquisitive beasts, buffalo tend to suddenly dash away when a vehicle approaches too close and then slowly return, when their curiosity gets the better of them, with their noses thrust forward, sniffing the breeze.

The rangers found that when they approached this particular herd, the elephant did exactly the same. It would dash off and then come back sniffing, trunk extended, among the buffalo.

This odd relationship became quite a feature of the Park as the buffalo and their orphan roamed wide across its southern region. Years later they were spotted in an extraordinary incident by a wildlife enthusiast on a visit to the Park.

The visitor was at a dam 12 km south of Balule early one afternoon watching eight young lions lazing about when the buffalo herd arrived with the jumbo in their midst. For some reason the buffalo took umbrage at the lions' presence and charged them.

The bewildered lions scattered in all directions, six fleeing into the veld and two climbing a tree in panic. And then, almost as if working to a plan, the

Left: This kudu, seen at the roadside near Lower Sabie, spent nearly an hour picking up and then dropping the fallen fruits of a sausage tree. It seemed content to suck at the fruit rather than bite through the tough skin.

buffalo slowly closed about the tree, the elephant with them trumpeting in anger.

The two lions tried to get higher in the tree and to hide behind its foliage. They tried to roar but could only manage anguished yelps. They were stuck there for some considerable time until the buffalo and their partner got bored and went off to the dam for a drink.

The lions stayed in the tree another quarter of an hour before they dared to come down and risk this formidable team. And then they left the scene in a great hurry.

The elephant was still with its buffalo family in 1986, still defending them against all comers.

Elephants have often been seen crossing rivers, either swimming or walking across the beds with just the tips of their trunks above the surface like snorkels. Young calves, however, cannot always manage the crossing and an incident on the Letaba river showed just how devoted their mothers are.

The mothers shepherded their calves into the water ahead of them and then each mother wrapped its trunk around its calf and held it up so it could breathe. Even when the mothers vanished beneath the surface, they still held their babies aloft above the water.

Elephant bulls are often seen wrapping their trunks around each other and jousting with their tusks – a normal part of growing up and a testing of strength. Sometimes bulls fight in earnest to achieve dominance in a herd, but rarely to the death because when the weaker knows it is being defeated, he gives way and runs.

At Nwamanzi lookout point near Letaba camp two bulls were seen fighting recently. They must have been at it for some time when discovered because the older of the two was tiring and trying to get away, but the younger would not let him.

"Eventually," the witness wrote later, "the older bull was forced on his haunches. By that time he was bleeding from various wounds inflicted by the younger one. He supported himself on his forelegs while the younger animal attempted to climb on his back to force him down.

"When this failed, he tried pushing him down with his head and hitting, but

This zebra with diagonal ears was sighted off the H1-7 road near Shingwedzi. Although a funny sight, it was probably suffering from an infection.

could still not do him enough damage.

"He then slowly circled the old bull, who was still in a kneeling position, and prodded him in the sides, probably in an attempt to roll him over. When he did not succeed the young bull charged at the old bull at a high speed over a distance of 2 to 3 m, turned his head sideways and thrust his left tusk into the older bull's neck just in front of the forelegs. With a terrible noise the assaulted bull slumped down and after a while rolled over on to his left side."

The young bull repeatedly tried to lift the old one and eventually had to be driven off by a game ranger. Despite a huge throat wound, the old bull was not dead and the ranger was forced to shoot him.

Crocodiles are creatures of the water but are surprisingly at home on dry land. In the 1960s the Orpen dam dried up and rangers patrolling the area found large crocodiles crossing dry, stony hill country in search of new pools. Few things could be more unnerving than finding a crocodile stalking the veld with its high-legged gait far from river or pool.

At the Luvuvhu river in the north crocodiles have been seen dragging carcasses of victims as far as 200 m from the water. Near Letaba some time ago crocodiles came a distance from the water to grab a buffalo killed by a pride of lions. The lions objected and killed one of the smaller crocodiles, but eventually the big armoured reptiles drove off the lions and dragged their booty back to the water.

Lions in love . . .

A truly fascinating display for visitors is the sight of lions mating. So intense is the attraction between the "honeymooning" pair that for several days they show no interest in either food or the carloads of onlookers nearby. The mating sequence usually begins with the female walking away for a few paces to be immediately followed by her mate. This triggers a display of "affection" – deep purring, nuzzling and grooming. The male often shows his teeth in a silent snarl, accompanied by a slow shaking of the head. Mating occurs every fifteen minutes over four or five days, on average.

The "no option" option

Since 4 July 1978, when the first wilderness trail hikers strode out from the Wolhuter camp in the south of the Park led by senior ranger Trevor Dearlove, more than 10 000 people from many corners of the world have enjoyed the unique experience of walking in the Kruger National Park (see p. 272).

It is the Park's proud record that in the estimated 50 000 hours people have spent exploring the veld, no visitor has been injured – although there have been moments of "intense excitement".

Trail rangers undergo exhaustive training, not only as ecologists so they can share their knowledge of the wild, but also as "safety officers" able to anticipate and thus avoid potentially dangerous situations in the bush.

If absolutely necessary, a ranger will protect his "clients" with his rifle. But it is a point of pride that since trails began eight years ago only three animals have had to be killed because they threatened danger. The first trail ranger who was forced to make that dreaded decision was Trevor Dearlove, when he and his group were charged by a rhino bull.

On 26 February 1983, he was escorting a party of four, two South Africans and a Canadian couple, on a walk from the Wolhuter base camp. It was a perfect summer morning and the visitors were well satisfied, having spotted many species of birds and a variety of game including wildebeest, zebra and three lions.

Trevor tells the story:

Senior trails ranger – Trevor Dearlove.

"We were approaching a bare brackish area below the Newu dam, having just come across three lions on a buffalo carcass and being charged by the lioness. The party, buoyed up by the 'adrenalin high' which this type of experience invariably produces, were quite naturally excited and talkative. As a matter of routine, I reminded them of the trail regulations, that they keep together, remain quiet and so on – this specifically because the bush that we were going into was dense, and I had experienced problems there on previous occasions. Two dry watercourses run through the area, and consequently the greener vegetation between them does not always burn together with the surrounding bush. To reach the vehicle we had to traverse the denser area.

"Suddenly a white rhino rose from behind an anthill 20 m away and without offering any warning, lowered its head and charged. There was no time for warning shots. I fired and the rhino's legs gave way beneath it. I looked back to see where the party and my assistant, corporal John Mangane was. The trail group had not moved and John had instinctively darted out to our right to obtain a clear line of fire, but fortunately, a backup shot was unnecessary. I walked

up to the rhino and fired an 'insurance' shot into the back of the head because experience has taught us that 'dead' animals can suddenly come to life with embarrassing consequences. But my .375 had performed as only a rifle of this calibre could and the second shot wasn't necessary.

"It is difficult to describe just how I felt at that moment. Mixed with the feelings of relief at having averted a possible tragedy, and having my party safe and alive, was a deep sadness at having to end the life of such a splendid animal. He was a bull in his prime, and killing him seemed truly futile – when the very *essence* of the wilderness experience which we were all sharing was the preservation of all the natural biological life forms to be found in the wilderness. I was most impressed with the calm reaction of my trail party during the whole incident. There was no question of panic. In fact one of the ladies calmly captured most of the action on film – pictures which are some of the more meaningful mementos of my stint as a trail ranger.

"During our campfire discussion that evening, it became apparent to me that the incident had made the party more aware of the intricacies attached to

One of a sequence of photographs taken by Canadian tourist Mrs Melanie Baker of the rhino shooting.

something of this nature. It made them more aware of the potential threat that man poses to nature, but at the same time, that modern man is nature's only hope of survival."

Both Trevor and corporal Mangane recall that is particular rhino had reacted aggressively in the past and on more than one occasion warning shots were used to turn a charge.

"In general one must expect big game to be predictably unpredictable," Trevor states, "and it is for those unpredictable situations that extensive precautionary measures are taken on all trails, and that all parties are accompanied by armed and trained rangers.

"I suppose something like this had to happen sooner or later, but this was the first time that an animal had had to be shot to protect the lives and safety of a trail group."

In later incidents Peter Davies had to destroy a charging buffalo on the Nyalaland Trail in the Park's north and Cleve Cheney had to shoot a white rhino on the Bushman Trail in the central area. There were no injuries to traillists or rangers.

Great tuskers of the Park

There is no creature in Africa quite as dramatic, as imposing or as regal as the elephant. He is the true king of beasts, the largest land mammal on earth, beside whom the lion is a mere courtier and before whom the powerful rhino, hippo and buffalo and all else give way. In his pure environment this majestic giant of up to six or seven tons has no natural enemies for he is not a predator and there is none large enough to challenge him. If there were music in the wild he would surely move to the strains of Beethoven, such is his ponderous grace and awesome presence.

Since recorded history the elephant has captured the imagination of man more firmly than any other animal – initially in the Far East, where he was trained as a friend and a worker, a formidable ally in battle or on the hunt, and a symbol of monarchy. Modern man later rediscovered the elephant in Africa, where he and early primitive peoples had existed in relatively peaceful parallel for aeons, and found him to be untrainable, unlike the Indian elephant, but a rich source of ivory. Thus man became his first and still is his greatest enemy. Lions are the only other because young elephants are sometimes killed by them.

Since then the existence of this extraordinarily gentle, sage species has been in steady decline. Only a few hundred years ago elephants inhabited the length and breadth of Africa, from Table Mountain in the Cape to the then grassy savannah of the Sahara, from the Maiombe forests in West Africa across the high slopes of the Ruwenzori Mountains in Uganda to the coral beaches of East Africa.

Today they have vanished entirely in vast tracts of the continent and within living memory have disappeared even in sanctuaries where they once roamed in tens of thousands. Their numbers have diminished in almost direct proportion to the spread of sophisticated weapons. In the Stone and Iron ages men took some toll of elephants for protection and provender but so rudimentary were their weapons that they made no impression on elephant populations. Then came the Arabs with gunpowder, followed by the European explorers, but their early guns were too inefficient to affect numbers.

The decline began in earnest with the arrival of colonialists and their guns which rapidly increased in efficiency from muzzle-loaders to high-powered weapons made famous by a host of today's hunters. Killing wild animals became an act of bravado in the Western world and the elephant rated higher as a prize than the lion, because of both the danger of the hunt and for the impressive trophy of the great, beautiful tusks, once also used for making billiard balls.

Adding to the pressure was the continuing population explosion in Africa. People needed more land, meat and money – making the elephant a prime target.

In the past two decades particularly the depredation of African elephants, and of almost all other animal life on the continent, has soared to unprecedented levels at a sharply accelerating pace. It is a tragic spin-off of the political turbulence shaking most of Africa.

Before the birth of the Sabie Game Reserve at the turn of the century the Lowveld elephant attracted only the ivory hunters. Today there are some 8 000 elephants in the Kruger National Park, about 2 000 more than the ideal population – a small number compared to the great herds which exist (or existed) in some other parts of the continent. But they are a stable, sheltered population and careful control of their numbers and habitat has regenerated characteristics that were typical before man began to plunder.

Most notable is the reappearance of the big tuskers. Because they were the

The biggest tusks on records, in fact, are a pair of 101,7 and 96,3 kg in the British Museum . . .

prime targets of the ivory hunters, these were the first elephants to decline in number and far fewer have been recorded in this century than in the last one. The biggest tusks on record, in fact, are a pair of 101,7 and 96,3 kg (226 and 214 lb) in the British Museum which were bagged by a hunter on the slopes of Mount Kilimanjaro, East Africa, in the late 1890s.

Now big tuskers can again be seen in the Park – not yet with tusks of that size

Mightly tuskers of the Park

Some of the big elephants that roamed the northern sector of the Park are shown on these two pages. **Above:** *Shingwedzi ranged the area around the rest camp of the same name. It died early in 1981.* **Right:** *Dzombo, a magnificent bull which lived to the south of Shingwedzi camp. It was shot by poachers.* **Upper right pair:** *Both these pictures are of João, the only survivor of the Magnificent Seven. Two views are included here because both its tusks were broken off and lost which means that João cannot take its place in the Park's Hall of Fame when it dies.*

Left: *Shawu, another elephant of the Shingwedzi district, died of infection probably caused by poachers' bullets.* **Above:** *An unnamed giant with just one very big tusk, usually seen north of Letaba.*

but nevertheless exceptionally impressive. Evidence of this has been gathered over a number of years by South Africa's leading authority on elephants, Dr Anthony Hall-Martin, a senior research officer in the Park, on whose observations most of the following facts are based.

Taking a census of the Park's elephants in 1980, using a helicopter, he and his team made a point of counting every bull elephant carrying at least one tusk estimated to weigh 45 kg (100 lb) or more. To their astonishment the final tally showed the Park had 41 bulls in this class and that at least eight had one tusk weighing an estimated 63 kg (140 lb). That put all those bulls in the trophy class, for which the minimum tusk qualification is 31 kg (70 lb).

Tusks and not body size have always been the bench-mark in assessing elephants, partly because the two are not necessarily related: a medium-sized elephant could have tusks of greater length and weight than a larger one. Also, it is the tusks which have always intrigued man, not just for their value as ivory, but for the elephants' versatile use of them as a weapon and a tool.

Tusks are modified incisor teeth composed almost entirely of dentine and grow by about 10 cm (4 in) a year throughout an elephant's life, which could be as long as 55 to 60 years. However, constant use as a digging and ripping tool wears them down. They are primarily weapons and Dr Hall-Martin states that recent research indicates that more elephant bulls are killed by other bulls in fighting for herd dominance and in breeding time than was previously suspected.

The massive bulls with the big tusks were found to be dominant in competition, as a general rule, although some could be bulls which simply avoided fighting. Tusks can grow to lengths of more than 3 m (over 10 ft) of which a third to three quarters is exposed – one hallmark of *Loxodonta africana*, the African elephant. In cross-section they are round or oval. They usually curve upwards and slightly outwards from the jaw, although some have been recorded which curve so sharply inward that they are more of a hindrance than a help to the owner.

Tusks are certainly used with deadly efficiency. Many a reckless or unwary hunter in Africa has been speared, smashed or thrown skywards by a tusk backed by the huge power of this gargantuan creature. Ground tusked over by elephants seeking tuber roots looks as if it has been harrowed by bulldozers. That long tooth can effortlessly split the bark from the toughest of trees. It is also commonly used as a kind of rack across which to drape a tired trunk.

The big tuskers are the kings of kings, the old monarchs who have passed the breeding and competitive stage and have hived off from the herds to become solitary wanderers in the wilderness. Solitary in a sense that they are not family members, but they frequently have the company of several younger bulls, like frigates around a battleship.

These are called *askaris* (a Swahili word for soldier) or *tsotsis* (a South African word for hell-raising teenagers) and are the youngsters who have been evicted from breeding herds on reaching puberty – a mechanism which prevents inbreeding, says Dr Hall-Martin.

"The young *askaris* learn a great deal from following the example of the old bulls – such as social manners, where to find water in the dry season, where choice stands of trees grow, and so on. The bonds between the big bulls and their young companions are not permanent. They move around much more than the older animals but return to them from time to time. The *askaris* are usually more alert and aggressive than the older animals. That is probably why they more often detect hunters and react to them (the wise older animals having learned that when man is around it is best to move off)."

For more than a decade public interest in the Kruger National Park's elephants has been dominated by The Magnificent Seven – seven behemoths with tusks well into the top trophy ranks although still short of the world record.

The names of the original Seven are:

João – the name is Portuguese for John and was probably given to mark the pioneer Lowveld settler of the 1840s, João Albasini. The elephant had a left tusk of about 50-55 kg (110-120 lb) and

A close-up of the giant tusker Shingwedzi, who died in January 1981. He was well known for his placid nature and seemed to enjoy the attention of tourists.

a right tusk of 41-45 kg (90-100 lb) before both were mysteriously broken off in mid-1984.

Ndlulamithi – means "taller than the trees", a name often given by the Shangaan people to very large elephants. His remains – tusks, tail and a few bones – were found in July, 1985, in the area north-west of Tshangwe viewpoint near Shingwedzi. He died of old age. His left tusk weighed 64,6 kg (142 lb) and the right 57,2 kg (126 lb).

Kambaku – meaning "big elephant" – had matched tusks of 63,6 and 64 kg (140 and 141 lb). He was crippled by a farmer's bullet in the knee while ranging outside the Park and had to be destroyed.

Mafunyane – means "the irritable one". The name was given to him not only because of his temperament: this bull was first noticed in the 1960s near a dam named Mafunyane because it was built by a ranger nicknamed Mafunyane by the local Tsonga people, thanks to his irritability. His matched tusks weighed 55,1 kg (121 lb 4 oz) each. Mafunyane died of natural causes in October 1983.

Dzombo – this elephant, named after the Dzombo river south of Shingwedzi where he roamed, had tusks of 55,5 and 56,8 kg (122 and 125 lb) and was shot by poachers in October 1983.

Shawu – also named after an area near Shingwedzi, this bull's left tusk measured 3,17 m (10 ft 4 in) and the right tusk 3,06 m (10 ft) – the longest recorded in the Park. The tusks weighed 52,6 and 50,8 kg (116 and 112 lb). Shawu died in October 1982 of natural causes possibly aggravated by bullet wounds caused by poachers.

Shingwedzi – named after the rest camp and river, he had tusks of 58 and 47,2 kg (128 and 104 lb). He died of natural causes in January 1981.

The Park has decided to let the Magnificent Seven rest in history. Their tusks and skulls will be preserved and will be displayed in an elephant museum when the Park can find the money to build one.

Each of the Seven was a character in his own right. The size and shape of their tusks varied widely. The most beautiful belonged to the legendary Mafunyane, a perfectly matched pair that scraped the ground as he walked. The heaviest, as a pair, were those of Kambaku and the longest was the left tusk of Shawu whose huge curving tusks earned him the name of *Groot Haaktand* (Big Hook Tooth).

The question is, who are their successors? There are several bulls who qualify, such as Mashakiri who lives near Klopperfontein in the north of the Park and has a long left and a short right tusk, or Phelewane, who ranges the Kingfisher spruit area near Orpen Gate and has one long and one broken tusk. And there is a bull yet to be named that roams a triangular area south of the

between 20 and 30 cm.

Kambaku was also well known and was photographed over a huge stretch of the Park from the Orpen road west of Satara to Crocodile Bridge. He suffered the fate of so many wild creatures that leave the sanctuary of the Park and venture into agricultural areas, where they are incompatible with the activities of man. Some unknown farmer shot him in one knee, so crippling him that the Park rangers had to put him down.

João is the sole survivor of the Magnificent Seven . . . he lost his tusks in mid-1984.

Sabie river bounded by the Kruger gate in the west, the Nkhulu picnic site in the east and the Transport dam in the south. He has very big bowed tusks and can be recognised by a large tear in his left ear. According to Dr Anthony Hall-Martin, it is between these last two elephants that the title of "monarch" rests.

João is the sole survivor of the Magnificent Seven and even he lives in comparative obscurity because he lost his tusks at the end of 1984, probably in a fight. Park experts believe that had he retained his tusks João could have become the king of them all because his relative youth would have allowed his already magnificent tusks to reach record size for the Park.

His home ground is Shingwedzi and he is often seen drinking from the Shingwedzi river a few hundred metres upstream from the camp, where he spends several hours a day. In 1981 he was drugged and fitted with a radio collar as part of Dr Hall-Martin's studies of elephant ecology, but somehow broke the transmitter from the collar. Fitted with another, he did the same again. He now has his fourth collar.

Dr Hall-Martin writes of him: "When seen feeding or drinking João appears to be a very placid elephant. Generally he is tolerant of vehicles – but he has charged on a few occasions. Visitors are advised to treat him with the utmost respect. In June 1982 João was wounded by poachers who hit him with four AK-47 bullets. Fortunately we were able to catch him and treat the wounds – all fairly superficial. They appear now to have healed up completely." Today João lives on, but with his great tusks mere stumps of

Several other bullets were found in him, including some of 0,22 calibre – pointless to fire at an elephant.

Apart from Shawu, about whose cause of death there is some doubt, and Kambaku, the only other of the Seven to have succumbed to bullets is Dzombo. He was killed by poachers from Mozambique using AK-47 assault rifles in October 1983, near the Dzombyane windmill in the Shingwedzi district of the Park. But for a coincidence, the poachers might have escaped with his superb tusks.

They were hacking the tusks from his body when they must have heard the sound of an approaching truck and fled. In the vehicle was senior ranger Ampie Espach making a routine check of windmills. When he reached Dzombyane he did not see the great corpse lying about 500 m away and there was nothing to tell him it was there, for it must have been shot barely 30 minutes before his arrival.

Two days later he returned and the vultures wheeling in the sky over the dead Dzombo told the story. In spite of the relative fortune the poachers could have got for the tusks on the Mozambique black market, they did not return. On the offchance that they might, an ambush was set up near the carcass but with no result.

The rangers drew Dzombo's tusks and these will be kept at Skukuza until they can be displayed in a Park museum.

Shingwedzi is testimony to the fact that elephants have much the same coronary problems as people; he died of a heart attack. So suddenly did death overtake this giant at the estimated age of 55 years that he collapsed on the spot in the bed of the Shingwedzi river a stone's throw downstream from the camp and his body was found still propped on its knees on 18 January 1981, with the long right tusk embedded deep in the soft river sand.

Mafunyane also died suddenly, in November 1983, on the banks of a small tribuary of the Zari river in the remote north-west of the Kruger National Park. The remains of this spectacular elephant who reigned supreme among The Magnificent Seven might never have been found had he not been wearing a radio collar – and how he got that is another story.

Mafunyane had long fascinated the Kruger National Park staff lucky enough to see him (for he preferred the remoter bush), for two reasons. One was his apparent size and his long, superbly matched tusks estimated at one stage to weigh possibly as much as 85 kg (187 lb) each. The other was the hole in his head.

*Right: After being darted from a helicopter, the mighty Mafunyane lies on his side as research staff prepare to take plaster casts of his tusks and attach a radio collar. **Below:** The elephant struggles to find his feet – with the ground crew's help. (Pictures by National Parks Board)*

The fist-sized hole was clearly visible, especially in photographs taken from the air: slightly to the right of centre on the very top of his head. How he got it was a complete mystery, as was how he lived with it, but of course nobody was going to get close enough to investigate.

So the bad-tempered old giant was left to go his own way in the vastness of the Park's north with his escort of young *askari* elephants, carrying before him those tusks so massive that he had to raise his head to keep them free of the ground when he walked. Sometimes he

and all the preparations and precautions, which included a front-end loader that was to save the day.

The spotter plane went up to look for Mafunyane and pinpointed him in a wild area of the north-west close to a fire-break road. The pilot radioed the position to the team and the Park's chief pilot, Hugo van Niekerk, went ahead with the helicopter and guided the ground crew closer by radio.

When all were positioned and ready, the helicopter moved in and regional ranger Bruce Bryden (now chief ranger)

fight, or by something like a dropspear – a weighted spear which falls down on to the victim when a trap is sprung.

When they had finished examining the elephant and had switched on his new radio, they gave him a heavy dose of antibiotics to counter some infection in the hole in his head, and then the antidote to the immobilising drug. They stood well back waiting for the fast-working antidote to do its job. Everyone worried about how Mafunyane would cope when rising to his feet with those heavy tusks.

Dr Hall-Martin describes the scene:

He crashed back in a cloud of dust, swung again . . . and almost managed it. He fell back, drained by the effort . . .

did not bother, and the polished tips were bevelled like chisels by the abrasion on the ground.

Then, on 8 June 1983, a small army of Park staff moved in on Mafunyane with four-wheel-drive bush vehicles carrying such dignitaries as Professor Fritz Eloff, chairman of the National Parks Board, the Park's warden, Dr "Tol" Pienaar, plus prominent visitors, photographers, a full veterinary team and Dr Hall-Martin and his fellow researchers, all backed up by a spotter plane and a helicopter. "Operation Mafunyane" was on.

The Park authorities had decided that partly to help keep an eye on the big elephant bulls to protect them against poachers, and partly to study their movements for research into elephant ecology, they should be fitted with collars carrying radios beeping out constant identifying signals. This way they could be monitored with ease instead of the scientists having to tramp or drive endless kilometres through trackless bush to find them. Collaring elephants was not new: the researchers had already done it to more than 50 of them. Because Mafunyane was such an extraordinary specimen, it was decided that he should be given this extra protection. He was, after all, considered an intrinsic asset to the Park, and his tusks of great aesthetic value; "like a work of art", as Dr Hall-Martin put it.

But in his case there were special problems. He was old and massive and if drugged he might injure himself falling. Hence the large team for the operation

leaned out and fired the drug-filled dart into Mafunyane's rump, then they swiftly lifted away to watch from a distance – ready, if necessary, to move in to shepherd him away from danger.

But, Dr Hall-Martin wrote later, the drug took effect within minutes and the ground crew quickly closed in on Mafunyane before he collapsed completely. There was a real danger that if his forelegs gave way too soon his body might be propped up by those great tusks, which could damage his neck or stop him breathing. The plan was to shove him onto his side if that seemed likely to happen, but there was no need – the king stood befuddled for a few moments and then crashed down on his right side.

The crew were onto him immediately, keeping a constant check of his heartbeat and breathing, measuring the huge tusks, making plaster casts so that if they did vanish in the bush some day there would be replicas, examining his other teeth to gauge his age, measuring the dimensions of the great body. The radio collar was fixed around his neck and Dr Hall-Martin and the others examined that curious hole in the top of his head.

To their astonishment they found that Mafunyane was breathing through it like a whale, as well as through his trunk. Clearly it went right in to his nasal cavity. They could not determine how the hole was made but theorised that it could have been by a charge from a muzzle loader fired downwards from a tree, or by the tusk of another bull in a

"When the antidote took effect and he woke up, he flapped one great ear forward – the usual preliminary before an elephant tries to get up. Then he swung his legs up into the air to rock his body and with the momentum of the swing tried to lift his head to pull himself over onto his brisket, get his legs under him and rise.

"On the first swing he almost made it, but the great weight of his tusks held his head too far down. He crashed back in a cloud of dust, swung again, straining to get his head up, and almost managed it. He fell back, drained by the effort, and then tried again.

"As it was the critical point in the change of balance required to roll his massive body over that could not be reached, the ground crew rushed in to help him by pushing against his shoulder and head as he came up once again. This was tried several times, but the elephant was rapidly tiring, or possibly sinking back into a drowsy state as a result of the drug immobilisation.

"It was also possible that because of his age he was not coping well with the drugs. Each effort was less vigorous and his great head fell back onto the ground as though in despair."

They put a rope around him, looped it over a forked tree and pulled on it with a bush vehicle, time and again pulled and pushed him by hand, and gave him more drugs and stimulants, but nothing helped and Mafunyane lay exhausted while gloom descended on the team. They decided to let him rest and keep him cool by dousing him constantly with water while they sent for reinforcements: the front-end loader.

When this came, driven by Francois "Fielies" Prinsloo, who maintained the vehicles in the Park's northern sector, it was decided to heap a mound of earth against Mafunyane's back so that his body would not pivot every time he tried to rise. They covered his head and breathing hole with a tarpaulin to keep out dirt and the job began.

The earth was piled against Mafunyane and he heaved himself up on the mound. Now Prinsloo tied a tree trunk and planks to the shovel of his loader to avoid injuring the elephant and moved in close. This time, when Mafunyane heaved his head and shoulders off the mound, Prinsloo drove the loader up so that when the elephant fell back, he was propped up by the loader's padded shovel.

Again Mafunyane tried to rise and

The moment of success in Operation Mafunyane – "Fielies" Prinsloo and his front-end loader shove the elephant to his feet. (Picture by National Parks Board)

now Prinsloo pushed forward with the loader and simultaneously lifted its shovel to virtually hoist the giant to his feet.

"The elephant rose out of the cloud of dust like a breaching whale and as his feet steadied on the ground, the old bull was instantly transformed into a raging monster. His first target was the loader, which he took head on. That seemed like a good time for 'Fielies' to get back to Shingwedzi, so he left the vehicle to the bull and departed.

"By now the helicopter, which Hugo had lifted off seconds before the bull rose to his feet, was on the scene to help 'Fielies' – stirring up even more dust and raising the noise level by so many decibels.

"Regional ranger (as he was then) Bruce Bryden, who was standing by with a rifle should anything go seriously

wrong, was moving in to see what had happened to 'Fielies'. Fortunately the downdraught from the helicopter's rotor blades caused the tarpaulin which had been over Mafunyane's head to start flapping and he spun around from the defeated front-end loader and attacked this new tormentor, driving his tusks through the canvas and soundly trampling it into the ground."

By then everyone else in the team was also departing for distant points on foot and in vehicles – aware of the sudden danger and of the disgrace of having to oblige ranger Bryden to shoot the bull to save a human life. The one exception was Professor Eloff, who calmly stayed at his vantage point atop a tall termite mound taking photographs.

Mafunyane, however, ignored them all. Having thoroughly destroyed the tarpaulin he lumbered off into the bush

like a galleon in full sail, "leaving the puny humans in total disarray – humbled by his style and grateful that he was back on his feet."

Mafunyane suffered some minor injury to his right rear ankle during the capture operation, so Dr Hall-Martin's staff decided to check on him every week. After a few weeks it became clear that he was fine and wandering about a large part of the north-west in search of food.

Early in November 1983, one of Dr Hall-Martin's researchers, Don English, picked him up by radio again but did not attempt to approach because the wind was blowing from him towards the unseen elephant, which made it dangerous. A week later Dr Hall-Martin joined English and they set out to find the old bull. Signals from Mafunyane's radio showed that he was in the same area where English had found him the week before – which was unusual because he moved about so much. Moreover, as they drove through the bush to get different radio cross-bearings, they discovered that he was in exactly the same spot, which meant one of two things: either the radio had fallen off and was lying on the ground, or Mafunyane was lying there.

The men moved in closer until the signals were obviously coming from very near. But they could not see him although the countryside was fairly open grass and mopane savannah. Then they saw a dead tree with white vulture droppings about it. They moved on and

had been chewed up and most of it carried off by hyenas. Only the transmitter and enough of its antenna to ensure a signal were left behind – and it was this signal, a small example of modern technology, that led us to the carcass."

Mafunyane's death had some profit. It taught the Park's staff a sobering lesson about judging the size of an elephant and its tusks. Far from weighing up to 85 kg, as was once estimated, Mafunyane's tusks each weighed 55,1 kg (121 lb), but were nevertheless a magnificent pair.

The elephant was actually rather smaller than was originally estimated, which gave the appearance that his tusks were bigger. Also, the tusks were oval, not perfectly round in cross-section, so that from certain angles they looked bigger than they really were.

Dr Hall-Martin was also able to examine, at last, the hole in Mafunyane's head in detail and came to the conclusion that it must have been caused by the tusk of another elephant, not by a bullet or spear. It was probably a very old wound inflicted early in the bull's roughly 57 years of life and, through infection, could have helped stunt his growth to a small extent – and have been the cause of his notorious ill temper.

Mafunyane's tusks will find a permanent place in an elephant exhibit planned for Letaba Camp. He was also one of the giants among The Magnificent Seven recorded by the nature

"The huge leg bones were dragged a hundred metres . . . and chewed, gnawed and crushed . . ."

finally found Mafunyane – the huge skull and scattered bones.

"Within days of his dying, lions, hyenas and vultures had torn apart his carcass and scattered it over a large area," Dr Hall-Martin wrote in his obituary report.

"The huge leg bones were dragged a hundred metres and more and chewed, gnawed and crushed by hyenas. The ribs were spread around, the pelvis had been gnawed away and one shoulder blade carried off so far that it could not be found. Only one piece of skin was intact, the rest was gone. It was as though the carnivora had determined that nobody would ever find the remains of the great bull. Even the radio collar

artist Paul Bosman, whose dramatic painting of him now hangs on the walls of many homes around the world and aids in drawing in the revenue the Kruger National Park needs to ensure that future Mafunyanes, Dzombos, Shingwedzis and Joãos will survive for the benefit of both elephant and mankind.

Dr Hall-Martin now has National Parks Board authority to preserve ten pairs of tusks, including six pairs of the Magnificent Seven who have died. João's will not feature because he has lost them. The remaining four pairs will be regularly replaced as larger pairs are brought in from other giants as they, too, die.

The white lions

Among the Kruger National Park's most internationally famous residents are its white lions. Basically they are like any other lions, often lazy, sometimes playful, gluttonous when they get the chance, and very powerful.

They are not completely albino, the freakish lack of any pigmentation which appears among a great many species of creatures from time to time, including humans. Like polar bears they are simply white. They have the same yellowish eyes that all lions have although sometimes, under certain light conditions, a blueish cast can be seen, but not the washed out blue-white of the true albino. Normal lions have black patches behind their ears and black tufts on the ends of their tails, but in white lions these patches and tufts are seldom darker than a light brown.

The white lions of the Lowveld caught public imagination world-wide in the 1970s after two were discovered in the private Timbavati game reserve adjoining the Kruger National Party by naturalist Chris McBride, who later wrote the best-selling book *The White Lions of Timbavati*.

Since then the number of white lions in that area has risen to more than a dozen and more have been born in captivity in the renowned Johannesburg Zoo. And these are not the first on record – white lions were seen in the Park itself as far back as 1928 and then again in 1959 and 1962.

But why white lions? According to ranger Pat Wolff of the Park's Tshokwane area where most of the mutants have been seen, the whiteness stems from an "autosomal recessive gene". Genes are the factors which determine the characteristics which creatures inherit and if two lions carrying this recessive gene meet and mate, whether they are white or not, the chances are that they can produce white as well as perfectly ordinary tawny cubs.

Wolff, who was born and bred in Kenya, has become fascinated by white lions since joining the Park staff. He states that if a white lion with white genes mates with a lion without them, none of the cubs can be white. However all the cubs will be carriers of the genes and therefore will be possible future breeders of the mutants.

If normally coloured parents are both carriers of white genes, they could produce a litter of entirely white cubs, or a litter of mixed white and ordinary cubs, or sometimes no white cubs at all.

"The reason that there are sudden 'crops' of white cubs is most probably based on the social structure of lion society," Wolff observes.

"Studies of these predators have indicated that the lionesses are the stable factor in a pride but that the males in the pride are continually changing . . . it has been noticed that groups of lionesses, or their direct female descendants, in places like the Serengeti National Park, have occupied the same pride area for many years and that these females are all related to one another.

"The males, on the other hand, are forced to emigrate from the parent group on becoming sub-adults. They wander around until they find a pride of lionesses without an adult male or they manage to chase the current male or males away. The new males then settle down with the pride until they die or, more usually, are themselves driven out by rival males.

"Thus it can be seen that if a carrier male moves into one group containing carrier females, there is a very good chance that white progeny will ensue. However, if the male that moves into this pride is not a carrier, there will be no white cubs but there still remains a chance that a number of the resulting offspring will themselves be carriers.

"Thus, the gene can be said to have 'gone underground' and it will remain hidden in the population until chance once again brings two carriers together."

Chris McBride and his family found the white lions at Timbavati when he returned to his father's game farm (part of the Timbavati reserve) after completing a wildlife management course at Humboldt State University, California, to make a study of lions.

They had been keeping close track of a breeding pair and when the litter of three eventually arrived, they were astonished to see that two were snow white, one male and the other female. These they named Temba and Tombi, the Zulu words for "hope" and "girl" respectively.

Some time later in a litter of about 10 cubs belonging to another pride they spotted a single white cub and named her Phuma, Zulu for "to stand out".

One of the famous White Lions of Timbavati photographed in Pretoria Zoo. They were taken there for their own safety.

Temba and Tombi were later captured and taken to the Pretoria Zoo. In 1981 Tombi died of liver cancer. This led to fears that these remarkable creatures might die out because then there were only three white lions left, as far as anybody knew.

Nature, however, stepped in and within a few more years there was a whole new "crop" of white lions in the Park's Tshokwane area.

Early in 1976 several tourists travelling through that part reported seeing a big white male lion and their sighting was confirmed by a Park policeman on 17 June of the same year. This lion is known to have been photographed at least twice, the first time by a tourist, Harold Resnik, and then again in August 1977 from a Park helicopter. Pat Wolff spotted it on 7 November that year and thereafter kept close records of its movements. This particular lion was last seen by ranger Wolff on 10 August 1981, "looking thin but fairly fit".

On 10 March 1979, an Australian visitor came up to a lion kill on the old Lower Sabie road close to Tshokwane and photographed three beautiful white cubs between six and eight weeks old in the company of several adults, all of normal tawny colour. Precisely what

happened to this trio thereafter is not known.

Just over two months later a previously unknown large male white lion was seen. He was recorded again in February 1980, and since then has disappeared.

Another pair of white cubs was found at the Shiloweni dam in August, 1979, 4 km south of Tshokwane. Wolff rushed to the spot but had to keep searching for more than a month before he saw the pair himself, and was then able to watch them for two consecutive days.

On the second day he was peering at them through binoculars when he noticed a third and smaller white cub, hardly more than four weeks old. There were two lionesses in this pride and one had two white and two normal cubs, and the other one white and three normal cubs.

The older pair, seen first at Shiloweni dam, did not live much longer (the mortality rate among lion cubs is as high as 75 per cent) but the third grew to become a fine lioness. Unfortunately, she crossed the border from the Park into Mozambique where she is known to have been photographed. She was apparently attracted by herded cattle, much easier prey than wild animals, and was snared and killed there.

A similar fate nearly claimed another white lion, a four-month-old wanderer discovered by Wolff in August 1980.

The magnificent sight of a full-maned white lion. He is the famous Temba, which means "hope" in Zulu, who lived out his life in Pretoria Zoo.

This male grew to full maturity and was also attracted by the cattle just across the border. But before the Mozambican ranchers could catch him, he slipped back across the border and in August 1983 is believed to have killed a giraffe within sight of ranger Wolff's post. This male was last seen in September 1983.

Since then, ranching on the Mozambique side of the border has virtually ceased because of the internal strife in that country so cattle are no longer an attraction for the Park's lions.

In June 1981, Mr Hans Taljaard, an honorary ranger, came across an unusual variation on the white lions. He was travelling on route S39, which follows the Timbavati river on the Park's western boundary near Satara, when he saw a white lioness with two ordinary-coloured cubs near the Leeubron windmill.

He took several photographs and these reveal a difference: the white mother had a black-tipped tail.

There seems to be no danger of the white strain among the Lowveld lions vanishing permanently. There have been at least nine recorded white lion sightings in the Park and seven in the Timbavati game reserve, which means the recessive gene must be well and truly spread through the local lion population.

In May 1984, a white lion was among three cubs born in the Johannesburg Zoo to Timba (not to be confused with Temba) and Suzie. Timba was brought to the zoo some years ago for surgery to remove the bullets after being wounded, probably by poachers, in the Timbavati reserve. Suzie was born in the zoo – so the white strain has travelled some distance from the Lowveld.

Conservationists are still in two minds as to whether the white colouration hinders lions in the wild. Author Chris

McBride speculates that the white recessive gene could have originated from lions which once roamed North Africa, where the white could have been a natural adaptation in the desert surroundings, until the Romans virtually eliminated them in their quest for arena performers.

White, he writes, makes a lion so obvious that it sticks out like a sore thumb and cannot stalk up close to its prey as do ordinary lions, whose tawny colour blends marvellously into the African bush. Ranger Wolff thinks this could be a reason why two white lions, at least, have chosen to hunt cattle in Mozambique, although he believes that they were probably in the company of other normally coloured lions.

The risk factor to the white lions was one of the reasons why Temba and Tombi were moved from Timbavati to the Pretoria Zoo. Ranger Wolff states that of the nine mutants recorded at Tshokwane a few years ago, five are presumed dead, one is missing but possibly alive and the other three have been seen alive and well, although not recently ... But, he adds, "this ratio is considered fairly normal as, in any population of wild lions, the mortality rate tends to be high."

By resolving to let their white lions roam free, the Park has determined that white lions can indeed survive in the wild. They watched one male for more than six years and a female for over two years.

"Lions are formidable animals and, on the occasion I saw the white female hunt, she melted into the bush as effectively as her normally coloured siblings," wrote Wolff.

The rangers did once have to step in to save white lions – the three which were found among two litters of eight altogether. In September 1979, they were watching these when they noticed that all eight cubs had a skin disease called "scab" which, if untreated, is invariably fatal in young lions.

They decided to do something about it. They darted the two mothers in the pride, scaring off the other adults, and while these two lay fast asleep they dashed about the bush catching the cubs by hand.

They then gave the whole lot a bath. The cubs were thoroughly washed from head to tail in disinfectant and when bathtime was over, the now sodden and hissing babies were put into a pen near the mothers in the bush to keep them together until the lionesses came out of their drugged sleep.

While two rangers stood guard, the rest of the pride – 11 lions – came back and rubbed cheeks against the sleeping pair. By pulling on a rope the rangers opened the door of the pen and the newly-scrubbed cubs joined the pride, which loafed about until the mothers awoke, when the whole pride moved off into the bush.

During 1985 there were confirmed sightings of two white lions in the Park. One was again seen at Leeubron windmill near the Timbavati river, this time by ranger Ted Whitfield of Houtbosrand, and another, a cub, was spotted slightly north of the Tshokwane picnic site, by Gert Erasmus, a Park information officer. The same cub was seen by ranger Pat Wolff near Tshokwane on 8 January 1986, and again by patrolling assistant rangers on 12 February. It was then confirmed to be a young male aged about 18 months. These latest sightings prove that the white gene is still active among the lions of the Kruger National Park.

Footnote: A number of other animal species with albinistic tendencies have been seen near Tshokwane, south of Satara, and at Letaba. A white kudu bull calf, in a herd of about 20, was seen for the first time by ranger Wolff on 9 November 1985 near the Mazithi windmill, 10 km north of Tshokwane. The same animal has been seen and photographed on several occasions since. In July 1985, a buffalo cow with white facial markings, grey body and black rear was photographed by Wolff near the Nkumbi mountain, south of Tshokwane, on the tarred road to Lower Sabie. This buffalo had a normal-coloured bull calf at its side. In January 1985 a state veterinarian captured a snow white buffalo calf near Letaba, but its colour gradually became normal as it grew to adulthood. An adult near-white giraffe (seen right) was well known in the area of the Sweni stream south of Satara and was photographed by the author in September 1982. It was usually in the company of several other giraffe and was extremely shy – moving away quickly to hide behind its companions when approached.

The "white" animals of Tshokwane
These are some of the "white" animals
photographed by ranger Pat Wolff of
Tshokwane. *Above:* A kudu calf. It is now
fully grown and wanders the area just north of
Tshokwane picnic site. *Right:* A buffalo cow
with white facial markings, grey body and
black rear which was seen near the main H10
tarred road to Lower Sabie near Nkumbe
mountain. *Top right:* A buffalo calf captured
near Letaba. Its colour became normal as it
matured. *Far right:* A white male cub at the
Shiloweni dam surrounded by normal cubs.
Centre right: The same lion photographed in
adulthood in September 1983. *Previous page:*
This extremely shy near-white giraffe was
photographed by the author near the Sweni
stream, south of Satara.

When nature takes its toll

During times of natural disaster, when fire, disease or serious drought strikes the Park – and turns nature ugly, life becomes a desperate struggle to which many animals and plants succumb. This happened in 1983 when serious drought gripped the Lowveld and many animals died, among them hippo, warthog, kudu and impala. These three pictures tell the story of a very young vervet monkey abandoned by its mother at that time. After he took these pathetic pictures, the author reported the monkey's plight to a ranger. However, it was decided that nature should take its course.

In a unique way the Kruger National Park can lay claim to running the world's largest, most diverse and most self-sufficient hotel. What other resort can accommodate more than half a million guests a year in its collection of 17 "wings" many kilometres apart from each other, most with their own restaurants, shops, snack bars and kitchens, together providing 3 180 beds in comfortable rooms served by 1 414 air conditioners and 1 159 refrigerators, plus laundries and an incomparable range of other facilities – including scores of camp sites?

And all this at prices which would make most hoteliers blanch but which include magnificent, 20 000 km² surroundings and a staff of well over 1 000?

The difference, of course, is that the Park is a non-profit organisation with no shareholders except, in the broad sense, every South African citizen.

Its financial turnover would be laughed at by the owners of the Hiltons, Intercontinentals and Sheratons: the record is just under R30 million a year, every cent of which goes back into running costs and improvements.

No other resort can match its extraordinary variety. Its 17 camps – many of them really small villages – range from the tiny Jock of the Bushveld with only 12 beds to big Skukuza, virtually a town, with 628 beds.

Each has its own distinct personality and atmosphere shaped by its situation, environment and age. One is in the middle of wide open savannah, another high on a ridge, yet another in riverine forest, one dominated by looming rock mountains, several beside rivers. There are also four small trail camps sited deep in the wild with simple amenities and only the most rudimentary of enclosures, and more camps of all kinds are being planned.

Left: Sunrise in the Kruger National Park –and a giraffe loftily looks down upon early-rising tourists leaving Satara camp.

Behind the scenes lies an astonishingly large and complex logistical organisation of which the following are examples:

- At Skukuza camp up to 17 000 sheets, pillow cases, tablecloths and similar items go through the laundry daily.
- Every month the Park restocks with an average of 10 000 2 kg bags of charcoal for campfires and barbecues.
- The Park's monthly electricity bill runs to over R60 000.
- The Park uses nearly 400 vehicles of all kinds, from road-building machines to motorcycles. Only three are ordinary saloon cars.
- The total number of meals enjoyed annually in Park restaurants now exceeds 500 000.
- The Park spends more than R80 000 a year carting firewood to camps and picnic sites.
- Air traffic to the Park's main airport at Skukuza reached a record 1 051 planes carrying 15 662 people in 1983, excluding official traffic. Since then traffic movements have declined slightly.

This section of practical information is designed to help the visitor gain a comprehensive insight into the extensive range of amenities available, the safety regulations, topography and climate, and health precautions in the Park – among many other things. There is also a section on understanding binoculars. This is followed by a chapter of hints for photographers and a major section describing each camp individually, the viewing routes around it and wilderness trails.

Topography and climate

The Park's topography is fairly uniform, like the rest of the Lowveld. Most of it is flat or gently undulating savannah which slopes gradually from the west down towards the Lebombo mountains, a long, low range which lies along almost the whole length of the boundary with Mozambique. Except in the far south there are no true mountains but the Park's surface is marked by many ridges

and koppies – huge, smooth, grey extrusions of granite which have withstood weathering, or hills topped by fractured stone which provide lairs for a whole range of large and small animals.

The Park's height above sea-level varies from extremes of 839 m on a mountain top in the far south to 122 m at the bottom of the Sabie river gorge, with most of it between 200 and 400 m. It is drained by six major rivers and bordered by the Limpopo in the north and the Crocodile in the south. There are also a number of often dry rivers which flow only when the rains come but retain water in pools for considerable periods.

The rains usually begin in September or October, reach their peak between December and February and dwindle in March and April. Most of the rain comes from erratic thunderstorms often accompanied by spectacular displays of lightning.

Generally the Park is hot by day, even in winter, and has recorded such extremes as 47 °C (116 °F) in January and 35 °C (95 °F) in July. But nights can be cool, in summer too, and at the other end of the scale temperatures have been recorded as low as 7 °C (44,6 °F) in January and -4,2 °C (24 °F) in July. But that the climate tends to be hot is shown by the averages: 30 °C (86 °F) in January and 23 °C (73 °F) in July.

The smiling salute of Judas Mashele has been welcoming visitors arriving at Numbi gate for more than 40 years.

Accommodation

So comfortable are the camps and so pleasant their gardens and surroundings that many visitors to the Kruger National Park choose to spend as much time in them as they do in their cars looking for game. There is always an amazing variety of birds to be seen and very often as many kinds of animals feeding just beyond the camp fences as outside on the roads. In the heat of high summer especially, the camps beckon with their cool rooms, restaurants with shaded patios and – at Berg-en-dal, Pretoriuskop and Shingwedzi – swimming pools.

Accommodation is usually in the form of the traditional thatched "rondavels" or rectangular chalets with shaded verandas, most with their own air-conditioning, refrigerators and showers. They range from luxurious multi-roomed guest cottages to simple, two-bed huts and tariffs range accordingly. The following are the different types of accommodation available:

Family cottage with kitchenette. Two double bedrooms, bathroom, toilet, kitchenette with gas stove, refrigerator, eating and cooking utensils, screened veranda.

Double family cottage (at Olifants camp only). Each half section consists of a bedroom, with three beds, bathroom or shower, toilet, a kitchenette with gas stove, refrigerator, eating and cooking utensils and a screened veranda.

Family cottage without kitchenette (at Pretoriuskop only). Two bedrooms with three beds each, bathroom, toilet, refrigerator.

Self-contained thatched hut with kitchenette. One room with two or three beds, shower, toilet, refrigerator, gas stove, eating and cooking utensils.

Ordinary thatched hut. Two, three, four or five beds, a hand basin with cold water. Shower and toilet not included. Ablution block nearby.

Thatched hut with bath and kitchenette (Punda Maria only). One room with four beds, bathroom, toilet, refrigerator, gas stove, eating and cooking utensils.

Guest cottages. In the larger camps there are a number of guest cottages which can be reserved three months in advance. They accommodate parties of from three to nine guests and have all facilities.

Private camps. Situated at selected sites throughout the Park are five mini-camps, each one being a complete living

Typical of the relaxing atmosphere at all the rest camps is this shady corner in Lower Sabie in the Park's south.

unit of several rooms with between 12 and 19 beds per camp. They have full facilities and are serviced by resident staff. They must be booked in their entirety – well in advance.

All types of accommodation are serviced daily to first-class hotel standards. Bedding, towels, glasses, soap and the like are provided, but not crockery, cutlery or cooking utensils, except where kitchenettes are included.

There is specially adapted accommodation available for paraplegics at Skukuza, Satara, Shingwedzi and Berg-en-dal, and their needs are also catered for in all the new campsite ablution buildings.

Restaurants

All main camps have restaurants with full facilities for three meals daily at remarkably low prices. Separate cafeterias are open from 7 a.m. to dusk for light meals, sandwiches, soft drinks, sweets and ice cream.

Visitors, especially those with families, who are not self-catering are reminded that the cafeterias close at 7 p.m. at all camps. After that time, because the shop is also closed, the only source of food is the main restaurant which is obviously more expensive. Too often newcomers to the Park, who are

anticipating take-away or cheaper veranda meals at a later hour, miss out on their supper. The only *à la carte* restaurant is in a beautifully restored old train next to the Sabie Bridge at Skukuza.

Main camps have communal cooking ranges where visitors can prepare their own meals and all have excellent barbecue facilities. Hot water for beverages is available at all hours in most camps.

Shops

All but the smaller camps have well-stocked self-service shops with frozen and canned goods, top grade fresh meat, some fresh vegetables, basic pharmaceutical supplies, film, books, binoculars, cooking and eating utensils, firewood, casual clothing and an extensive variety of other goods at everyday prices. This includes cold beer and soft drinks, a small range of good South African wines and several kinds of "hard" liquor. Under South African law no liquor may be sold across the counter before 9 a.m., or on Sundays and other religious holidays. It may, however, be served with meals, and on any day of the week one may order drinks on camp verandas and in lounges during lunch and dinner times.

Right: Rows of potted Lowveld plants are on sale at the Skukuza nursery. They are also sold by most Park shops and at main entrance gates.

The following are the shop opening times at all main camps:

Monday to Saturday
March, September, October.
  08h00 to 13h00 and 15h30 to 18h30
November to February incl.
  08h00 to 13h00 and 16h00 to 19h00
April to August incl.
  08h00 to 13h00 and 15h00 to 18h00

Sunday and religious holidays
March, September, October
  08h30 to 11h30 and 15h30 to 18h30
November to February incl.
  08h30 to 11h30 and 16h00 to 19h00
April to August incl.
  08h30 to 11h30 and 15h00 to 18h00

Other amenities

All camps have basic laundry facilities and some have modern coin-operated washing machines and tumble dryers, all in the camp ablution/laundry blocks. Camp domestic staff may be hired only in their off-duty hours to wash clothes or dishes for visitors.

For people who fly into the Park – and those who come in their own cars, but may prefer a minibus for game-viewing – there is an Avis car hire service at Skukuza. Bookings should be made well in advance. Visitors can also fly by regular commercial air services to Nelspruit or Phalaborwa, where they can hire cars.

The only bank in the Park is at Skukuza and is open during the usual banking hours. Reception offices at all main camps will cash travellers' cheques, but shops will not, except in emergencies or for purchases. All generally recognised credit cards are accepted for making purchases at shops and paying for restaurant meals and accommodation.

Conference facilities are available at Berg-en-dal camp and by arrangement the camp staff will organise tours in the vicinity for spouses and children while the conferences are in progress. An increasingly popular practice for people who do not want full conference amenities is to hire in entirety one of the small "private" camps which accommodate up to 19 people.

Only Skukuza and Pretoriuskop have post offices with telephone links to the outside world, but emergency messages will be relayed at all hours from Skukuza to any camp. All camp reception offices will accept mail.

All main camps have electricity supplied through the Escom grid, except Shingwedzi and Punda Maria, which rely on diesel-powered generators. Power is available 24 hours a day, except at Punda Maria where the generator is shut down from ten o'clock at night to just before dawn. Although candles are supplied to all rooms there, it is advisable to keep a torch within reach during the night.

Tap water is safe to drink at all main camps, as each has its own purification plant. All water is drawn from Park rivers except at Punda Maria, which has a borehole. Lengthy pipelines supply two camps: Satara is 40 km from its pump station on the Olifants river (just west of Balule), and water to Pretoriuskop is piped from the Sabie river, 17 km to the north of the camp.

The authorities warn that malaria is endemic throughout the Park.

Plants indigenous to the Lowveld can be bought at the nursery about 3 km from Skukuza on the Paul Kruger Gate road. Qualified staff in attendance will give advice on the selection and care of plants. Some of the more popular varieties of shrubs and trees, like the impala lily, are also on sale at camp shops and Park gates.

There are full information bureaux at Berg-en-dal, Skukuza, Satara, Olifants and Shingwedzi camps and more are planned. Camp managers will also gladly help visitors with any queries about where to go and what to look for.

All camps regularly screen films on conservation at outdoor cinemas and whenever possible, a Parks information officer is made available to answer questions after the show. Queries from children are especially welcomed.

Your health

The authorities warn that malaria is endemic throughout the Park. This mosquito-borne disease is debilitating but not in itself fatal. However, it can lead to fatal complications. It is therefore essential that visitors take one of the several recognised courses of prophylactic tablets which they must begin before entering the Park, then continue during and after their visits for the prescribed periods. The tablets are commonly available from pharmacies in South Africa and from all Park shops.

The elderly and people with heart problems are advised not to exert themselves in the heat, which in summer in the Park can become extreme, and to relax in the shade during the middle of the day. The same advice goes for people who have recently had or still have viral infections like influenza.

Children under 10 are also particularly vulnerable to heat discomfort and it is sensible for them too to stay in camp in the shade during the hot hours. For those taking advantage of the swimming pools, special care should be taken against sunburn. For that an array of barrier creams is available at camp shops. Hats and long-sleeved shirts should be worn if necessary and drivers are warned to protect their "window" arms from the sun.

Do not use salt tablets to counter the effects of heat or sunburn. The main effect of heat is the loss of fluids, not of body salts. Therefore, drink plenty of water or fruit juice, or eat fresh fruit.

There are modern state hospitals just outside the Park – at Nelspruit in the south and Phalaborwa in the mid-west, and a mission hospital near Punda Maria in the far north. There is a resident doctor in the Park with consulting rooms at Skukuza and all camp managers are trained in first aid. Anyone needing medical attention should call the camp manager who will contact the Park doctor by radio if necessary.

Emergency aid can be summoned at any time by driving to the main gate of any camp and sounding the car horn until help comes.

Ambulances are based at Skukuza and Letaba to take patients to hospital, but in emergencies they will be flown out by helicopter.

How to get there

By air. There are regular commercial flights from Jan Smuts airport, Johannesburg, to Skukuza and also to Nelspruit and Phalaborwa just outside the Park, as well as charter services. At present private planes may land only at Skukuza. Pilots wishing to make use of the new all-weather airstrip at Punda Maria, in the Park's north, may only do so by special arrangement with the Park's chief warden. Overseas visitors are advised that all aircraft entering the Park require special clearance and details of how to obtain this are available from any major airport or air charter company.

By road. The main route from Johannesburg is to Nelspruit and from there to Malelane Gate or Crocodile Bridge on the Park's southern border, or to the Numbi or Paul Kruger Gate entrances on the south-west side. Some people prefer to travel further north along the Park's western side through Acornhoek to enter at Orpen Gate.

Another way is to travel due north from Johannesburg to Pietersburg and then turn east, passing Tzaneen and Gravelotte and entering the Park at Phalaborwa Gate in the mid-west.

By going further north to Louis Trichardt and turning east, visitors can drive through the Venda Republic and enter at Punda Maria Gate.

The northernmost entrance to the Park is Pafuri Gate a few kilometers south of the Limpopo river border between South Africa and Zimbabwe, and north of the Luvuvhu river. It is most easily reached from the town of Messina, the main South Africa/Zimbabwe border post, and increases the Park's accessibility to Zimbabweans.

All these routes are scenic and the excellent network of modern highways in the Transvaal provides a wide diversity of routes to the Park, such as along the magnificent Long Tom Pass from Lydenburg down to Sabie and then to Numbi or Paul Kruger Gate, or through the spectacular Blyde River Canyon further north and from there to Phalaborwa.

Travellers should ensure that they give themselves enough time not only to reach the Park entrance gate, but to drive from it to their destination camp – bearing in mind the internal speed limits and the closing time of the camp gates (see table on p. 143).

While in the Park, visitors should plan their trips out from camps to leave themselves enough time to get back within the speed limits before gates close.

The motorist

Filling stations are open during business hours at all main camps and provide the normal range of fuels, oils and other requirements for the motorist – with one exception: automatic gearbox fluid is sold only by the camp shops. Diesel fuel is available at all camps except Crocodile Bridge.

The Automobile Association provides breakdown services from Skukuza, Satara, and Letaba. They can cope with routine and emergency repairs but not major mechanical troubles, and they stock limited supplies of items such as fanbelts, batteries, tyres, tubes, spark plugs and radiator hoses.

The AA is an organisation formed to serve only its members, but in the Park it serves all motorists at its usual rates. Only members, however, can make use of its free towing service. Non-members

The popular à la carte Railway coach restaurant at Skukuza is renowned for its venison dishes – buffalo fillet in particular.

are charged the full rate unless they immediately join the AA, when it is reduced by half.

Visitors should have their cars serviced and tested before entering the Park and in particular ensure that their tyres, including spares, are in good condition.

If a motorist driving in the Park experiences minor mechanical problems, such as a punctured tyre, broken fanbelt or split radiator hose – something that can be fixed immediately at the roadside – one or two people may leave the vehicle to carry out repairs, but strictly at their own risk. Should the breakdown be a major one requiring the assistance of a mechanic, the motorist should not under any circumstances attempt to walk away to seek help. Flag down a passing car and send a message to either the nearest camp manager or AA service centre and then *stay in your vehicle* until help arrives – however long it takes.

All are warned that law enforcement officers using speed traps strictly enforce the speed limits of 50 km/h on tarred roads and 40 km/h on gravel.

Travelling hours

One of the strictest of the Park rules, for the welfare of the visitors, is: *get back to your rest camp on time.* The insistence that people be back in their camps at closing time is not idle when one considers that the Park staff are responsible for the safety of thousands of people wandering all over a wilderness full of potentially dangerous animals and pitfalls for the unwary.

To make that easier, they have set standard closing and opening times for all camp gates as well as Park entry gates. They are:

	OPEN	OPEN	CLOSE
		Entry	Camps
	Camps	gates	& gates
January	05h00	05h30	18h30
February	05h30	05h30	18h30
March	05h30	05h30	18h00
April	06h00	06h00	17h30
May-August incl.	06h30	06h30	17h30
September	06h00	06h00	18h00
October	05h30	05h30	18h00
November-December incl.	04h30	05h30	18h30

Note: The reception offices at all Park entrance gates are *closed* from 13h00-13h30 and camp reception offices are *closed* from 13h00-14h00 for lunch.

Caravanners and campers

The least expensive way of visiting the Park is to take one's own caravan or camping equipment and set up home in one of the many allocated camp sites, all of which are wooded and some grassed. So popular are caravanning and camping that some 240 000 of the Park's annual flock of visitors choose to do it.

All the main Park camps have caravan and camp sites except Olifants, which is only 10 km from the planned Ngotsamond caravan camp on the south bank of the Olifants river immediately west of the low level bridge. The opening of this camp has been delayed because of

All tarred roads in the Park are periodically sealed with a spread of coarse sand.

recent important archaeological discoveries there. The Park authorities are eager to allow scientists to finish their work before the camp is completed.

Meanwhile caravanners are welcome to use one of the 10 sites available at the small, delightful Balule camp nearby, but they must check in at Olifants camp before going there.

Another exclusive camp for caravanners is the Maroela camp situated a stone's throw away from Orpen Gate in the central area of the Park. Although it is most often used by late-comers unable

143

to get to Satara before the gates close at night, it has become a favourite among traditionalists who enjoy true bush isolation.

Camping and caravanning areas have been expanded in most of the main camps and bathroom facilities have been extended and modernised throughout the Park. Many now have coin-operated laundries. There are also gas or coal-fired cooking ranges available and hot water is available on tap at all hours.

Caravanners are reminded that, as a general rule, they may not tow their vehicles off the tarred roads in the Park.

Roads and by-ways

If rivers are the arteries of the Park, the road network is its nervous system. Its prime function is to carry the tourist traffic, but roads also serve as firebreaks and as the main communication and service channels in the running of the Park.

Stevenson-Hamilton's few rough tracks have grown into more than 4 000 km of roads, of which 1 255 km with gravel surfaces and 790 km tarred are

open to the public. In addition there are a further 1 000 km of patrol roads and more than 900 km of access roads used only by Park staff, plus – although strictly speaking they are not proper roads – nearly 4 000 km of firebreaks.

Maintaining all this and also 24 high-level bridges and thousands of culverts (whose large diameter pipes are widely used as economy housing by warthogs and breeding hyenas) is a large roads department using a fleet of more than 100 specialised vehicles, from trucks and trailers to giant rollers and graders.

For years one of their more exasperating tasks has been to re-erect or repair some 500 "no entry" signs used to keep visitors off service roads. Elephants seem to take a particular dislike to these and are always uprooting or simply flattening them. Now the department is designing new signs of stone or concrete – similar to the route markers at every intersection in the Park – which are short, pyramidical and solid so elephants will be disinclined to push them about.

Tarred roads might seem incongruous in a wilderness area and caused wide-

Elephants and tourist traffic meet south of Letaba camp.

spread controversy when construction began on them in 1965 between Numbi Gate and Skukuza. In fact they have proved their worth.

Critics protested that tarred roads were a contradiction of the principle of keeping the wilderness pure and natural, that they were an intrusion of civilisation, that animals would avoid them, and that motorists would drive too fast on them.

Not so, said the protagonists. Any roads including gravel disturbed a nature area, but they were a necessary evil if people and nature were to be brought together. Gravel roads, moreover, needed much more costly maintenance and were the source of a huge dust problem even when motorists travelled slowly. That this is true is proven by the fact that on a still day dust clouds can be seen hanging above some gravel routes. Worse, the dust thickly coats vegetation on both sides of dirt roads for as much as 20 to 30 m away, turning it to

144

uniform grey or brown and making it unpalatable to browsers and grazers, whereas alongside tarred roads the vegetation is clean and green.

Today most of the main arterial routes between camps are tarred and the surfaces are regularly spread with bitumen and river sand to help seal the road and soften its dead black colour. Most other roads will remain gravel, not only because of the expense of surfacing them, but for their intrinsic character and because they wind and twist deep into the bush.

Far from avoiding them, the animals have accepted tarred roads as readily as they did the others. Elephants like to use all roads as unencumbered "freeways" – as is obvious from their droppings – and lions loaf on them in the cool of the mornings and evenings as people will on a patio. Moisture condensation and run-off on tarred roads makes the grass grow greener along the verges and attracts the grazers, especially in drier months.

The Park is divided into three sections for road maintenance purposes, each one manned by 30 workmen under a regional foreman.

Other than the major task of road sealing (every metre of tarred surface is covered on average once every ten years), they are responsible for new road building and for repairs to culverts, river causeways, bridges and road verges – as well as grading of dirt roads and patching of tarred ones.

Tarred roads have to be cleared of elephant droppings and especially of buffalo dung – a real headache because if it is not swept off immediately, it causes tarred surfaces to break up and lift in patches. Maintenance crews are also responsible for the collection and delivery of firewood to all camps.

The authorities are now commencing a long-term scheme to expand the road network to 4 600 km to cope with a planned increase in visitors – but still at the same average flow of traffic that it carries now: 75 cars to each 100 km.

It will, however, impose a greater burden on the rangers and traffic control officers who constantly have to curb drivers selfish enough to spoil things for others.

What the officers cannot control are

A section of the main building complex at the new Berg-en-dal camp.

the animals themselves. A pride of lions might choose to take a nap and spread themselves right across a road, blocking traffic for kilometres. Elephant sometimes appear to deliberately do the same thing, standing in roads and placidly waving their ears at tourists. In one instance wild dogs gave a traffic officer an unusual problem.

The officer had laid a speed trap device across a road near Skukuza: a set of twin pneumatic tubes which, when crossed by a car, registers its speed on a gatsometer. He was watching the dials of the instrument from his hiding place behind a screen of bush when they suddenly registered a vehicle passing at high speed – but he neither heard nor saw any vehicle. Puzzled, he peered out and saw a group of wild dogs playfully chewing up the tubes and switches. He ran out, roaring in protest, and the dogs fled – taking the apparatus with them. It was found the next day some 400 m away.

There are several law enforcement officers based at Skukuza and Letaba and their strictness in enforcing speed limits is justified by the fact that in a recent period of six months more than 30 impala and bushbuck were killed on one short stretch of road near Skukuza.

For safety's sake nobody is allowed to enter the Park on a motorcycle or in a soft-topped car or open vehicle. For the same reason dirt roads are closed when rain is heavy, to avoid motorists becoming stuck in mud or suddenly rising

streams, and also to prevent the road surface from being churned up and damaged. With a few exceptions all gravel roads are also closed to towed caravans.

Rules of the Park

Rules and regulations in the Park have been kept to an absolute minimum. The prime aim of the Park is that people should be able to relax and commune with the environment without the harassment of a multitude of petty restrictions. The only code of behaviour is inherent in common sense, but since some people ignore even that, a few basic common-sense rules have to be unambiguously spelt out.

Do not leave your vehicle except at camps, lookout points, picnic sites and other places where signboards specifically indicate it is permitted to do so. This is not an idle rule: people and animals in open confrontation spells danger for both. For the same reason it is a breach of rules to lean out of your vehicle with head, shoulders and arms protruding. The only exception to this rule is in the event of a flat tyre or breakdown, when people may leave their cars but do so entirely at their own risk.

Do not feed any animal. This rule is strict, for the sake of the animals as well as the tourists. Animals are frequently fed by tourists who are ignorant of their natural diets, sometimes seriously harming or even killing them. Some animals, such

as baboons, become so accustomed to being fed that they lose their fear of man, which makes them extremely dangerous.

Do not exceed speed limits. It is extremely easy to hit a wild animal which knows nothing of road rules. Hitting an elephant, rhino, buffalo or kudu can be fatal to the speeding driver and his passengers. Remember: animals *always* have the right of way.

Stay on designated roads. Those "no entry" signs are there for good reason. Tourists who ignore them might find themselves stuck along a rough road designed for official four-wheel drive vehicles and it could take several days before they are rescued.

Do not disturb the wildlife. It is totally contrary to the spirit of the Park to intrude upon its natural activities. The tourist is there as a privileged outsider. Shouting or hooting at animals – or, as some deviates have done, playing tape recordings of animal sounds to them – is severely disruptive.

Do not light fires – except in designated places like camp and picnic site barbecue areas. There is *nothing* more devastating in a national park than a runaway bushfire. Never toss cigarette butts or "spent" matches out of car windows: it could cost thousands of lives.

Do not remove anything from the Park which you are not entitled to. That covers everything from seed pods to elephant tusks, unless they are bought at a Park shop. Taking unauthorised items out breaches the strict health control regulations without which the Park cannot survive and could also land visitors in trouble under veterinary laws.

Bring no pets. Only rangers may have pets within the Park, and then only with special permission.

Do not drop litter. Animals have been known to choke to death on pieces of plastic or other rubbish discarded by unthinking visitors. Litter bags are handed out to all vehicles entering the Park and rubbish bins are copiously available wherever tourists may go.

Get back to your rest camp on time. Nobody is allowed to stay the night outside a rest camp and every new overnight visitor must report his arrival to the camp reception office.

Do not play radios, tape recorders or musical instruments so loudly that you might disturb others. The beauty of the camps at night can only be properly appreciated in the silence of the surrounding wilderness. Visitors often retire to bed early and rise

to see the bush at its best in the early morning. Therefore the rule of thumb is: no noise from 9 p.m. to 6 a.m.

Do not harm trees or Park property. The most unwanted visitor is the kind who carves his initials on a tree trunk or scrawls his name on a bridge, rock or hut wall.

Declare all firearms. They can be brought into the Park but only if they are sealed at the point of entry. If the seal is broken when the owner leaves, he has some serious explaining to do.

The Park staff and system are not a bureaucracy. They are devoted to caring for their magnificent and priceless ward, the wilderness, while at the same time making it possible for the rest of the world to enjoy this treasure with them – an exceptionally difficult balance of motivations.

Yet if anyone breaks their simple set of rules, they can and will be severe, in so far as the law allows. The penalty for littering and for feeding animals is the same: a paltry R25. For the more serious offences such as speeding or disturbing the animals, fines range from R25 to R150.

That the fines are so low is a measure of the trust placed in the public.

DON'T JUST STAND THERE — FLAG HIM DOWN!

Tourism in the future

The Park authorities are committed to a long-term R80 million programme to further develop its facilities under the guidance of a planning committee consisting of its leading scientists, engineers and tourist officers, headed by chief warden Pienaar and advised by landscape architect Willem van Riet, a renowned "wild water" canoeist and conservationist.

Their brief when the National Parks Board approved the project in 1980 was to expand the Park's capacity to a maximum of 700 000 visitors a year (believed to be the top limit), which means providing more than 1 000 additional beds, extending the 4 017 km road network by another 600 km, building more camps and opening up more picnic sites and hides from which people can study animals at close quarters.

To do that they have designed five new medium-sized and five new small camps – although whether and when they will go ahead with these will be determined by current circumstances. Of the new medium-sized camps the R6 million Berg-en-dal in the far south is already open to tourists.

The other four on the drawing board are a *de luxe* camp to be named Narina sited just 3 km north-west of Skukuza, Doornplaat on the Nwaswitsontso river between Orpen Gate and Tshokwane, Pioneer on a koppie overlooking the Pioneer dam between Letaba and Shingwedzi, and Lanner Gorge on the northern bank of the beautiful Kowakuly gorge on the Luvuvhu river.

Four of the new small camps have been completed: Jock of the Bushveld on the main road between Malelane and Skukuza, Roodewal camp on the bank of the Timbavati river near Satara, Malelane camp in the extreme south which was converted from the much larger camp that used to exist there, and the unusual Boulders camp 10 km south-west on the Mooiplaas picnic site north of Letaba, whose luxury huts are raised on stilts and built into and around a picturesque, boulder-strewn koppie, so it needs no fences.

Like the existing ones, the new camps will have to be reserved in their entirety. They are the Lonely Bull camp in a newly opened wilderness area northeast of Phalaborwa Gate and Silverfish

The new Boulders mini-camp north of Letaba is built on stilts so no fences are needed.

camp on the Shingwedzi river very close to the Tsange lookout point in the Park's far north.

The Ngotsamond caravan camp will replace the historic little camp of Balule on the banks of the Olifants river, and Crocodile Bridge camp is to be bulldozed in two stages (the old square huts will go first) and completely rebuilt.

Existing camps are being upgraded at a cost of more than a third of the development budget. Filling stations and other "commercial" services are being resited so that the constant flow of travellers in and out of camps does not disrupt the tranquillity for people staying in them. Information services are being expanded and will include a film unit to make documentaries for screening in the Park as well as nationally and internationally. Most huts are being given *en suite* bathrooms, swimming pools will eventually be built in all camps, and special ramps and other aids are being provided for people confined to wheelchairs.

A freshwater aquarium is envisaged for Letaba or Skukuza, and an elephant Hall of Fame.

Another part of the development project is to create greater freedom for visitors to leave their vehicles and stretch their legs. All the existing picnic sites are being carefully modernised, to include such luxuries as running water, waterborne sewerage and mini-shops, without damaging their aesthetic appeal. Eventually there will be more than 30 picnic

and view sites scattered all over the Park, with modern toilets, thatched shelters and barbecue facilities where people can relax along the roads between camps.

All the new large camps and some of the older ones will have mini-trails within the actual camp boundaries on which walkers will be escorted by guides or can learn what they want to by using pushbutton audio-information points along the routes. The new camps, like the old, will be positioned to give ready access to different parts of the Park's ecological structure. However, great tracts of its wilderness will be kept closed to tourists and accessible only to Park staff to remain undisturbed reservoirs of wildlife.

In response to fears expressed about over-modernising and overcrowding the Park, chief warden Pienaar wrote recently:

"The [National Parks] Board is very conscious of the fact that the Park has a finite capacity for tourist development and that a saturation level will be reached in the not too distant future. With this in mind the necessary wilderness areas, which will not be available for tourist development (except wilderness trails), have been demarcated and confirmed in the master zonation plan.

"Roads, trails and other facilities for tourists should be planned in such a manner that they serve as showcases of

Binoculars are a must for successful bird-watching.

the natural landscapes and the splendour of the wildlife indigenous to the area. The facilities provided should in no way constitute a threat to the protected environment or detract from the natural character of the area."

Dr Pienaar said recently that the latest assessments indicated that the maximum annual tourist capacity the Park could maintain without harming its character was roughly 700 000 people, and that figure could be reached only after new camps had been built to provide 1 000 additional beds, bringing the total to 4 000 beds (the maximum permitted by the National Parks Board) plus additional camping facilities. Only small and medium-sized camps would be planned in future.

Visitors were already restricted to no more than 10 days at a time in the Park and during peak periods such as public holidays and long weekends the number of day visitors was limited by the issue of permits.

The Park was receiving a growing number of black and brown visitors from nearby African homelands and from as far afield as Cape Town, and it was essential to encourage this trend, Dr Pienaar said, to spread the gospel of conservation. It was particularly important to promote visits by parties of schoolchildren, so additional facilities had to be provided for them.

Responses to questionnaires given to visitors had shown them to be generally satisfied with the range of amenities the Park provided, from trail camps to private camps to luxury camps like Berg-en-dal.

But to further enhance the Park's natural attraction he intended, he said, to build more vantage points and hides from which to view wildlife. "Rather bring the features of the Park and animal life to the tourist than let him ride around wasting petrol."

Understanding binoculars

A good pair of binoculars is the wildlife enthusiast's greatest asset. I truly regret the years that I wasted insisting stoutly that "there's nothing wrong with my eyes," until I was given a pair of compact 9 x 25s. Now even the shortest trip away from home becomes frustrating if I have forgotten to take them along.

Without binoculars, for instance, it can be difficult to tell whether that bird in shade in a tree 30 m away is a long-tailed widow bird or a longtailed shrike.

Binoculars are an excellent investment because they require virtually no maintenance beyond cleaning and, unlike most precision instruments, will give a lifetime of trouble-free service. Yet it amazes me how many people select theirs with no more than "a quick squint down the street," a shake of the head at the price tag and little forethought about what they wish to use them for, and in what conditions.

This cavalier attitude is a pity because today's manufacturers are producing an ever-wider range from which to select and personal preferences are no less important than in choosing a good camera.

The following is a brief insight into the lore of binoculars.

Each is a pair of identical twins – two perfectly matched optical systems (if the binoculars are of a good make) set in absolute parallel and linked by a mechanical focusing device. The advantage of binoculars over the monocular (such as a telescope) is that by giving both eyes an identical view, normal three-dimensional vision is retained.

Engraved on most binoculars is a set of figures which state its performance, for instance 7 x 50 7,3° or 9 x 25 5,6°.

Let us examine the first set of figures. The symbol 7 x indicates the magnification strength of the optics and in this case the image shown by the binoculars will be seven times larger than that seen by the naked eye. An interesting experiment to prove this is to look at a distant object with one eye staring through one barrel of the glasses, leaving the other eye free to look at the subject naturally. The double image will show the subject seven times larger through the binocular lens than the object seen by the unaided eye.

The figure 50 gives the diameter of the front element, or object lens, in

millimetres. Generally speaking, the bigger this element the more light it will gather, which means that binoculars with larger front elements are more efficient in weak light – and also proportionately heavier to carry. The size of the rear element, the one put to the eye, can be calculated by dividing the magnification figure (7) into the diameter of the front element (50) – so in this case the rear element's diameter will be 7,1 mm. (It should be noted that with some binocular designs, this calculation does not apply.)

If you consider that the pupil of the human eye contracts in bright light to about 2 mm and dilates to about 7 mm in poor light, it becomes obvious that the large 7,1 mm rear element means that these binoculars perform well in weak light. However they are no less useful in very bright light because obviously the pupils of the eyes contract automatically to compensate for the extra light intensity.

While the big waterproof kind might be convenient in the Kruger National Park for looking at ducks in the rain, they are very cumbersome.

Zooms boast ranges of magnification between as much as 8 and 24 degrees so that the subject being looked at can be brought instantly closer by a mere flick of a thumb lever and, in effect, give you a whole range of binoculars in one. The disadvantage of most zooms is that they are not close-focusing.

The compact "roof-prism" binoculars are small enough to fit in the pocket and light enough to wear on a strap without giving one the proverbial pain in the neck. Because of their size, they are very popular with rangers and trail hikers in the Park and have the advantage of focusing to as close as 4 m – most useful to bird watchers.

I must add a word of caution to people contemplating buying the more powerful types, like those of 16 magnification and more. Although excellent for long-range viewing, they require extra steadiness and it can become very tiring holding their weight up to the eyes. Ideally they should be firmly supported or fitted on a tripod, like a very long camera lens. The advantage of 16 times magnification is often destroyed by the fact that even the slightest tremor of the hand is also magnified 16 times, causing a very shaky image – and eye strain.

Finally, when choosing a pair of binoculars, beware of inferior makes. If the lens barrels are not perfectly in parallel, or if there is any play between them, you will see a double image. If the lenses are poorly matched you will see coloured edges around the object you are viewing. Inferior binoculars are also unlikely to be properly proofed against moisture and dust – which will soon give you unhappy viewing, and very tired eyes.

The figure 7,3° is the field of view as seen through the binoculars – the width in degrees of the scene as you see it. Each degree represents a lateral width of about 17,5 m at a distance of 1 000 m from your viewpoint; therefore at 1 000 m the width of the area seen through the binoculars will be about 128 m. For a simplified example of a field of view of this size, stretch out your arm at full length with the palm of your hand facing outwards. Then separate your forefinger from your middle finger as wide as possible in a V-for-victory sign. Now sight a distant object between the fingers. The field of view between your entended fingertips will be roughly 7°.

The glass surfaces of binoculars, like camera lenses, are "bloomed" and should be cleaned with extreme care (see p. 157). The majority of those sent in for repair require the re-alignment of the delicate glass prisms which have shifted in their seats thus causing double vision – usually a result of having been dropped or heavily knocked.

Binoculars come in a wide range of sizes and shapes, from the tiny opera glasses favoured by ladies of fashion, through the very compact folding models now produced by leading Japanese manufacturers, to the big night glasses favoured by mariners, which are so powerful and heavy that they have to be mounted on gimbals.

Much favoured by followers of horse-racing and other fast action sports are the beautifully computed wide-angle binoculars which give fields of view of as much as 12° without loss of magnification strength. Other variations include those which are rubberised to make them waterproof and resistant to rough treatment, those that zoom through a range of magnification strengths and the compact "roof-prism" types.

Right: *A summer sunset near Pafuri in the Park's far north.*

It is man's instinct to hunt. Whether he kills to satisfy his belly, his greed or his vanity, he hunts. Be he a sinewy little Bushman, armed with bow and arrow running his prey to ground across a dry, scrubby desert, or a poacher, merciless with his cruel wire snare or automatic weapon of war, or an aristocrat who, in the good name of sport, flushes grouse from the heath to gun them on the wing, he hunts.

Today there is a new breed of hunter, equally skilled with eye and trigger who boasts of his "shooting" not in talk of guns, powder, tracking and kills, but in a language of shutter speeds, f-stops, film ratings and filter factors. He too displays his precious hunting gear with pride and some prejudice – everything from humble Instamatics and clockwork Super-8s to modern auto-focus and multi-lensed electronic masterpieces of video and microchip with pure gold price tags.

These are the photographers.

Today most of the thousands of visitors to the Kruger National Park are equipped with cameras. With varying degrees of enthusiasm they click away, blindly confident that their faithful old Kodak or shiny new jewel of optical perfection will keep to a minimum the ratio of that great bogey of photography – "the ones that didn't come out".

All too often that ratio is much too high because they are intimidated by the so-called mysteries of photography, they bash away with the hit-and-miss attitude of a defeatist, despite the near perfect conditions of weather, light, and subject matter usually available in the Park.

The formula for better wildlife photography is simple: patience, pre-planning and a good basic understanding of the ways of the wild – in other words, self discipline, keen observation and well-rehearsed camera know-how.

Not only will picture taking then become more fun, but you will develop a much keener awareness of the Park's life and environment. You will learn to look not only at the more obvious birds or animals, but also at the other things around them: the varied textures of leaves; the flowers, often tiny and easily missed; the delicate grasses and their seeds; or the bark of trees. Nature offers an endless selection of patterns, rhythms, planes and colours for those who care to see – like the incredible designs of the leadwood bark, the fan of fine leaves of the sickle bush, or the almost computer-perfect diagrams in a spider's web.

After patience and perseverance, what a wildlife photographer needs most is a sense of positive thinking. I am sufficiently old-fashioned to believe that if you want a picture badly enough, and you really try for it, you are very likely to be rewarded. Lady Luck has much to do with it, of course, but you can give her all kinds of help towards ensuring better pictures.

Hints for better wildlife photography

Understand how your camera works. I realise that many camera boffins will heave a sigh of indignation at my daring to suggest that they do not know how to use their gear to the fullest. To them I apologise, and ask that they give a thought to the many eager "amateurs" whose failures and frustrations stem from ignorance of how to make the fullest use of the often excellent equipment they own. Remember the legions of good pictures that have been lost through lack of attention to that most basic of photographic operations: loading a film. Bad loading usually means torn perforations, film not winding on – and no pictures. Consider too the person with auto-focusing, auto-exposure camera kit worth thousands who, at the critical moment, finds it useless because the battery is flat.

Then there is the camera fiddler who knows all the "tricks" his equipment is capable of in theory, but is ill-rehearsed in how to exploit them. This is the one who keeps his wedding group stewing in the sunshine, or his prettied child staring woodenly at him while he fiddles with film speed, shutter and aperture settings, filters, flash, tripod and the other gadgetry – and then later wonders why his pictures are lifeless and boring.

In the Kruger National Park, where the four-footed and feathered residents are probably easier to photograph than wildlife anywhere in the world, being completely familiar with your equipment will take you most of the way towards an album of fine, alert pictures. The fumbler will be left with a collection of views showing empty bush, blurs or departing backsides.

Obviously, the essence of good photography is also to understand the limitations of one's equipment and I hasten to confirm that you do not have to be an expert armed with exotic paraphernalia to collect good pictures. People with simpler equipment can get excellent results, only their range of picture opportunities will be narrower.

It is vitally important to be familiar with the perspective limits as "seen" by a camera lens. Let us take for example the standard, non-interchangeable, short focus lens fitted to simple cameras – the kind carried by most visitors to the Park.

Such a camera is designed for good pictures of subjects not more than about 5 m distant, like the wife and kids posing for a family group. Too often in the Park, the photographer using the same camera does not realise that the magnificent sable antelope bull standing perfectly posed 50 m away, sharply visible to the naked eye (which pinpoints his view in three-dimension), will fill only a very small part of the scene encompassed by his camera lens. He clicks excitedly, and is terribly disappointed later when he collects his 35 mm colour slide or postcard print to find that the sable is an insignificant speck lost in a wide landscape.

The answer is to determine by experience just how far away an animal of a

Left: Elephants are marvellously photogenic and generally easy to photograph. However, they should be treated with the utmost respect and it is unwise to get too close – especially to breeding herds with calves. This "wide-angle" photograph of an advancing "convoy" was taken under the guidance of a ranger.

given size should be to produce an acceptable image and then, if necessary, try to complete the picture by composing it in the viewfinder – taking advantage of a framework of trees or storm clouds, or the simple starkness of a grass plain to create surprisingly pleasing results.

The technique also applies in reverse. There is the camera buff with such powerful telephoto equipment that he can zoom in to the fly on an elephant's eye, which may make a fine picture, but over-use of the long lens will leave him few photographs showing the animals in their natural surroundings. Movie and video makers are often guilty of this, by constantly shooting with their lenses zoomed to the longest focal length. The finished results show an array of close-ups that could have been taken at a zoo. Therefore, for greater variety, zoom back for the wider view as well as forwards for the close-up.

In a nutshell, the photographer who is usually pleased with his picture results has mastered the functions and understands the limitations of his camera, however simple or sophisticated it might be, to a point where he shoots automatically. His whole concentration is geared to seeing and creating the photograph and not to the mechanics of his equipment. He then becomes an "artist", with the camera simply the tool of his creativity.

Make your pictures "come alive". A picture of an impala is a "so what" picture if it is just standing there staring towards the camera or into the middle distance. Individuality begins by seeking to photograph an animal doing something, even if it is merely feeding, otherwise it tends to look like a stuffed museum specimen stuck out in the veld. The trick is to put patience before camera and to study animal behaviour so that you can anticipate the actions of different species.

Observation teaches us such animal idiosyncracies as the insatiable curiosity of the buffalo; he may well run away, but he will usually turn back again because he just cannot resist another look at you. Kudu, waterbuck and nyala will usually offer only one chance – their curiosity will hold for a moment and then they move off without so much as a backward glance. If a resting lion yawns once he (or she) will nearly always do it again – so be patient. An antelope or bovine chewing the cud appears hyp-

Tree squirrels, common throughout the Park, are fun to photograph and are quite tame in the camps – Letaba and Shingwedzi in particular.

Ignoring cars and pedestrians, this glossy starling dutifully ferried food to its nest in a hole in a tree in the car park opposite the bank at Skukuza.

notised as he does so; he swallows and regurgitates dreamily, over and over again, seemingly oblivious of a working camera. Birds preening and impala grooming offer easy close-up views, as do waterbuck when lying down to escape a chilling August wind. Note the shake of their heads, the flick of their ears or the grimace of a mouth as they rid themselves of worrying flies. Within the zebra herds, stallions may suddenly fly into a frenzy of rearing, slashing, nipping and kicking – a challenge of dominance that is usually in fun, but sometimes in deadly earnest.

There is that moment just as the fish eagle spreads its broad wings for take-off – easy to anticipate because this bird almost always defecates just before launching itself.

There is the mother baboon using fingertips to groom her diminutive, big-eyed baby; the brilliant kingfisher cocking an eye at the water the moment before it dives from its perch; the weaver clinging upside down to the bottom of its dangling grass nest as it flutters and chatters to attract a mate; the impala rams facing each other with lowered heads just before they clash horns; the elephants sucking up dust to blow over their bodies, or stretching up with their trunks to reach the juiciest seed pods;

the extraordinary lips and tongue of a giraffe folding around a branch bristling with thorns to strip it clean of foliage; the oxpecker almost vanishing into the buffalo's ear.

During the afternoon you may notice at the roadside one or more hyenas that appear reluctant to move off as you approach. Stay with them, or make a point of returning to the spot towards dusk. They will probably have cubs hidden close by (often in culverts under the road) and the little fellows are certain to show themselves just before sundown.

Be prepared to spend the better part of a day at a site overlooking a waterhole because almost all animals and most birds drink daily. In thick bush the giraffe is tricky to photograph because the trees, dappled shadow and excellent camouflage diffuse its shape. But at the waterhole, where it must spread its forelegs and drop that long neck to drink, it is an easy and excellent target. Giraffes lift their heads often while drinking to look about them and the moment to shoot is when the lips leave the water, with sunbright droplets flying from the snout.

All the photographer must do is be sure of his exposure and focus, pick his target like a hunter of old, and wait for just the right moment, the right action – a swing of the head to shake off a pesky

oxpecker bird, a scratching of the neck against a tree trunk, or any one of the multitude of other nuances of animal life. The scope is vast and, in its own way, can provide as much excitement as a lion's kill or a battle for superiority between two elephant bulls.

The Park is constantly alive, always in motion, and the better pictures are there for the patient and the observant.

A wildlife photographer should always be alert and prepared for the unexpected.

When a cloud moves across the sun, or you drive through an area of predominant shade, or simply the day begins to draw to an end, be aware of those changes and try to keep the camera settings adjusted accordingly (not a problem, of course, if your camera is blessed with an automatic mode). You can then have your camera up and shooting with only the focusing to concentrate on. Even if your exposure is not absolutely perfect, you will have got the picture in the bag. If your subject then stays put, you can continue shooting having made finer exposure adjustments.

When approaching a scene close to the roadside which is likely to be disturbed by your arrival, such as the shy cheetah feeding on a kill, you should prepare your camera in advance; stop the car some distance away, pre-focus on a tree or log that is about the same distance from you as the cheetah is from the point where you are likely to stop, and then approach slowly, smoothly and quietly with the camera poised at the ready.

A common fault is trying to do too much. It is tempting to keep driving to see as much as possible; some visitors feel cheated if they have not covered at least 100 km in one day. When you come across a herd of buffalo (which can number many hundreds) or of elephant (be specially cautious if there are youngsters with them), then stay with them: pull up at the roadside and be prepared to watch them for as long as they are within camera range, which could be an hour or two. That way you will see many things that people who are constantly on the move will miss, and you will be offered a wealth of picture opportunities.

Take full advantage of camps, picnic sites and lookout points. I can understand the frustration of photographers who are keen on close-up or macro work but are confined to a car. They should remember that rest camps, trail camps, picnic sites and lookout points are also part of the Park and give access to a rich variety of little creatures and plants which can be photographed easily, especially just inside the perimeter fences of the camps. Exploring these areas can be very rewarding.

Birds are particularly easy to photograph, because they are prolific and comparatively tame. At the Timbavati picnic site I was able to follow the whole slow, ponderous process of a hornbill luring a female into a hole in a hollow tree trunk by first wooing her on a branch, then attracting her to the hole by dropping inside delectable items of food. Had I been able to stay long enough, it might have been possible to photograph him walling her in with mud to lay her eggs and eventually raise their brood while he fed her.

One of my favourite pictures, of a glossy starling taking food to its young, was taken at the nest right opposite the door of the bank at Skukuza.

All the camps play host to flocks of impertinent glossy starlings, hornbills and redbilled wood hoopoes which hunt for scraps around campfire sites, and also to brilliantly hued agama lizards. Shy nocturnal animals like the furry civet can be captured on film along camp fences after dark and at Balule one is a regular night visitor. If you hear the crash of refuse bins being overturned in the dead of night at Satara or Shingwedzi camps, it is almost certainly a honey badger raiding for scraps of tourists' leftovers – a dogged animal easily followed and photographed but out of whose direct path it is wisest to stay. A metre or two outside the fence in front of the main building at Satara is a mud wallow beloved by passing warthogs who often disport themselves there in the heat of the day.

Camps and picnic sites are also a cornucopia for photographers of flowers, especially in springtime after the early rains. Nwanedzi picnic site and Satara's new camping area are special favourites. But the ideal for this purpose is to take one of the Park's trails – walks through the bush with a ranger from a small base camp during which the camera can be aimed at anything from dung beetles to elephants.

A courting hornbill offers a tasty tit-bit to woo a reluctant female into a nest at the Timbavati picnic site near Satara.

Why is the colour disappointing? It is not always, of course, especially in the spring and summer when the veld comes alive (if the rains have been normal) with flowers and greenery beneath cobalt blue skies. But the most popular months for visiting the Park are the cooler ones from May to September, the dry time when animals rely heavily on waterholes. Much of the bush is then drab except for the green along rivers, the sky is made grey by the smoke of distant bushfires and the vegetation beside most roads is brown from the dust raised by cars. An exception at this time is the north of the Park where the colours of the mopane that dominates the countryside are splendid rusty reds and gold and the magnificent crimson blossoms of the flame creepers are in full bloom.

It is often possible for a photographer to add colour by carefully positioning a car, for example, so that a picture can be taken through a selected framework of heavily out of focus red blossoms or green leaves.

Exciting effects can be had by deliberately adding colour to an otherwise dull sky or foreground, by using one (or more) of the huge range of graduating filters which are so much in vogue today. They can be specially useful to give a more even overall exposure by toning down a comparatively over-bright sky –

especially in overcast conditions – and at the same time adding colour. However, these filters should be used sparingly and always with definite creative forethought.

The veld undergoes its own colour transformation from dawn to dusk. At sunrise it is full of rich reds and golds and then, too, the animals are often at their most photogenic with their coats glistening from the morning dew, and the birds are like pompoms with their feathers puffed out against the early chill.

Sunset can be just as rewarding, especially at the edges of dams and large waterholes, with herons, hamerkops and storks feeding in the shallows, making clean silhouettes against the reflection of sun on water as it sets.

Shooting in rain or mist can give beautifully delicate colour hues and although a faster film might have to be used, the heavier grain often enhances the mood. If desired, the blue cast of a cloudy day can be "warmed" by using a suitable filter.

Who drives the car? This is a matter of personal preference. The advantage of having someone else drive in the Park is that it leaves you free to concentrate entirely on the readiness of your equipment for fast use and on watching the bush all around instead of mainly the road. However, I personally prefer to drive myself as I can stop and position the car exactly where I want it, change direction quickly and accurately when searching for an angle and also, frankly, because I prefer to trust my own judgement, should it become necessary to make a hasty but dignified retreat.

The ideal is to have a companion who, just as the car stops, can stretch across to turn off the ignition (to kill vibration) and pull the handbrake, leaving you free to have your camera up and shooting without delay – and then, when necessary, pass you the additional combinations of cameras and lenses at the moments required.

On the subject of driving, I have found it very worthwhile to pause along the road and talk to other photographers. Being fellows of the same breed, they will happily exchange information about what they have seen and what is available to be photographed elsewhere. Park rangers are also very helpful and ready to direct you to scenes of interest. If you see one on patrol, flag him down and ask.

Protecting equipment. Photography can be a very expensive business, even after the initial outlay on cameras and lenses. There are many ways to save on costs such as taking advantage of the film, processing and printing bargains which are often available, by loading your own cassette from bulk stock (if the amount of film you use warrants it) and, above all, shooting selectively so that each frame counts, instead of using the machine-gun approach of shooting off a whole roll of film on one dozing rhino.

The biggest expense can be the maintenance and repair of equipment. New camera values have more than doubled in four years, the price of spare parts has soared equally, and today so much modern technology is being incorporated that repairmen have to be better qualified, camera testing equipment is more sophisticated and repair costs have gone up accordingly.

Taking good care of your equipment demands some special measures and attention in the bush because of the harsh environment and climate. The factors that take most toll are dust, heat, moisture and vibration.

Dust is the real camera killer – and the most difficult to control because no matter how careful you are, it always seeps in somehow, especially during the dry winter months which happen to be the best for wildlife photography.

Cameras with fixed lenses have the advantage of usually having fitted cases which can be kept closed most of the time affording at least some protection against dust. The more advanced camera with interchangeable lenses becomes a problem because of its bulk and the keener or more proficient photographer, in any case, does not keep his camera in the case because of the inconvenience of constantly removing it to reload film.

When using several cameras with different lenses, I find the answer is simply to wrap each combination in a clean square of old white linen or towelling – old so that it no longer gives off fluff (which can be *as* harmful as dust) and white simply to reflect the heat.

Lenses need special care because, for obvious reasons, dust will inhibit the sharp transmission of light if allowed to

Times of drought, although often ugly, still offer sensitive picture-story opportunities – such as these hunger scenes of plant (above) and animal – in this case a bushbuck searching for food at the roadside near Lower Sabie (opposite page).

collect on the glass (but remember that a few specks of dust have very little ill effect).

A more sinister problem lies in the manner in which dust is cleaned off a lens. A lens should never be just wiped clean, even with a chamois leather, because inevitably extremely fine particles of grit will be dragged across the lens surface and, especially if any pressure is applied, they will damage the "bloom" of the glass with very fine scratches hardly visible to the naked eye. Repeated damage like this will eventually make the lens lose its optical contrast and produce "soft" images, which then means sending it back to the factory for expensive re-blooming and polishing.

The best way to remove dust is to blow it off – but *not* with your own breath. Spittle spray can also be harmful. Rather use one of those inexpensive, small rubber bulb-brush combinations or, even better, a mini-canister of compressed air designed for cleaning precision instruments, which is available from most photographic dealers. However, special care should be observed not to aim the nozzle directly at the camera's reflex mirror or shutter curtain because the surprisingly powerful air jet may cause damage to these delicate parts. More stubborn dust particles can be

flicked off with a fine-haired brush and if the dirt persists, such as a fingerprint on a lens, then very gently wipe it away with a lens cleaning cloth or a well-worn clean cotton handkerchief which can be very slightly dampened with pure water if necessary.

Remember that wiping and brushing can generate static electricity which simply attracts dust right back, especially in winter, so keep them to an absolute minimum.

The best barrier against lens dust is to keep an ultraviolet filter permanently fitted to each lens. If that becomes damaged it can be replaced at a fraction of the cost of the lens it protects.

Heat is another insidious enemy of photographers: it attacks both cameras and film. Most cameras are finished in heat-absorbing black and, if left to lie in the hot sun, both mechanism and lenses may be damaged (especially the modern electronic kind). Excessive heat may also

reticulate the emulsion of a film. If equipment must be locked in a car, make sure the car is parked where it will remain in shade. Too often we forget that the shade moves as the sun "moves". Remember that the temperature in a car standing in the sun rises within minutes to double the temperature outside.

Moisture can cause a camera to malfunction and again the electronic kinds are most vulnerable. Guard against humidity, morning mist, condensation in damp weather and especially perspiration from hands and face in hot weather. Photographers going on trail in the Park should take particular care in the rainy summer months when the weather can change suddenly. The best is to wrap equipment in a small towel and then enclose the bundle in a plastic bag should there be a shower. Be wary of using plastic bags on their own because during temperature changes they may cause water to condense

A "wider" view to a picture can be most satisfying.

within. Also avoid trying to protect a camera by wearing it inside your shirt against your perspiring bare skin.

Vibration. As with any precision instruments, vibration can cause damage in cameras – like settings going awry, wear and tear on mountings, scratches, and screws coming loose. This is especially so when travelling on rough gravel roads. Try to ensure that camera bodies and lenses do not rub together, damaging enamel, glass coatings and the equipment's value. If the various pieces are not contained in separate protective compartments within a camera bag, then keep each major item in its own drawstring pouch of soft cloth or chamois leather. Do not travel with the camera bag on the floor of the car, where vibration is worst. If there is no room on a seat then bed it on blankets or pillows on the floor.

158

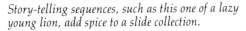
Story-telling sequences, such as this one of a lazy young lion, add spice to a slide collection.

A few other hints that might help towards greater picture fulfilment in the Park – and also prevent a tear or two of frustration – are:

- Whether you are a serious wildlife photographer or just an enthusiast who collects good pictures to show proudly to family, or friends at the office, you can enjoy peace of mind by having your camera checked by a reputable photo-repairman before departing for the Park. The cost is very reasonable.

- Modern electronic cameras are almost completely reliant on their own electric power – one or two tiny coin-like silver-oxide or mercury batteries. Although some cameras have a mechanical back-up system, should there be a power failure most of these cameras become virtually useless. The only answer is to *always* have at least one set of spare batteries for each camera with you. Before inserting the new batteries, it is a good idea to polish them and the terminals inside the camera with a clean, dry handkerchief.

- If you are fortunate enough to have an extended stay in the Park or have been using your camera extensively elsewhere before your visit, it is advisable to spend a little time check-

ing the workings of each item of your equipment at least once a week. Check the exposure meter for accuracy by comparing readings with those of at least two other cameras (or hand meters). Also check that the lens stop-down mechanism is working smoothly (if your camera is of the reflex type) and also visually check the operation of the shutter by holding the camera against the light and firing it at a variety of speeds while looking through the open back.

This lesson I learned the hard way. A hair, caught just in front of the shutter mechanism, remained undetected in one of my cameras and the result was a sharply defined hair imprint across every frame of seven rolls of film. It was only by chance that I checked the camera and removed it, otherwise many more films would have been ruined.

- A tripod is ideal when using long lenses, but is almost impossible to use in a car. However, one can buy or even make a special clamp which screws into the camera's base plate and then slots over the edge of a half-raised window. This works well with lenses of focal length up to 200 mm, but with bigger lenses swinging the camera to change direction becomes awkward.

- With longer lenses, I would suggest using an ordinary bean bag or small

sandbag draped over the edge of the car window as a dead rest to keep the lens steady. The bag can easily slide back and forth along the length of the glass, or be quickly transferred from one window to another.

- To protect the lens barrel in case it slips off the bag and crashes onto the glass edge in a moment of carelessness, wrap it in a padded sleeve. This can either be a homemade item of towelling and elastic sewn to fit snugly around the barrel or, for shorter lenses, one of those elasticised wrist sweat bands, sold at most sports shops.

- To keep camera shake to a minimum, always use the highest possible shutter speed within the limitations of the light available in relation to your film speed – especially when using telephoto lenses.

- Many fine wildlife photographs are spoilt by foreground or background clutter. This phenomenon, the result of inexperience, usually remains undetected while the picture is actually being taken because one simply doesn't see it as visually offensive at the time. The reason is that we are blessed with two eyes that see three-dimensionally and which are "trained" to spotlight the subject we are looking at and isolate it from the surrounds. The answer to this problem is to get into the habit of seeing as the camera

The photogenic zebra
Zebra herds within camera range usually offer excellent chances for action pictures. Amusing antics such as rolling in the dust, mock combat and a range of amazing facial expressions are common and will reward the photographer with patience and quick reflexes.

The doughty buffalo is inquisitive. When a vehicle approaches it will run off but usually turn back for another look – a perfect moment for the photographer.

does – by studying one's subject through just one eye for a moment. With the third dimension gone, it will be surprising just how those "defects" behind and in front of the subject immediately become obvious. One then simply moves into a position that is more pleasing.

- With regard to film, always keep it as cool as possible before and after use. In fact, unused film is best kept in a fridge, but then remember to take it out at least an hour before opening the container, because of possible condensation. In addition to your usual favourite stock, carry a few rolls of fast film just in case of bad weather or sudden action occurring early or late in the day when the light is low.

For those who are macro-photography enthusiasts, I would also offer a few suggestions that may make life a little easier.

With extreme close-up work, there are always problems with the lack of depth of field. I always carry two of the cheapest available flash units (one for back lighting) which, although not very powerful for general use, will deliver just the right amount of light at a distance of about 20 cm at an aperture of f22 or f32 with 100 ASA film.

Another valuable light source for close-up work is an ordinary mirror. One is often confronted with the problem of photographing a flower, spider's nest or the like situated in deep shade or with the sun at the wrong angle, or the subject may be inaccessible, so that a telephoto lens is required. A flashgun is not always the answer because there may be problems with the power-to-distance ratio or the matching of exposure between flash and natural light, especially at the usual synchronising shutter speed of one-sixtieth of a second (unless you are lucky enough to own a camera with higher synchronising speeds).

The answer is a mirror. Any kind will do, even a silver tea tray or the silver bottom of a biscuit tin, but the size should not be less than about 20 cm². (I

use a wing mirror unscrewed from my Volkswagen minibus.)

Simply have a helper reflect the sun's rays with the mirror onto the subject – but take care to position the mirror fairly close to the camera and on the opposite side of it to that of the sun. Excellent results can be achieved by using the mirror to provide frontal light and the sun itself as a backlight, but obviously avoid having the sun shine directly onto the camera lens because of "flare".

Because of the disturbance caused by the intense, flashing reflected light, this system should be avoided when photographing animals or birds.

While on the subject of reflectors, another valuable tool is a cake board – the flat base found under most bought cakes with hard icing such as a wedding or Christmas cake. These boards have a silver-grey matt finish which gives off excellent fill-in reflected light, softening harsh sunlight when doing macro work. They also double as useful wind shields when photographing flowers waving about in the breeze.

A final suggestion is to invest in a pair of elbow pads of the type used by rifle shottists. Macro photography often means lying flat on your belly, which can result in very tattered elbows if they are unprotected.

Favourite picture places

In many ways taking photographs – getting good pictures, that is – demands all the nuances of skill and the finesse of trout angling. You should be as fully familiar with your camera as with rod and tackle to be able to use it fluently by instinct, without having to think. You should know exactly which film and lenses to use, as you would know lines and flies. You have to use great patience, stalk with care and strike at the right instant. And precisely as the accomplished angler seeks out the pools where the trout lurk instead of haphazardly flaying a whole stream, so the photographer must know where to look, and what for.

To help the photographer who is a newcomer to the Park, I suggest the following places where patience may be well rewarded, places which I have found to be the best for variety and accessibility of wildlife to the camera lens.

The might of the moorhen

Two moorhens jealously guard their chick against intruders. When a flight of much larger whitefaced whistling ducks landed close by, the adult moorhens immediately skidded across, squawking and stabbing at the intruders until they moved off. This sequence was taken at the Sweni dam, just south of Nwanedzi picnic site on the eastern boundary of the Park.

163

Punda Maria. The S64 and S63 roads at Pafuri that run east and west along the south bank of the Luvuvhu river from where the main north road reaches it. The variety of subjects – from scenic views to birds, from hippo to crocodiles – is very wide. The Maritubi dam on the Mahonie loop which circles west of the camp, mainly for bird life. The S58 loop road where it runs immediately below the Dzundwini hill, between the H13-1 and the H1-7 main roads where a water seep attracts buffalo, lion, leopard, kudu, roan antelope and especially nyala.

Shingwedzi. All of the S56 Mphongolo river road which runs to the west of the main road north of the camp, from its southern junction with the H1-7 to where the Mphongolo river meets the Phugwane – mainly for elephant, also antelope, bush squirrels and birds. The Kanniedood dam road from the river crossing just east of the camp all the way past the dam wall to the 23 km road sign where the road swings southwards away from the river – for waterbirds, especially rare kinds; also crocodile, hippo, leopard, bushbuck, nyala and elephant. The concrete causeway that crosses the Shingwedzi river at the extreme west end of the S52 loop road which leads to the Tshange lookout point – for birds and buffalo.

Letaba. The whole of the S47 road which leads from the high-level bridge just north of the camp to the Mingerhout dam – for big herds of buffalo, waterbuck, giraffe and especially leopard. Take the S46 east of Letaba: about 2 km from where it turns off from the S94, there is a short side road of about 400 m which runs along the south bank of the Engelhard dam. Here you are very likely to see big herds of impala, a variety of waterbirds and elephant.

Olifants. Take the S92 south shortly after leaving camp. That part of the road which closely follows the Olifants river, especially at the low-level bridge leading across to Balule camp, is an excellent area for lion kills; also bateleur eagles, giraffe and waterbirds.

Satara. Leeubron windmill on the S39 Timbavati river road to the west of the camp – for very big zebra, impala and wildebeest herds and good pool reflect-

Animal behaviour such as grooming adds interest and charm to photographs and usually the subjects are so absorbed that they will allow a vehicle to approach to within a few metres.

ions of drinking animals, bird life and lions. The Ngirivane windmill on the S12 loop off the S40 west of the camp is probably the best waterhole in the Park for a variety of game including large herds of giraffe and waterbuck – and especially bathing elephants. The whole of the Nwanedzi river road running east from the camp offers riverine forest on one side, open veld on the other, with sightings of big impala herds, waterbuck, giraffe, warthog, leopard, cheetah and lion. The S41 which runs north-south east of the camp, near the border, where it crosses the Nwanedzi river – for water birds, crocodiles and leguaans. Just south of Nwanedzi camp where the Sweni river crosses the S37, the road forms a dam about 100 m long. Crocodiles bask on the banks and bird life includes whitefaced whistling ducks and breeding moorhens.

Skukuza. The Manzimhlope waterhole and trough on the S36 road which runs north from the main road between Tshokwane and Skukuza can be spectacular in spring. Herds of sable antelope, waterbuck, many giraffe and big buffalo herds can be seen, and very often lions resting in the shade of the concrete water tank. The best known viewing road in the Park for all kinds of animals that frequent riverine forests is the H4-1, which runs from the old low-level bridge just north of the camp eastwards to where it joins the H12 at the high-level bridge. Cross over the high-level bridge going north on the H12; turn east to follow the north bank of the Sabie river where you are likely to see elephant, hyena, kudu, giraffe; travel to just beyond the Nwatindlopfu causeway where a loop to the right takes you down to a river pool, a likely place to see the drama of a crocodile kill. Travel in the area around the Nkhuhlu picnic site on the H4-1 road to Lower Sabie for spectacular riverine forest, lion, leopard, elephant, bushbuck, duiker and nyala. The Nwaswitshaka waterhole south-east of Skukuza on the S65 road linking the S1 Doispan and H1-1 Nahpe roads is well known for both black and white rhino and lion.

Lower Sabie. All of the H4-1 Lower Sabie road north-west of the camp to the Nkhuhlu picnic site, for elephant, bushbuck, raptors (including a pair of martial

eagles), whitefronted bee-eaters and warthogs. The emergency water supply dam just west of the camp next to the H4-1, for waterbirds and superb reflected sunsets. The entire Gomondwane road from the camp to Crocodile Bridge, particularly near the Gomondwane windmills – where big herds of zebra and wildebeest gather. This road can also be very good for warthog, white rhino, cheetah, wild dog and lion.

Pretoriuskop. The entire network of roads immediately to the east of the camp

Although high-speed colour films have coarse grain and are more expensive, they "save the day" in low light situations such as this.

which includes the Pretoriuskop and Manungu hills – for sable antelope, kudu and the very pretty scenery.

Berg-en-dal. The dam immediately in front of the main camp buildings – for bird life and animals drinking. Also, the loop road swinging north from the camp past the Matjulu lookout point – for elephant and kudu, and magnificent scenery, especially in stormy weather.

165

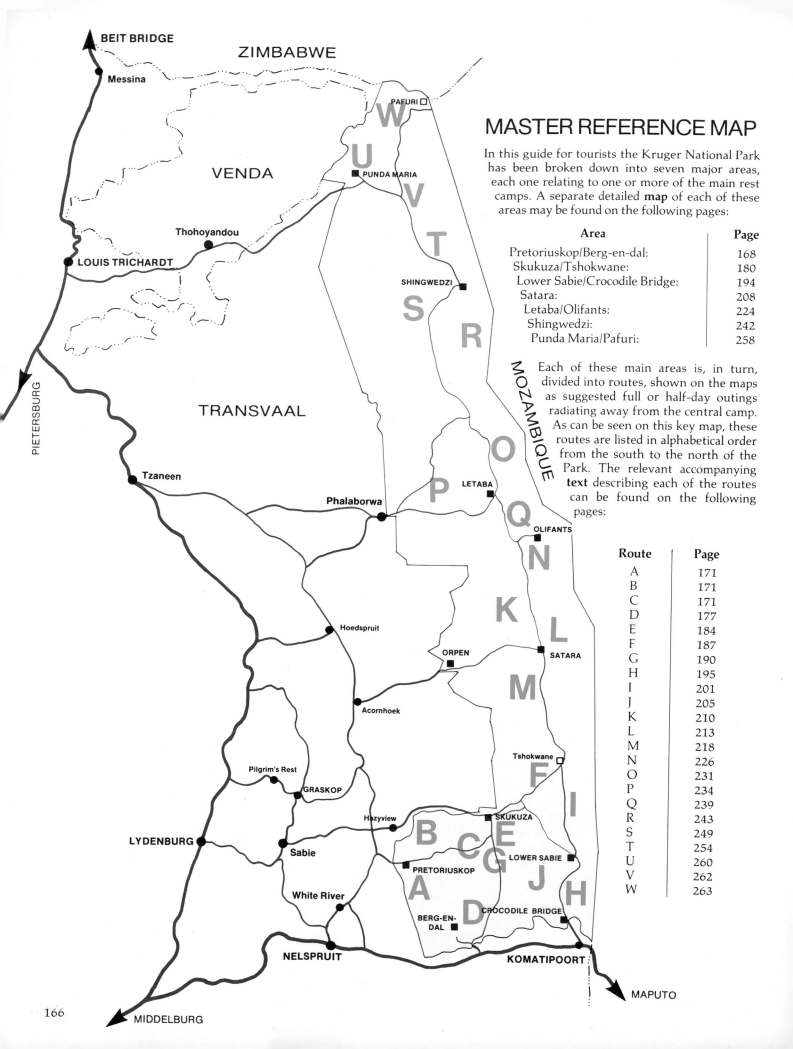

MASTER REFERENCE MAP

In this guide for tourists the Kruger National Park has been broken down into seven major areas, each one relating to one or more of the main rest camps. A separate detailed **map** of each of these areas may be found on the following pages:

Area	Page
Pretoriuskop/Berg-en-dal:	168
Skukuza/Tshokwane:	180
Lower Sabie/Crocodile Bridge:	194
Satara:	208
Letaba/Olifants:	224
Shingwedzi:	242
Punda Maria/Pafuri:	258

Each of these main areas is, in turn, divided into routes, shown on the maps as suggested full or half-day outings radiating away from the central camp. As can be seen on this key map, these routes are listed in alphabetical order from the south to the north of the Park. The relevant accompanying **text** describing each of the routes can be found on the following pages:

Route	Page
A	171
B	171
C	171
D	177
E	184
F	187
G	190
H	195
I	201
J	205
K	210
L	213
M	218
N	226
O	231
P	234
Q	239
R	243
S	249
T	254
U	260
V	262
W	263

This section of the book is a comprehensive guide for tourists. It will help them whether they are at home planning a lengthy trip to the Park, or deciding on a single day's outing from a particular camp, or simply as an armchair reminder of happy memories afterwards.

It describes each main camp in the Park separately and in detail – its character, the amenities on offer and then, where to go and what to see when out exploring the network of roads around it.

It is not intended to be a rigid guide. Its purpose is to assist you to plan ahead by studying the text, in conjunction with the accompanying maps, to give you an idea of what to expect from different areas of the Park – with each camp as a centre point. This should be especially useful to the uninitiated. It is important that first-timers to the Park should avoid wasting time and fuel by aimlessly rushing about from dawn to dusk desperately trying to cover as much ground as possible. Result: they see little, unknowingly pass by much and end the day hot, tired and frustrated. With a little planning that can be avoided.

In a reserve of such size and wildlife variety, visitors should bear in mind that it is impossible in a single trip, of even as much as two weeks, to see and experience everything. Therefore consider your time and your preferences carefully. Do you prefer to see big game such as lion, cheetah, rhino or hippo? Are birds your priority? Do you prefer spending a full day at a single waterhole, or viewpoint, or hide, or in the shade beside a river? Are you fascinated by little things, remembering that insects are wildlife too? Or do you simply want to stay in camp and relax as one can do only in this kind of wilderness, sleep through the heat of the day in air conditioned comfort, enjoy a good book, or watch the bird life and spectacular sunsets from your own veranda?

Whatever you choose, the following notes will help you to make the most out of whichever camps you have booked into and, when you venture forth, to keep yourself and those in the car with you – particularly if they are children, who get bored quickly – interested and entertained.

Having made your choice, exploit the detailed route notes and map concerning your camp and work out a basic schedule for your day, leaving yourself time for some alternatives. If you intend to be out on the road when the sun is high and hot, for instance, when the animals tend to seek cover, pinpoint from the notes a cool and shady picnic site or viewpoint where you can rest a while. It is surprising how much one can see by staying in one spot for an hour or two.

Do not ignore the small things and travel slowly, for speed blurs and opportunities are lost. There can be as much fun in watching a lilac-breasted roller trying to grab a scorpion as in watching lions feeding; there is no less pleasure in seeing a couple of hyena cubs rocking and rolling through the grass tufts behind their mother than there is in seeing a pair of young elephant bulls wrestling in mock combat; a martial eagle cooling its behind in the river shallows is quite as amusing as ostriches doing their mating dance.

Because it is not possible to list in this book the full details of the variety of wildlife in the Park, people who need more information about a particular mammal, bird, reptile, tree or other aspect should study the list of books in the bibliography, many of which are available at camp shops.

And please, always remember that it is illegal to feed animals. It can be extremely dangerous. The authors believe, however, that there should be one specific exception to this – the flocks of birds of all sorts which scamper up to be fed in camps and when visitors pause at picnic sites.

For children in particular this close proximity to wild birds is an exciting and elevating experience and there is no real harm in feeding them. With the help of parents and the use of a bird reference book, this is an ideal way to encourage children to learn about birds, their habits and characteristics, and their names. But again a word of warning: *do not feed birds if there are monkeys or baboons in the vicinity.* In some areas they have become aggressively "tame" and will not hesitate to snatch food or climb into a car to seek it and if resisted, they will bite.

Some final points about the notes which follow:

- Please bear in mind that the description of each game-viewing road, waterhole and viewpoint is based on personal experience over a number of years and is described at its best. Within broad parameters, the Park's population is constantly shifting and altering and an area that might be rich in animal life today could be empty tomorrow. Half the fun is the element of luck and expectation.

- The time of year is also very significant and affects game viewing. During wet weather waterholes are often abandoned by game for the sweeter rainwater pools scattered throughout the veld. In the spring migratory birds start arriving, in the summer the Park is usually at its scenic best and the autumn heralds the mating season for many species of antelope.

- Roads described here may be closed from time to time, for maintenance or other Park operations, and new roads may be opened.

- Crocodile Bridge camp has not been described because it is due to be completely rebuilt soon.

- Balule camp has been mentioned but it is likely to be closed soon. The Ngotsamond caravan park nearby will be opened when archaeologists have finished their work there.

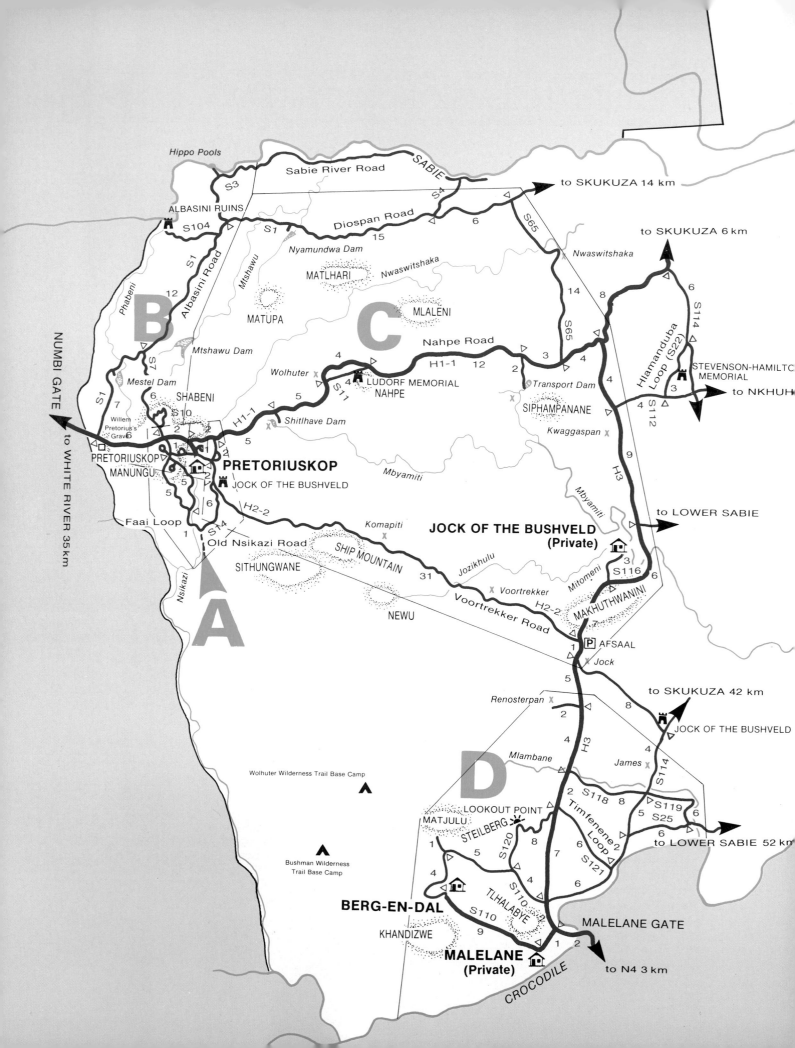

PRETORIUSKOP/BERG-EN-DAL AREA

Pretoriuskop camp

By far the greatest volume of traffic entering the Kruger National Park does so through the Numbi Gate and for many weary travellers their first taste of camp life is nearby Pretoriuskop – known affectionately to many as the "friendly camp".

With its spick and span whitewashed and thatched chalets and pretty gardens it has an air of quiet calm and efficiency, borne out by the fact that since the inception of the coveted jade-ivory-and-silver Yvonne Knobel trophy awarded to the Camp of the Year it has been won four times out of seven by Pretoriuskop.

Its history stretches far back to before the Park was born. It was through here that the swashbuckling transport riders and their creaking freight wagons passed on their way between the Lydenburg goldfields and the coast.

The first tourist to enter the Park came through the Numbi Gate in 1927, just a year after its proclamation as a national park, and two years later Pretoriuskop camp had grown to eight tourist huts.

Today there are 145 huts formed roughly into three big circles – named Maroela, Moerdyk and Wolhuter. The latter comprises quaint little mushroom-shaped rondavels and is reserved for visiting groups of schoolchildren. The most luxurious huts line the open parkland leading to the "human waterhole" – the swimming pool built into a granite outcrop, donated to the Park by the Rembrandt Tobacco Corporation in 1954. These front huts are being renovated into six-bedded family units. Other huts are being given bathrooms or showers but the planners are determined to maintain the rustic atmosphere of the past. The delightful thatched restaurant and shop are also due to be remodelled and a new reception area will be built soon.

Pretoriuskop has a post office (the only non-thatched building in the camp) and public telephones.

A unique aspect of Pretoriuskop is that it is the only camp to break the Park's strict "indigenous-trees-only" rule. The whole camp was once part of the local ranger's garden in the days before any real rules applied (except those against poaching) so today the red flamboyants and maroon bougainvillaeas planted then still bloom proudly. The rangers themselves believe these old trees should be allowed to live out their lives for historic reasons. However the jacarandas are doomed since that much-loved but misunderstood tree was declared a pest plant in the Lowveld. It has "invaded" much of the area, causing untold damage to banks of rivers and streams.

The parkland that surrounds the swimming pool abounds with tall, wide-branched indigenous trees: marula, sycamore fig and Natal mahogany among the many. One particular fig tree trunk has totally enveloped the stem of a silver cluster leaf tree (*Terminalia sericea*), giving the impression that the branches of the latter are sprouting from the fig tree like a strange hybrid.

Pretoriuskop is reputed to be not the best area for game viewing because of the thick vegetation, tall grass and unpalatable sour veld. The most dominant tree is the silver cluster leaf which is generally inedible to wildlife. However, the greatest concentration of kudu is found here and if one is lucky, small herds of eland, sable and roan antelope may be spotted quite near the camp, as well as a few of the usual predators plus impala, reedbuck, giraffe, zebra and wildebeest. Five rare black rhino have settled in the Faai loop area, just south of the camp.

It is rhino, the white variety, that Pretoriuskop is most famous for. It is generally accepted that the square-lipped (white) rhinoceros became extinct in the area as far back as 1896 – thanks to ceaseless persecution by hunters. Sixty-five years later, on 13 October 1961, two bulls and two cows were captured in Natal and released into a specially built "boma" near the camp, so once again these giants walk the Lowveld plains.

Since then more than 340 white rhino have been brought from Natal and today the Park's population is firmly established, numbering over 900.

Other less dramatic but equally important animals that help to complete the ecological picture of the area have been "imported" and are thriving. They include oribi, red duiker (to be seen north of the camp towards Shabeni koppie) and grey rhebok (vaalribbok) in the wild terrain south of Stolznek dam well away from tourist routes.

Visitors are sometimes surprised to come across groups of labourers scything long grass from vleis and watercourses around Pretoriuskop. This is because most of the thatching used in the Park is collected from this area.

Within minutes of leaving the reception office at Numbi Gate the visitor will become aware that game viewing in this area may be difficult: the veld grass is tall and the bush thick, but browsers thrive in this part of the Park and the tourist is almost certain to get a "tally ho" from the white fluff under the tail of a fleeing kudu.

A late afternoon view of an impala herd grazing near Pretoriuskop camp on the S14 Faai loop.

Close to the roadside, 3 km from the entrance gate, is the grave of Voortrekker Willem Pretorius. He died in 1845 and was buried by the pioneer Portuguese trader João Albasini. Behind this grave Pretoriuskop hill can be seen.

A little further on is the Faai loop turnoff where a keen eye will notice a neat line of metal discs planted at the roadside on the right like mining claims. They are markers for a long-past grass-growing experiment.

A kilometre further on a main road branches away to the left to Skukuza and straight ahead, exactly 9 km from Numbi Gate, are the welcoming gates of Pretoriuskop camp.

ROUTE A

By far the most popular area for touring from Pretoriuskop is the network of roads that explore the rocky outcrops and koppies to the immediate west of the camp. Turning left off the main tarred road 1 km from the gates, one immediately discovers the first of the area's many small ring roads – to the right. Herds of impala frequent this little loop, especially after the first rains when the granite rock formation offers pockets of cool, clear water and consequently a backdrop of grey stone and blue sky for photographers as the animals climb up to drink.

Continuing westwards after the loop, a crossroad is reached with another ring road straight ahead. The right hand road leads up to the tarred Numbi Gate road while a sortie straight on will take you around Pretoriuskop hill with its fine vista over the surrounding veld. Kudu browse in the thick bush and it is one of only two places in the Park where I have spotted numerous Barberton daisies growing wild (the other was near Berg-en-dal).

Taking the left turn southwards, however, will bring you to a fork and both its roads skirt around Manungu hill before stretching away southwards for 5 km when they join again. By driving along the left one of these branches you can make a right turn to travel up and around Manungu hill to enjoy excellent views not only across the Lowveld but also down towards Pretoriuskop camp.

After circling this loop and resuming your drive south, you will notice a long line of sturdy game fencing on the left. This is the "boma" into which the rhino from Natal were released. Look carefully and you will almost certainly spot the great beasts quietly feeding.

Continuing southwards (you are on the S14 Faai loop), the road swings suddenly to the east at an intersection with a "no entry" firebreak. Old-timers to the Park will remember this as the old Nsikazi road that once took visitors further to the south from this point.

The S14 soon loops northward, dipping through a stream drift (note the natural waterhole upstream) before joining the main gravel H2-2 Voortrekker road leading back to camp.

At this intersection there is a monument to Jock of the Bushveld: a reminder that it was here that transport riders, including the famous dog's owner, Sir Percy Fitzpatrick, trekked to and from the Mozambique coast late in the last century.

ROUTE B

A pleasant and leisurely day's drive is to explore the area to the north of the H1-1 Numbi Gate road, either along the S1 past the Mestel dam or on the parallel S7, both of which join up to form the Hippo Pool road.

The S7 passes close to the dramatic Shabeni rock formation and you have the option of turning off on the S10 to circle this "mountain" (you may travel anti-clockwise only), said to be a favourite lair of leopards. Scan the undergrowth carefully for the rarely seen red duiker and oribi and keep an eye on the sky too, because a variety of raptors soar in the upcurrents from the rocky slopes.

Sixteen kilometres further north is the S104 turnoff to the historic Albasini ruins, the remains of a homestead and store built by the Portuguese pioneer trader, João Albasini on the site known as Magashulaskraal (see p. 42).

Tourists may leave their vehicles to view the site and a small thatched shelter exhibiting old photographs and a selection of ancient bottles, nails, chinaware, spent cartridges and a model of what the trading post looked like when it was first built in 1846. The site is well shaded by giant marula and jackal-berry trees and is excellent for birdwatching.

Another nearby spot where you can leave your vehicle is at the hippo pools, slightly further north on the Sabie river. Visitors may sit along the shady bank and watch the rotund river residents snorting and blowing loudly. Picnickers are warned to be wary of a troop of vervet monkeys that make a habit of boldly raiding hampers and even enter cars in search of food.

ROUTE C

Another excellent run of medium length (about 90 km – a full morning) is a round trip taking the H2-2 Voortrekker road south-east to where it meets the main H3 Skukuza-Malelane Gate road, then north to the H1-1 Skukuza-Pretoriuskop road junction where you swing westwards for the final leg back to camp.

Turn right immediately after leaving Pretoriuskop and then right again at the next intersection to travel southwards for 2 km, where the Jock of the Bushveld plaque signals the beginning of the Voortrekker road.

Exactly 8 km beyond the long rein-

One of the white rhinos transferred from Natal, photographed with her calf in the rhino "boma" south of Pretoriuskop.

A place for elephants . . .
Although there is no shade to counter the midday heat, a visit to the Transport (Vervoer) dam, off the H1-1 Nahpe road, can be very rewarding during the hot hours for those who enjoy the antics of elephants bathing.

forced game fence of the roan antelope holding camp is the historic landmark of Ship mountain, so named because it roughly resembles an inverted ship's hull. Further away from the road to the right is another landmark: the conical-shaped Sithungwane hill.

The road winds its way onwards for another 16 km past two well-situated waterholes – first the Komapiti windmill and then the Voortrekker windmill, both worthy of a pause, especially between 8 and 10 a.m. when game can be plentiful.

Just before the Voortrekker windmill there is a short side road turning off left, which takes you to a place which was once a favoured outspan of the transport riders in the 1880s. About 40 m upstream from the usually dry weir near the site, a large leadwood tree still stands which was used for target practice.

On reaching the tarred H3 road to Malelane Gate, turn northwards towards Skukuza, but keep a sharp lookout at that intersection for white rhino, which are frequently seen in this area. Slightly to the south, on the left of the tarred road, is the new Afsaal (off-saddle) picnic site, a good stop for a leg-stretch and a cup of tea. A shop there sells soft drinks, sweets and basic provisions and there are toilets and gas cooking facilities in a solid, modern complex which sadly fails to blend with its surroundings.

Travelling north again on the tarred road, past a series of beautiful, broken granite koppies, you will notice, on the left, the entrance to the private Jock of the Bushveld camp which was opened for block bookings by tourists in October 1983. Better known simply as the Jock camp, it was sponsored and largely planned and designed by Jack and Cecily Mackie-Niven of the Eastern Cape. Mrs Mackie-Niven is the daughter of Sir Percy Fitzpatrick, author of *Jock of the Bushveld.*

The camp boasts its own natural waterhole and consists of two huts, each with two bedrooms and a veranda, a family cottage and a kitchen/dining room complex overlooking the waterhole which is shaded by a wide-spreading jackal-berry tree. The accommodation is for a maximum of 12 people. Sir Percy's granddaughter-in-law, Marina Mackie-Niven, was mainly responsible for the interior decoration and she has included such memorabilia as pictures, antique porcelain, rifles, gin flasks, old brake-

The Shitlhave dam, east of Pretoriuskop on the H1-1 Nahpe road, is usually home to a family of reedbuck.

shoes and many other artefacts of the colourful era of the transport rider.

After 6 km you drive past the S113 link road to the eastern sector of the Park (described under Route J, Lower Sabie camp), and after another 9 km the S112 turns off to the right. A detour may be taken down this road as far as the S22 Hlamanduba loop, which turns off to the north. Off this loop is a short side road that winds up Shirimantanga hill, the final resting place of Colonel Stevenson-Hamilton, the Park's first warden, and his wife Hilda.

On reaching the main T-junction with the tarred H1-1 Nahpe road from Skukuza, you turn west on the final leg towards Pretoriuskop.

After 4 km, a turning to the right offers a lengthy alternative circular route back to camp via the Nwaswitshaka waterhole, the S1 Doispan road going west to the Albasini ruins, and then southwards to Pretoriuskop. If you should consider making this detour, at

least an hour's added travelling time should be allowed.

However, by keeping to the tar you soon reach the Transport dam (Vervoerdam) on the left, where during the heat of the day elephant love to congregate, dipping and wallowing in the water to cool off – much to the disgust of a lone, large crocodile, which is compelled to evacuate up the far bank for safety.

There are two further worthwhile stops before returning to camp. They are the Nahpe boulders on the S11 loop on the left where a plaque has been erected to the memory of Joseph Francis Ludorf, a past chairman of the National Parks Board who died on 21 December 1958, and then 5 km further on is the Shitlhave dam, where the surrounding tall grasses offer sanctuary to a family of southern reedbuck.

While in the area close to the camp, a sharp lookout may reveal sable or roan antelope and eland, which range quite close to Pretoriuskop camp in small herds, but they may be difficult to see because of the heavily wooded terrain.

Herds of three of the rarer species of antelope – eland, roan and sable – can seen in the vicinity of Pretoriuskop. Th two sable were part of a herd feeding on new grass sprouting after a burn near Shitlhave dam. On the left is a bull, w above is the more graceful cow.

JOCK OF THE BUSHVELD

The most graphic account of the wild and free life of the transport rider comes from a man who later rose to prominence in South African politics and administration but was then a young novice seeking adventure and fortune: Percy Fitzpatrick, later Sir Percy when he was knighted for services to South Africa.

His story is told in *Jock of the Bushveld*, now a South African classic, the biography of a dog, a runt-of-the-litter born in the Lowveld in May 1885 of a bull-terrier mother and a father of unrecorded lineage.

Fitzpatrick regularly plied the wagon roads from the Lydenburg and Barberton goldfields to and from Delagoa Bay, risking tsetse fly and the *Anopheles* mosquito to ferry supplies in, routinely making the hair-raising descent down the perilous Great Escarpment tracks.

Except for the brief stops of a few days at each end of a trip, he lived in the Lowveld bush, camping alone at night with his dog and his Zulu helper, or sometimes with other transport riders at established stopping places. On these occasions, there was always a campfire, plus the tales around it and often the eyes of hyenas, jackals and sometimes lions reflecting the flickering firelight. And the hunt, always the hunt for meat for the pot, although many transport riders also shot for trophies and for profit.

In *Jock of the Bushveld*, Jock is a hunting dog *par excellence* who instinctively knows how to tackle and bring down a wounded buck, or distract it from a charge, or keep an angry lion at bay until the hunter can safely fire the killing shot. It is a true story of great courage and high drama in which the beauty and deep lure of the Lowveld come through strongly. So firmly has Jock become entrenched in South African folklore that all over the Lowveld and Great Escarpment today there are markers to show spots where Jock passed or camped. Some of these are inside the Kruger National Park, reminders of that brief but fascinating passage in the history of Southern Africa.

The old route trekked by Jock, his master and other transport riders – and the very early settlers before them – was, however, lost soon after 1892 when the new railway from Pretoria to Lourenco Marques made it obsolete. It was simply swallowed up by the bush. Colonel James Stevenson-Hamilton and others tried to retrace it but discovered only parts.

But in the past few years that part of the transport route which traversed the present-day Park has been almost fully rediscovered by senior ranger Thys Mostert of Pretoriuskop, aided by rangers Charlie Nkuna, July Mona, Ngwela Mangane, Salmão Mongoe and Million Zitha. It runs roughly along a line from Pretoriuskop to the south-east, crossing the Park boundary about halfway between Malelane Gate and Crocodile Bridge, and in part following the present H2-2 Pretoriuskop-Malelane tourist road.

The one thing they have not found is Jock of the Bushveld's birthplace, although Fitzpatrick gave a fair description of it and they have narrowed down the region to a matter of a few kilometres.

The transport road diverted north of Ship mountain, which lies south-east of Pretoriuskop, so the transport riders could avoid the boulders and clinging clay which lie at its foot. Somewhere there – "under a big tree on the bank of a little stream . . . it was the tree under which Soltke prayed and died," Fitzpatrick wrote – a bull terrier bitch named Jess bore six pups, of which Jock was the runt.

Jock, the Lowveld's most famous dog.

Thys Mostert's only firm clue was that phrase ". . . it was the tree under which Soltke prayed and died". Soltke was one of those odd characters who flicker briefly across the screen of African history, a German of about 23 whom Fitzpatrick and some other transport riders came across on the way from Lourenco Marques to the hinterland. They found him marching along the transport road wearing a suit and a bowler hat and shading himself from the sun with an umbrella. He spoke little English, but the transport riders felt sorry for him and took him along.

They were camped under the big tree on the bank of the little stream when Soltke saw a bird he wanted to collect – a roller. In his excitement he jumped onto a wagon to get his shotgun. As he jumped down, both barrels went off and the charges hit him in the right leg, shattering it.

One of the transport riders immediately rode more than 100 km to fetch a doctor but when he returned two days later, all he could bring was a chemist named Doc Munroe who was so drunk he could do nothing for the unfortunate Soltke, who by now was saying his prayers. After another two days a doctor arrived from Mac Mac on the Escarpment by horse and immediately amputated Soltke's now badly infected leg.

But blood poisoning killed the young German early the next day. The transport riders buried him beneath the tree where he had prayed. A few months later Jock was born there.

That grave is now Thys Mostert's target. A grave has been found on the bank of the Samarhole creek where the old transport road crosses it between the Komapite windmill and Pretoriuskop and the remains are being examined to try to determine whether it is that of Soltke or of some other, unknown, victim of the Lowveld.

If it is Soltke's, then the last important link will have been completed along one of the most fascinating historic "roads" in South Africa.

Near Berg-en-dal on the H3 north road, an elephant bull shakes earth from a trunkfull of grass before eating it *(left)*, and a slender mongoose poses at the roadside *(below)*.

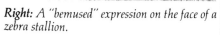

Right: A "bemused" expression on the face of a zebra stallion.

Berg-en-dal camp

The new Berg-en-dal ("hill and dale") camp, officially opened on 25 February 1984, is 12 km from the imposing new Malelane Gate entrance to the Park on its southernmost border, marked by a new high-level bridge across the Crocodile river.

The opening of the camp on time was little short of a miracle because less than a month before the ceremony, on 28 January, a fire completely gutted the entire kitchen complex in the camp's main building. The fire, which started in the roof due to an electrical fault, caused R100 000 damage as every item of the as yet unused super-modern catering equipment was destroyed. However with invitations already posted and preparations nearly completed for the opening ceremony, the show had to go on, so Park staff worked day and night to repair and re-equip the kitchens for the big day.

Berg-en-dal is unique in several respects in the Park: it is the only camp with a rugged mountain environment, it was built on the site of a prehistoric settlement, it has full conference facilities and its style is modern, quite different from the traditional thatch-and-white-washed-walls-character of the others.

Set in a bowl amid the steep, rocky Malelane mountains whose flanks are lightly covered by trees and scrub, it has neat chalets of raw brick beneath thatch which are placed in carefully landscaped, undulating terrain with limited green lawns and totally indigenous trees, in such a way that each chalet is private and the whole camp fits unobtrusively into the environment.

It has 69 three-bed chalets, 23 with six beds, and two guest houses, one with eight and the other with six beds. The smaller chalets have low-walled outdoor brick patios with barbecue facilities and also electric hotplates indoors, refrigerators, glass-doored showers, sculleries and stylishly designed lounges with the beds around the walls. The larger family chalets are similar in concept but have full bathrooms and two separate bedrooms.

Caravanners and campers have a treat in store for them with a park that must rank with the best in the country. There are 70 beautifully shaded and grassed sites for up to six people each, served by three superbly designed ablution and kitchen units. Full facilities for paraplegics have been incorporated in these buildings.

The core of the camp is an ultra-modern complex of facebrick buildings on the west bank of a small dam which is home to several crocodiles and in the mornings and evenings hosts a wide variety of bird life. At the complex is a restaurant, a cafeteria, a shop and three separate conference units with full audio-visual facilities which may be hired singly or together. Nearby is a swimming pool and, on the road out of the camp, a filling station; these blend with the rest of the camp.

Also in the central complex is an information centre designed to house an exhibition depicting the region's ecological history from Stone Age times, when Bushmen lived in complete harmony with the environment, through the advent of Iron Age man and his impact upon it, up to the harsher impact of modern man in recent history, and then the return to pristine environment through the creation of the Park as a sanctuary.

When the camp was being built, the swimming pool dug and trenches excavated for electricity and other services to the chalets, the Park staff found traces of ancient man. Archaeological core samples taken at various places in the camp's vicinity and test trenches produced some bone material and potsherds. The bone material had been too changed by soil acidity to make accurate scientific dating possible, but the designs on the potsherds indicated the presence of an African culture in the Iron Age.

As a further educational device a Rhino Trail, marked by rhino spoor, is being created along which visitors can make a 20 minute walk to see points of geological, archaeological and ecological interest within the camp.

From the main complex, looking across the Mowebeni dam, visitors have a view of a distant sugar mill in the fertile farmland outside the Park in the south-eastern Lowveld.

Nine kilometres south-east of Berg-en-dal on the tarred road to Malelane Gate is the short turnoff to the old Malelane camp, now a private camp which must be hired in its entirety and is a favourite of traditionalists because of its old-style huts and verdant tropical growth.

ROUTE D

The routes around Berg-en-dal offer splendid scenery while the game is at times rather sparse – although the area does contain rhino, kudu, impala, giraffe, some elephant, reedbuck, roan antelope, klipspringer, grey rhebok and warthog.

Leaving the camp through its brick-pillared gates beside a thatched gate-house, the most impressive view is of the bulk of Khandizwe mountain looming to the south.

A few metres from the gate a gravel S110 road turns off the tarred road (also the S110) to the right and loops for 4 km northwards, past a turnoff to the north-east, to the Matjulu lookout point overlooking a windmill, a worthwhile place to stay for a short time to see birds and quite often rhino coming down to drink. The road continues but access is barred.

Returning to the turnoff (still the S110), now turning left to the east, the road twists and turns through hilly country for 6 km to another turnoff, to the north, the S120. The S110 continues eastwards for 4 km to the main north-south tarred road. The S120 climbs north higher into the mountains to the top of the Steilberg (Steep mountain) and after 5,7 km reaches a viewpoint on the top, off the road to the right which gives excellent vistas of the Malelane mountains to the south-west and the broad bushveld to the east and north.

The Steilberg road continues for another 2 km to the main north-south H3 tarred road between Malelane and Skukuza. A short distance north of this junction along the tarred road, another gravel road, the S121, turns off east, named the Timfenene loop. This leads out of the mountain country back on to the Lowveld plains.

It follows the north bank of a dry creek for 6 km with two very short loops to the right leading to the creek's edge. It then reaches the S114, the old gravel main route between Malelane and Skukuza. The S114 goes north for 2 km where it is joined by the S25 Crocodile river road which runs eastwards along the river, although the river itself is out of sight. Take the S25.

After 5 km the Crocodile river road reaches a turnoff northwards, which is the S119 road following the Mlambane river. It is worth going just past the turnoff for about 30 m to reach a steep dip through the bed of the Mlambane, where one can park in the shade of big, spreading roadside trees and watch birds and animals coming to drink at water beside the causeway. On occasion unusually large groups of grey louries gather at the water, and also impala and kudu.

To get back to Berg-en-dal, return to

As regular as clockwork, the last week in November signals the lambing season and within days the Park proudly boasts hundreds of mini-impalas.

the S119, the turnoff just before the river, to follow the road along the south bank of the Mlambane (note that the road sign at the turnoff gives the Mlambane road as S118 when it should be S119, as on the official map).

The Mlambane river road travels for 6,3 km north and west before it reaches the old Malelane-Skukuza road, the S114. Two good river views are given by loops to the right. After 2,7 km a mud wallow appears on the left which is much enjoyed by elephant, rhino, warthog and, when there is enough water, scores of terrapins. Just before the S119 reaches the S114 there is a fine river view loop to the right with excellent shade.

Turn north onto the S114. After 100 m the Mlambane south bank road leads off the S114 to the west (now the S118). This continues for 8 km, still close to the river, and has another four river views, the third of which offers a fine grove of *Euphorbia*.

Where the S118 joins the H3 tarred Malelane-Skukuza road, one can either return to Berg-en-dal or travel north on the H3 for 4 km to a turnoff west to the Renosterpan waterhole, a good place to wait for rhino, giraffe, kudu, impala, lion, various birds and particularly for lone elephants or elephants in small groups.

For tree enthusiasts, Berg-en-dal has a variety not found elsewhere in the Park because of certain kinds which grow only in mountain environments, such as the mountain seringa, wonderboom fig, wild olive, Lowveld chestnut and sweet thorn.

During the late spring and early summer, the display of wild flowers in this area is outstanding, especially the brilliant reds and pinks of the Pride of De Kaap, which often dominates the riverine scrub.

A brand new member of the comical wildebeest clan at play and at rest on the H3 road just north of the Jock of the Bushveld camp.

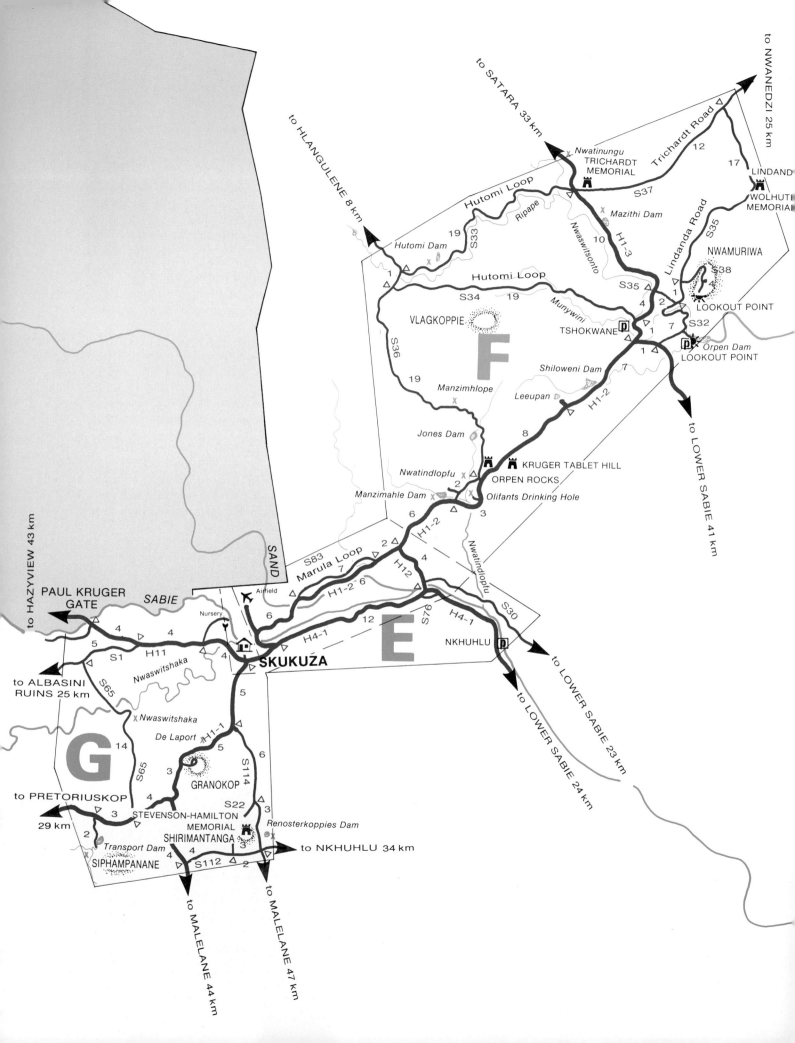

to NWANEDZI 25 km

to SATARA 33 km

to HLANGULENE 8 km

Trichardt Road

Nwatinungu TRICHARDT MEMORIAL

12

17

LINDAND

WOLHUT MEMORIAL

Hutomi Loop

S37

Lindanda Road

S35

Ripape

Mazithi Dam

Nwaswitsonto

10

H1-3

19

S33

NWAMURIWA

S38

Hutomi Dam

1

Hutomi Loop

S34

19

Munywini

S35

4

2

1

LOOKOUT POINT

VLAGKOPPIE

F

TSHOKWANE

1

7

S32

S36

1

7

Shiloweni Dam

Orpen Dam LOOKOUT POINT

19

Manzimhlope

Leeupan

H1-2

to LOWER SABIE 41 km

Jones Dam

8

Nwatindlopfu

2

KRUGER TABLET HILL

ORPEN ROCKS

Manzimahle Dam

Olifants Drinking Hole

6

H1-2

3

SAND

S83

Marula Loop

2

H1-2

4

Nwatindlopfu

S30

SABIE

7

H12

Airfield

H1-2

6

S76

H4-1

PAUL KRUGER GATE

Nursery

6

12

E

NKHUHLU

to HAZYVIEW 43 km

4

4

H4-1

to LOWER SABIE 23 km

5

S1

H11

4

SKUKUZA

to ALBASINI RUINS 25 km

S65

Nwaswitshaka

5

to LOWER SABIE 24 km

Nwaswitshaka

H1-1

De Laport

G

14

5

6

S65

3

S114

GRANOKOP

to PRETORIUSKOP

4

S22

3

29 km

3

STEVENSON-HAMILTON MEMORIAL

SHIRIMANTANGA

Renosterkoppies Dam

to NKHUHLU 34 km

2

4

3

Transport Dam

4

S112

2

SIPHAMPANANE

to MALELANE 44 km

to MALELANE 47 km

SKUKUZA/TSHOKWANE AREA

Skukuza camp

Skukuza is the Tsonga word for "he who sweeps clean" – the African name given to the short, tough and determined colonel of the British Army, James Stevenson-Hamilton, who was the first warden and the legendary champion of the Kruger National Park.

Initially called Sabie Bridge, or simply Reserve, Skukuza was rechristened in 1936 to mark Stevenson-Hamilton's 30th anniversary in the Park and as a permanent memorial to him.

Today Skukuza ranks as one of the biggest towns in the Lowveld (including the revolving population of tourists) with its ever-busy shop, bank, petrol station, post office, restaurants, library, doctor's rooms, café, Automobile Association workshop, plant nursery, information centre, police station and airport. It has more than 200 tourist huts, various family cottages, several luxury cottages and a rest house for VIPs. There are also conference facilities for seminars on nature conservation.

Behind the scenes, away from tourist routes, are the Park's nature conservation and research complex, state veterinarian, laundry (one of the largest in the southern hemisphere), motor and roads maintenance workshops, refrigeration workshops, abattoir and by-products factory, weather station, church, primary school, sports complex for rugby, cricket, squash, swimming, tennis and golf, and staff villages that house about 500 whites and 3 000 blacks.

The window of the chief warden's office overlooks a neat garden with a tree in the centre of a lawn. In circles around the tree are more than 50 stone and cement tablets marking the graves of rangers' dogs over many years, all of which served their masters well and some gallantly. Leaning against the tree is an upright headstone engraved with the words "G. R. Healy – in loving memory of Mary, died 14.11.07 aged 4 years." When this headstone was first found in thick undergrowth near the Skukuza boundary fence it was thought to mark the grave of a child. But old documents later revealed it to be the grave of ranger Healy's dog.

Built on the banks of the Sabie river, Skukuza is steeped in colourful history. Dominating the scene, just east of the camp, is a big steel bridge, the last relic of the old Selati railway line that once cut through the Park and which was closed to traffic in 1973.

On the site of the original railway siding there is now a restaurant and lounge cum bar in beautifully refurbished old railway saloons. Cocktails and *à la carte* meals are served in the three-car lounge, dining saloon and kitchen combination attached to an old British-built 24 Class steam locomotive (No 3638).

The dining car, named Lundi, was built in Durban in 1924 at a cost of £10 225, complete with its own separate tandem kitchen/staff accommodation car. It was retired in June 1978 after 54 years of service throughout South Africa. The lounge car, of Belgian origin, entered service in 1929 and was subsequently converted to a State Funeral Coach. It was re-converted before being donated to the National Parks Board.

The station has also been built in the old style and it is said to have been modelled on the original station at Komatipoort. It boasts a hand-painted name board "Skukuza – 875 ft above sea level" and now houses the Selati Railway Museum.

The oldest hut in the camp, numbered S1, is now also a museum. Known as the Campbell 1929 Hut Museum (after Mr W. A. Campbell of Mount Edgecombe, Natal, a founder member of the

National Parks Board of Trustees who donated money towards the project), it has been restored to its original simple state, including a ventilation hole in the door. On display within are original rawhide thong beds with reed-matting "mattresses", paraffin boxes used as bedside tables and also cupboards, old maps, lamps, entry permits, books, crockery and many other items of historic interest.

Outside the hut is one of the original red Pegasus petrol pumps that were used in the Park during that period and behind the hut is the smallest rondavel in the Park, now used as a broom room.

Nearby, in front of hut B80, is the last of the concrete blocks which anchored the cables of the pontoon ferries that carried vehicles, horses and people across the Sabie river between 1928 and 1937. They were eventually replaced by causeways and later by high-level bridges.

Just inside Skukuza's entrance gates, opposite the garden of lawns and ponds, is a clock erected to the memory of Herbert Boshoff Papenfus, once Member of Parliament for Hospital Hill, who was a driving force for the proclamation of the Park and a member of the first National Parks Board. He died in 1937.

One of Skukuza's main attractions – for both scientists and tourists – is the Stevenson-Hamilton Memorial Library built in 1961 chiefly with money donated by the Wildlife Society of Southern Africa. A statue of the man, who was as small in physical stature as he was great in achievement, dominates the entrance hall, with his favourite pipe in one hand and a fly whisk (a wildebeest tail) in the other.

The building boasts many facilities of which the most valuable is the library, recognised internationally as housing one of the finest collections of ecological reference material. In it, too, are several notable wildlife paintings by artists such as Paul Bosman, Simon Hodge, Gabriel de Jongh and David Shepherd.

The building also houses fascinating Park memorabilia, including Stevenson-Hamilton's rifle, saddle, binoculars, water bottle and other travelling kit displayed in a mock-up camp, his wife's guns and those of ranger Harry Wolhuter. There is the skin of the lion stabbed to death by Wolhuter and the knife with which he did it, poachers' guns, historic maps, Stone Age implements and the like. The

Park's plan is to hold periodic exhibitions by wildlife artists and photographers on the premises.

Behind the library is the information and environmental education centre with its modern 150-seat auditorium where trained officers arrange films, lectures and audio-visual programmes for both adults and children.

Children are specially catered for at Skukuza and are welcomed in organised school groups to the centre. A dormitory complex with dining-room and staff accommodation is sited adjacent to the caravan park. The children enjoy field trips into the Park by bus and lectures by Park information officers and other experts on wildlife. At least 100 bus loads visit the Park each year.

During school vacations, when there are no organised child tour groups, children holidaying in the Park are invited to attend courses at the centre given at both primary and senior school levels. There is also an outdoor "touch" display specially designed for younger people where they can handle and closely examine animal skulls, horns, skins, and nests.

Campers and caravanners have an extended area reserved for them with two modern ablution blocks and laundries with coin-operated washing machines and tumble dryers.

Skukuza airport is fast becoming one of the busiest in South Africa. In 1983, civil aircraft movements reached a record 1 051 with 15 662 passengers passing through the thatched terminal building.

Within the next few years, the camp is to undergo a transformation: the existing reception/shop/restaurant building on the Sabie river bank is to be renovated to house the restaurant, shop and snack bar only, and a new complex will be built near the main camp gates which will include reception, bank, post office, carhire, AA services and petrol station. A state guest house is also planned and the present post office will be moved. A swimming pool is to be built on a site beyond the caravan park near the train restaurant.

In spite of all the activity, Skukuza still maintains the atmosphere of a bushveld camp. Impala, warthog and other game often wander nonchalantly between the huts on their way to feed on the camp's green lawns. One of the "resident" warthogs can be seen on most

After wading the Sabie river, an elephant ambles down the H4-1 towards Skukuza, oblivious of tourist traffic (right) until a roadside mud-wallow becomes the new attraction (series above).

nights snoozing happily, squeezed tightly inside a storm water pipe which can be found a few metres across the road from the outdoor cinema in the direction of the river.

A favourite evening pastime is to sit on the benches overlooking the Sabie river to watch the chattering weavers prepare for sleep in their nests dangling from the huge wild fig trees overhead, the variety of animals that frequent the opposite bank to drink and the crocodiles swimming slowly by.

Visitors there were given a special Christmas treat in 1982 – a rare insight into the ways of the wild. A leopard

Left: A mother baboon from one of the large troops that frequent the H4-1 Sabie river road enjoys the attentions of her infant. Above: The same youngster shares a "toy" with a playmate.

killed a young bushbuck soon after sundown and dragged its prey 10 m up into a fork of the giant sycamore fig tree right in front of the camp restaurant. A keen-eyed waiter first noticed the perfectly camouflaged cat and its prey. Although the leopard was frightened off, many visitors abandoned their beds hoping to catch a glimpse of it returning to complete its meal.

ROUTE E

Tourists visiting the Park for the first time usually choose Skukuza as their first base camp and to many it becomes a firm favourite for subsequent visits. In the early mornings when the gates swing open it is a fair bet that most people will drive to the H4-1 route, or "Piccadilly Highway" as it is sometimes known because of its popularity. This stretch of road, which runs along the south bank of the Sabie river between the low-level bridge a stone's throw from Skukuza and the new high-level bridge just over 12 km downstream towards Lower Sabie camp, has by far the highest volume of tourist traffic in the Park – and understandably so.

Between the two bridges there are 16 loop roads and riverside viewpoints to explore and the magnificent riverine forest is a haven for elephant, buffalo (sometimes in very large herds), lion,

impala, giraffe, bushbuck, kudu, warthog by the score and, if one is lucky, nyala – all so accustomed to traffic that they scarcely look up as visitors drive by.

Bird life too is prolific. Many of the kingfisher species are busy over the river, the rowdy hadedas with their short legs and curved beaks scuff for grubs and brownheaded parrots shriek from the treetops. There are green pigeons and hornbills galore, including the mournful wailing trumpeter variety and the huge ground hornbills with their bright red dewlaps – sometimes so tame that they strut up to cars to beg.

The antics of the large troops of vervet monkeys and chacma baboons are constantly amusing, especially for younger visitors. Sadly, many of these animals are now spoilt and potentially dangerous because people continue to illegally feed them, in spite of warnings.

Also often seen are the secretive leopard (look carefully on the lower boughs of trees in the early mornings and late afternoons) and smaller animals such as the mongoose (especially the banded variety), the leguaan and in the evening the African civet.

The river itself is the home of many crocodiles and hippos. A huge crocodile, one of the Park's biggest, can often be seen snoozing on the bank opposite the roadside resting place just 4 km from the T-junction where the Skukuza link road meets the H4-1. A busy hippo pool

Fully alert for lurking predators, herds of giraffe and zebra share a corner of the Manzimhlope waterhole on the S36 west of Tshokwane picnic site. During the dryer months of early spring, this waterhole is one of the best for game viewing.

lies 8 km further on, or 1,5 km before the H12 high-level bridge.

From this bridge the motorist has the choice of several alternative routes. One is to continue on the H4-1 towards Lower Sabie – and I recommend a stop at Nkhuhlu picnic site where there is shade, hot water, barbecue facilities and a refreshment kiosk. The bird life there can be spectacular.

Other alternatives are to turn back towards Skukuza – simply because one can discover a whole new vista of wildlife in the same places after a short time – or to travel north across the bridge for 4 km to join the Park's "great north road" (H1-2). Travellers can then return

to camp following the usually dry Sand river on the left, or take the parallel S83 Marula loop, a worthwhile gravel road alternative covering the same distance.

However, from the Sabie river high-level bridge, I always favour travelling east for a few kilometres, down the gravel S30 Salitje road before turning back along the same route. It turns off to the right immediately after the bridge and follows the north bank of the river, and can be very worthwhile.

There are three short, deep-shaded riverside loops off this road which are especially pleasant on hot days. The third of these, exactly 5 km from the turnoff, can be richly rewarding for

people with the patience to sit quietly for some time. The entry to the loop is easily identified by a concrete causeway across a dry stream close to a giant fig tree on the right. You can drive to the water's edge and park in the shade of two jackal-berry trees overlooking a still pool beside a stretch of sandy beach. It is here that several photographers have captured the real life drama of a crocodile kill as animals come down to drink.

*Left: A hyena lying in the middle of the S30 Salitje road becomes agitated during a roadside feeding frenzy of several hyenas around a buffalo kill. **Above:** Another hyena dashes off with a chunk of the carcass.*

ROUTE F

A good day trip from Skukuza is to explore the area around Tshokwane tea room 40 km away to the north-east. It is advisable to leave fairly early and to allow a full day for this tour.

From the Skukuza gates, make your way to the H4-1 Sabie river road, where you may take either the H1-2 road via the two low-level bridges over the Sabie and Sand rivers (also past the airport), or more directly, by following the Sabie along the well-trodden H4-1 to the H12 link road that crosses the river over the high-level bridge. Both ways are equidistant.

Make a point of diverting along the short loop to the left to the Manzimahle dam and the Nwatindlopfu windmill, and also down the short road to the Olifants waterhole. This area is well known for its prides of lions, large herds of waterbuck and sable antelope.

At this point, I would also recommend a drive along the gravel S36 road off to the left for exactly 7 km, past the Jones dam to the Manzimhlope windmill – a beautiful waterhole right next to the road on the left. You should also keep an eye on the concrete trough away to the right of the road.

In my view this is one of the finest waterholes for game variety in the Park, especially during the dry winter months and early spring. During one August morning in 1982 I watched herds of 15 giraffe, more than 30 sable, over 100 buffalo and numerous wildebeest and zebra there, plus a pair of reedbuck, a mother cheetah and two cubs, four tsessebe, a pair of hyena that sniffed the air from a distance and decided the place was too crowded, and a lone honey badger which waddled by without stopping. The highlight of the morning came when two sable bulls decided to settle their territorial rights by fighting it out right in the middle of the pond. A rare sight indeed.

Returning to the H1-2 main road and continuing northwards, one soon arrives at two memorial tablets set into giant granite boulders at the roadside. The first on the left honours Mrs Eileen Orpen for her donation to the Park of seven farms situated on the western border of the Park between Orpen Gate and the Olifants river – a total area of 28 633 morgen (24 500 hectares). An amazingly tame family of klipspringers lives among these rocks.

The second tablet, 2 km further on the right, commemorates the institution of national parks in the Union of South Africa, the proclamation of the Sabie Game Reserve by President Kruger in 1898 and the National Parks Act introduced in Parliament by the then Minister of Lands, Mr P. G. W. Grobler, in 1926.

Moving on, one passes the attractive Leeupan and the parklike setting of the

Shiloweni dam. Both these waterholes are 1 km off the main road on the left and are worthwhile detours.

The next turnoff to the right is the tarred H10, the new direct road to Lower Sabie camp. If you turn down this road, almost immediately to the left is the gravel S32 road to the Orpen dam viewpoint.

It is situated on the Nwaswitsontso river and visitors may park their vehicles under a reed-thatched carport and walk to the tiered viewing shelter which overlooks the dam lying in the deep valley below. Binoculars are necessary to watch the abundance of waterbirds that feed there. It is a favourite place for thirsty animals and is the home of several very impressive crocodiles.

Instead of returning directly to the H1-2 main north road, continue northwards on the gravel S32 – a pretty drive along the river with rugged rocky cliff faces on the far bank – to the Lindanda road (S35) intersection. By now you cannot help but notice the Nwamuriwa mountain rising hugely on the right.

Turn right on the S35 and then, after 4 km, right again on the S38 for a short drive to the summit of this range for spectacular views in all directions across the Park.

Returning to the S35 and continuing north, you reach another historic monument: three plaques that graphically map the sequence of ranger Harry Wolhuter's famed encounter with a full-grown lion. The trunk of the tree that the desperately wounded Wolhuter climbed after stabbing the lion to death still stands there with its base wrapped in a pyramid of concrete to protect it from marauding termites (see p. 107).

At the next T-junction, the S37 Trichardt road intersection, turn left and travel west for 12 km to meet the main H1-3 north road 14 km north of Tshokwane. At this intersection is another memorial plaque: it records Louis Trichardt's historic wagon trek from Soutpansberg in the hinterland to Lourenco Marques – now Maputo – on the coast. He is said to have passed this exact spot in March 1838.

The choice now remains whether to drive south directly along the tarred road to Tshokwane and Skukuza, or to continue westwards along the banks of the usually dry Ripape river along the S33 Hutomi loop.

Opposite page: Several species of game gather together to enjoy a drink at the Manzimhlope waterhole. Above: A crowned plover and her chick scrape at the roadside gravel nearby for grub.

Other than the lucky chance of seeing a large herd of buffalo on this road, the main attraction is the Hutomi dam.

A windmill fills a trough which is out of direct sight and the dam itself is sometimes dry before the onset of the rains. In the heat buffalo love to wallow in its muddy bed, and when the sun begins to sink the sight of wet elephants dusting themselves in silhouette against the red sky is memorable.

The southern S34 Hutomi loop road (see map, p. 180) back towards Tshokwane has little to offer, but the direct route to Skukuza down the S36 can be most rewarding in the late afternoon – and you can enjoy another visit to the Manzimhlope windmill *en route*.

Visitors exploring the roads around Tshokwane should be aware that this is the area of the "white" animals (see pp. 128-33).

The last confirmed sightings of a white lion was of an 18-month-old male seen by ranger Pat Wolff and his assitants on two occasions during January and February, 1986.

During 1985 there were several sightings of a snow white kudu bull in the same area. It has been seen mostly north of Tshokwane feeding with a herd along the banks of the Nwaswitsontso river between the Baobab Tree landmark and the S33 Hutomi loop.

A buffalo with peculiar white facial markings was seen in a herd in the Nkumbe area, about 6 km to the southeast of Tshokwane on the H10 main road to Lower Sabie, also during 1985.

The ranger at Tshokwane is eager to keep track of these rare animals and would welcome details of any definite sightings from visitors including the date, time of day, and exact location.

ROUTE G

The 12 km stretch of main road between Skukuza camp and Paul Kruger Gate is an ideal short drive for the visitor with an hour or so to spare.

Skukuza has always been known for its hyena population and they are very common all around the camp, particularly along this road in the early morning. Impala herds and kudu are plentiful and very tame. Giraffe and rhino may also be seen crossing the tar heading for the Sabie river to drink.

On the left soon after the turnoff to Paul Kruger Gate, there are a number of stark, dead trees which are favourite marabou roosts. The whitewash colouring is the result of years of bird droppings. The "homecoming" of these giant storks in the evenings is a circus: each new arrival is welcomed with a noisy clacking of beaks as it clings to a branch, teetering desperately after a clumsy, flapping landing which rocks the whole tree, threatening to dislodge those already settled.

This road also offers a good alternative route to Pretoriuskop camp or the Numbi Gate via the upper Sabie river hippo pools and the Albasini ruins, along the S3 Sabie river gravel road and the S1 Doispan road, both of which offer reedbuck and sable antelope in addition to other species.

Another option from Skukuza is to take the H1-1 Nahpe road, the main tarred route to Pretoriuskop. After passing the old gravel S114' turnoff to Malelane Gate on the left, continue for exactly 1 km and at that point a (usually) dry stream bed passes under the road through a culvert. This is a favourite hyena breeding den, and if the vicinity shows fresh tracks use a little patience to see the occupants, often including cubs which may come out to play around at dusk.

On the left 4 km further is the Granokop turnoff. A ride to the top of this solid rock massif reveals beautiful views in all directions.

Further along, one passes the new H3 tarred road to Berg-en-dal in the south, and then travels another 7 km through attractive rocky outcrops (where leopards are often seen) to the Transport (Vervoer) dam. Although there is no shade, it is well worth staying at this dam for an hour or two, as it is the

An excellent place for lions is just south of the Orpen rocks in the area of the Nwatindlopfu bridge and the Manzimahle dam. The picture of the lioness nuzzling her cub (above) and the mother transporting one of her litter (right) were both taken in that vicinity.

haunt of many animal and bird species and a favoured bathing place in the heat of the day of buffalo and elephant, whose antics are a delight to watch.

Returning towards Skukuza, one can turn northwards off the tar up the S65 to the Nwaswitshaka windmills; donated to the Park by Alex and Netta Wrede of Durban. Although the waterhole is rather unattractive with steep, bare banks, it is excellent for game viewing and is a popular watering place for rhino. There is also a concrete water trough nearby which attracts large herds of impala, zebra and giraffe.

Alternatively you can continue towards camp back along the Nahpe road, but then make a point of detouring southwards down the gravel S114 Malelane road for exactly 6 km, where you take a right fork. After a short distance, another turnoff to the left winds up Shirimantanga hill to a quiet, shady place high above the plains. Here a simple bronze plaque set into a giant granite boulder marks the last resting place of James Stevenson-Hamilton, who gave 44 years of his life to the Park. The ashes of his beloved wife Hilda were also scattered there.

A picture parade
All the pictures of these two pages were taken one August morning at the Manzimhlope waterhole north of Skukuza. The parade included impala, giraffe, kudu, a big herd of buffalo . . .

. . . and also waterbuck, zebra, sable, (including
two bulls that fought each other for herd
supremacy (see pp. 112-3) and a cheetah and
her two "teenager" cubs who lay in the shade
nearby for several hours before also coming down
to drink.

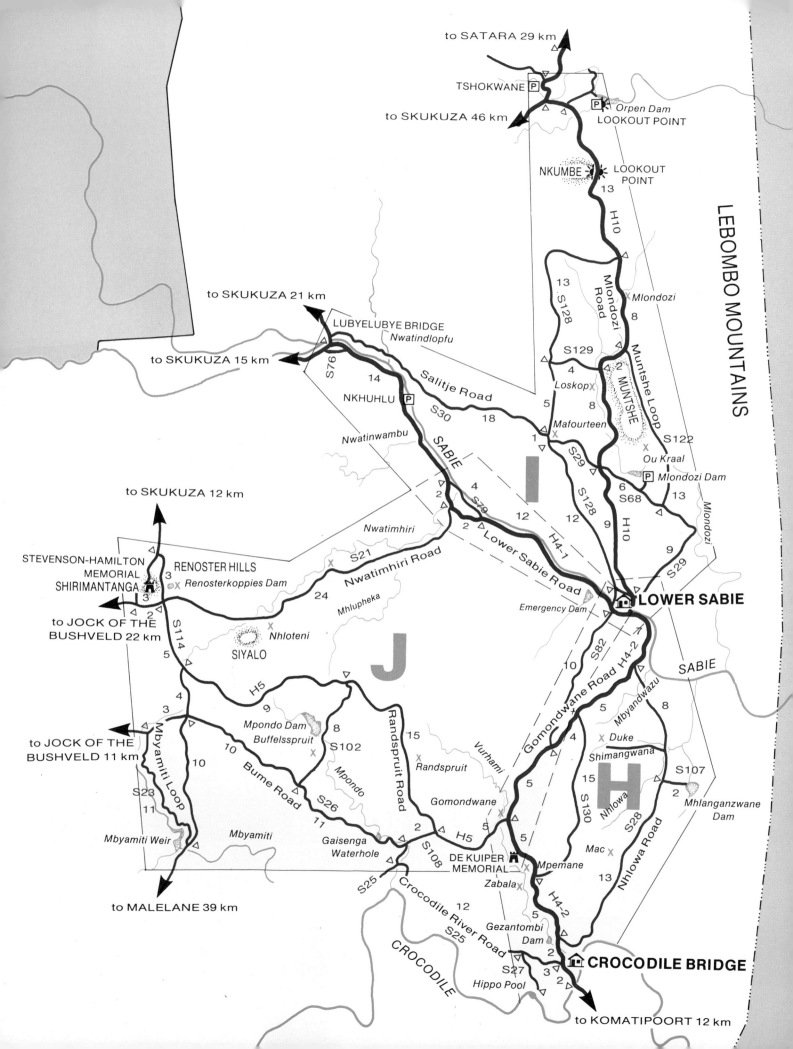

LOWER SABIE/CROCODILE BRIDGE AREA

Lower Sabie camp

Situated just 11 km from the Mozambique border on the banks of the Sabie river in the south-eastern corner of the Park is Lower Sabie camp.

Second only to Letaba for its giant trees and well-shaded lawns, this camp is a firm favourite among regular visitors to the Park. The wide river banks hug the camp's northern boundary and visitors can spread themselves on the green lawns in the shade of marulas, huge figs, knobthorn and Natal mahogany trees to watch elephant, giraffe and buffalo feeding along the riverside.

Bird life is spectacular: many kinds flutter in the spray of the lawn sprinklers and almost daily a pair of purplecrested louries can be seen feeding high in the trees, showing off their brilliant colouring. Fever trees are festooned with the nests of chattering weavers.

For some distance before reaching Lower Sabie, the green and gold flag of the National Parks Board marks one's approach. The entrance is dominated by low, stone walls that flank the timber gates. On entering the camp, one swings past the flagpole to park in the shade of a huge jackal-berry tree that dominates the entrance of the reception area. The very compact administrative block includes the camp's restaurant, shop, snack bar and reception office, all situated around a covered patio with rustic leadwood supports and a gnarled old weeping boer-bean tree that pushes its way through the roof.

Afternoon tea is often shared with cheeky lesser masked weavers that quickly move in to perch on milk jugs, sipping leftovers.

The long-term development plan for the Park includes a new, much more spacious main building to be built on the site of the present one.

The caravan and camping areas are on the west and east sides of the camp and have been neatly divided into asphalt parking bays interspaced with lawns. The camp's two ablution blocks have recently been completely renovated and modernised for the convenience of caravanners.

Lower Sabie is also renowned for its unique clusters of "clover-leaf" huts and the three-sided block of cottage-style rooms, 18 in all, with circular rondavels on their corners which are much favoured by larger families. However, much of this characteristic accommodation is soon to be either demolished to make way for new huts or refurbished to include bathrooms or showers.

The variety of game to be seen along the roads surrounding Lower Sabie is unequalled in the Park. It is an area noted for lion, cheetah, elephant, hippo, rhino, herds of buffalo numbering at times between 500 and 800, and a very large population of warthog.

Visitors would do well to spend at least half an hour in the evenings before the gates close at the emergency water supply dam a mere half kilometre from camp on the Skukuza road. At sunset the silhouettes of waterbirds feeding and animals drinking are spectacular to see and photograph.

ROUTE H

Many loyal old-timers, who regularly travel to the Park, swear that the Gomondwane road, the tarred H4-2 from Lower Sabie to Crocodile Bridge, travels through the best stretch of game-viewing country in the Lowveld.

In fact many believe that for total game-viewing satisfaction, they need do no more than continuously repeat the round trip between Lower Sabie and Crocodile Bridge, driving from either camp in one direction along the H4-2 for the full 35 km and then returning by the gravel S28 Nhlowa road in the other direction.

To explore this area, you leave Lower Sabie, then turn left at the main T-junction. Almost immediately to the right is the gravel S82 which runs roughly parallel to the H4-2 Gomondwane road and is also an alternative, slightly shorter route southwards.

However, keeping to the main road which follows the Sabie river on your left for a short while, visitors are virtually guaranteed sightings of elephant, fine kudu, impala and warthog crossing the road to and from the river. The view to the east is often uninterrupted and early rising photographers will be rewarded with excellent pictures of the sun lifting out of the distant Lebombo foothills bordering Mozambique, possibly with kudu and giraffe posing conveniently in the foreground.

After another 7 km one passes on the left the gravel intersection of the S28 Nhlowa road; it is this area (and a little further south) that seems to be favoured by white rhino.

After 9 km the S82 comes in from the right. From here the forest on both sides of the main road is quite dense and slower, more alert driving is required.

Soon the Vurhami river "depression" appears on the right side and occasional open vleis between this and the road are haunts of cheetah taking advantage of the long grass and dappled shade to watch for passing prey. Stalking cheetah have been known to use moving or stationary vehicles on the road as cover in their attack.

Past the halfway point to Crocodile Bridge, the windmills, concrete trough and waterhole of Gomondwane can be seen across well-trodden grassland on the right. A criss-cross of paths brings caravans of zebra, wildebeest and buffalo to this fine drinking place. If only there was a road closer to this spectacle, it could be a paradise for photographers. At present animals at the trough can be viewed satisfactorily only through binoculars.

At this waterhole a major gravel road (H5) turns away to the right (to Berg-en-dal) where, by moving a short distance along it, excellent shade can be found for more comfortable watching and waiting should the weather be hot.

Continuing southwards, a short loop to the right circles a splendid example of a common wild fig tree which is well worth photographing. A kilometre further on is a roadside brass plaque marking the spot where, in 1725, explorer Francois de Kuiper and 30 of his followers travelling from the coast into the interior were ambushed by Chief Dawano's impis (warriors), forcing them to retreat to Delagoa Bay (now Maputo).

Lone elephant bulls and sometimes small herds of elephant feed on the reeds and grasses of the usually dry Vurhami riverbed, which continues to wind its way beside the road to the Gezantombi dam just beyond the Nhlowa road intersection. You should turn off from the main road (after the bridge over the Vurhami) and park at the viewpoint overlooking this pretty stretch of water to watch waterbirds or just enjoy the shade and the view with a cup of tea or lunch.

Two kilometres further on is the final turning to the right before reaching Crocodile Bridge camp. This is the S25 Crocodile river road that follows the Park's southern boundary to Malelane Gate. After a short 3 km down this road, turn left for another 3 km and you will find hippo pools where you can leave your car and, accompanied by an armed assistant ranger, watch hippo blowing and splashing in the river. Bushman paintings can be seen here too.

The final approach into Crocodile Bridge usually offers surprisingly good game viewing: very tame giraffe, numer-

Two cheetahs keep an eye on game heading for the Gomondwane waterhole . . .

ous warthog and both banded and dwarf mongooses are often seen in the area.

Returning directly to Lower Sabie along the tarred road can offer a whole new vista of animal and bird life, but an excellent alternative is to take the S28 Nhlowa road, once the original main road to the south. It is especially worthwhile during peak season when heavy traffic along the tarred road forces game to find more peaceful sanctuary further to the east.

The vegetation is knobthorn savannah and consequently the veld is fairly open and game is easier to see. Look out for large herds of buffalo and for lion and cheetah, which often rest in the shade near the reservoir and trough at the Mac windmill, the first waterhole on the left. Because of tall grass alongside and a heavy camber to some sections of this old road, it is sometimes difficult to see out across the veld from the windows of a low, modern car.

There is only one turning off this road – the short leg eastwards to the Mhlanganzwane dam, a popular drinking place for game and home to several hippo, but the water is too far away from the car park for even the best equipped photographer. Where the road turns off to the dam there always seem to be two or three elderly wildebeest

. . . an impala ram on the alert nearby.

A kudu at sunrise – a view looking out across the open veld towards Mozambique, just east of Lower Sabie camp.

Above and left: The last few minutes of the day, before the camp gates close, can be well spent at the emergency water supply dam immediately west of Lower Sabie. The reflected sunsets are spectacular.

bulls ready and willing to show off their humorous antics for a camera. (Note: the official Park map shows a second road, to the Duke windmill, situated slightly to the north-west of this point, but it was not open to tourist traffic at the time of going to print.)

Before joining the main tarred road again, one crosses the low-lying vleis of the Shimangwana and Nhlowa streams, frequented by feeding elephants, giraffe, waterbuck and buffalo. Then, immediately before reaching the main road, there are the roadside pools of the Mbyandwazu river, an excellent spot for raptors and other easy-to-photograph bird life. It is also a favourite mud-wallow of warthog and rhino.

ROUTE I

Another worthwhile morning drive begins by turning right immediately after leaving camp and going north to cross the Sabie river over the low-level bridge. It may be advisable, especially in the drier winter months, to travel slowly and quietly over the causeway so as not to disturb the abundance of river life feeding in and around the shallows.

It is a good idea to turn right just 1 km after the bridge, taking the S29 Mlondozi road to the east. Cross over the new H10 tarred road and continue through open knobthorn countryside, past the S122 Muntshe loop to the

north, until you reach the Mlondozi dam lookout point high on a hilltop, at the end of the short S68 turnoff. Pleasantly cool, even on the hottest of days, this open-sided thatched shelter was originally erected by members of the Wildlife Society and recently renovated by Park engineers. There are nearby toilet facilities and cold drinks are sold by the attendant.

Although the 1982/83 drought took a heavy toll on hippo and crocodile at Mlondozi dam, which dried up completely, it has recovered well and these two species are again in residence.

A good pair of binoculars will highlight an abundance of waterbirds feeding in the shallows: a great variety of storks – yellowbills, saddlebills, sometimes African openbills, woollynecked and black storks, herons, hamerkops, and even an occasional flamingo. In May and June flocks of spurwinged geese, sometimes numbering over 100, gather to feed. A diversity of animals drink there including elephant, kudu, several waterbuck bachelors and, in the winter months, often spectacular herds of wildebeest and zebra.

Returning to the S29, you can turn right and drive for 4 km before joining the new H10 tarred road between Lower Sabie and Tshokwane. A run north up this road can be productive with a strong possibility of seeing cheetah and ostrich, but the highlight of this drive is the beautiful view from the top of the Nkumbe mountain across the Park. Two parking areas have been provided, one facing west and the other looking out

towards the eastern boundary of the Park. Visitors lucky enough to be there at the right time may see the spectacular mass migrations of zebra, wildebeest and buffalo moving to their seasonal grazing areas. Also along this high ridge, visitors who enjoy plant life can compare two species of *Euphorbia* found growing side by side there – the common *cooperi* variety which is found throughout the Park and the rarer *confinalis*, found mainly along the Lebombo range that borders Mozambique. The latter is easily identified by its more delicate "leaf" structure. It should be remembered that the milky sap from all these trees is poisonous, causing painful blisters when touched. It may also cause blindness or impair sight if it comes in contact with the eyes.

Motorists should take heed of the lower 30 km/h speed limit in force along the top of Nkumbe mountain. It is imposed because of sharp, blind bends along the road and because of the likelihood of traffic swinging suddenly to and from the lookout points.

Depending on the time available, you may either complete the northwards drive, visiting Tshokwane tea room, or turn back to travel south again and explore the recently opened S122 Muntshe loop which skirts the eastern slopes of the giant koppie of the same name. The pretty, rolling, rather open countryside stretching away towards the Lebombo range offers kudu, ostrich, zebra, wildebeest and cheetah. During the wet season there is usually an excellent display of wild flowers.

Wattled plovers are regular visitors to the emergency water supply dam at Lower Sabie.

Below and right: Warthogs are very common around Lower Sabie and Crocodile Bridge and in the spring one of the most absorbing sights is the sheer exuberance of the piglets that appear soon after the first rains.

Top: The gregarious, fast-traveling wild dogs, although generally rare in the Park, are known to range the area around the S26 Bume road near the Mpondo dam. *Above:* After the first spring showers, leopard tortoises trundle on to the tarred roads to drink from puddles. *Right:* There are always kudu to be seen along the Gomondwane roadside.

Other Lower Sabie residents include *(left)* a usually nocturnal scrub hare feeding on grass shoots sprouting beside the tarred road, *(centre)* a yellow-throated longclaw cocking an eye at the camera and *(right)* a black-backed jackal shaking its head for a reason only it knows.

You may also wish to try your luck along the S128 loop to the west of the tarred road. Incidentally, much of this loop includes stretches of the old gravel H10 road to Tshokwane. I have always found this area to be rather disappointing. However, make a point of stopping at the Mafourteen windmill situated just to the east of the S128/S29 intersection. Dedicated to the memory of Edgar Hansen "who loved all animals," this waterhole, although not clearly visible from the road, is a haunt of tsessebe, cheetah (I once saw nine, including four cubs) and an impressive lone sable bull that is often seen in the company of zebra.

Again if time allows, a very pleasant alternative route back to camp is to drive westwards from Mafourteen windmill (the intersection is less than 1 km to the north), travelling on the S30 Salitje road along the north bank of the Sabie river to the H12 high-level bridge over the Sabie and then back to camp along the main H4-1 Skukuza-Lower Sabie road. However, allow at least three hours to make the best of this 50 km loop.

The stretch of H4-1 road east of the bridge up to the Nkhuhlu picnic site and then beyond for a further 8,3 km to where the Nwatimhiri road turns away to the south, is heavily wooded and can be hazardous because surprisingly large elephant breeding herds cross the road daily, as do buffalo and even hippo heading for the river.

While travelling this road, you should look carefully into the shadows of the huge riverside trees for the more secretive animals that live there, such as leopard, bushbuck, nyala and, in the early evening, a civet. I have also seen hippo grazing at the roadside in broad

The usually shy and rather secretive bushbuck is bolder and thus easy to photograph along the H4-1 Lower Sabie road. The ewe is on the left and the magnificent ram on the right.

daylight – a very unusual sight indeed.

Make a point of stopping at the newly extended and refurbished Nkhuhlu picnic site. It is beautifully shaded and right on the bank of the Sabie river. A kiosk supplies refreshments. Birdlovers should be well satisfied with the variety of feathered visitors: greenbacked and purple herons stalk the river along a collapsed tree; trumpeter hornbills can usually be heard if not seen and many of the kingfisher species and paradise flycatchers are often in the area. A party of groundscraper thrushes sometimes scrapes and chatters among the tables and chairs, darting between the feet of visitors.

ROUTE J

This is a large area to explore and a morning, or even a day, should be set aside for it. It is advisable to take with you sufficient food and drink, as there are no refreshment kiosks along the way – unless a visit to the new Afsaal picnic site, situated on the main H3 road between Skukuza and Malelane Gate is planned as a halfway stop.

Leaving camp, you turn right along the main H4-1 road to Skukuza. The first section of this tour is on the smooth luxury of asphalt, but do not be tempted to drive quickly because this section will reward the slower, alert traveller.

On the left, just half a kilometre from Lower Sabie, is the emergency water supply dam, a gathering place of resting marabou storks and feeding waterbirds. A further 4 km on is the Lubyelubye bridge where, on the right, there is a rocky outcrop of cave sandstone of geological interest because it is of the same type of formation found nearly 500 km away in the Golden Gate National Park in the Orange Free State province.

Two kilometres further on the right is a fine specimen of sausage tree. In its shade bushbuck and kudu feed, gathering the fallen fruit and blossoms virtually within touching distance of a quietly parked vehicle.

Left: Elephants can be a hazard when they suddenly appear out of the thick roadside bush along the H4-1 Lower Sabie road near the Nkhuhlu picnic site. Above: A pair of yellowbilled storks paddling in the Mbyandzawu stream on the S28 Nhlowa road.

After another 2 km there is another dry river crossing and of special interest are the vertical earth banks (on the right) peppered with the nesting holes of the beautiful whitefronted bee-eaters. Although these photogenic birds breed in the spring, they often return to their nests to roost at other times of the year.

This entire stretch of road is un-equalled for its bird life: green pigeons fluttering noisily among leaves in search of wild fruits; the scarlet flash of a purplecrested lourie in flight; the haunting cries of the fish eagle; the arrogance of a martial eagle with crest held high and yellow eyes staring back at you imperiously.

The thick riverside forest is home to leopards and you should again keep a wary eye on the dense bush alongside the road for one of the many elephant breeding herds that frequent the area.

A right-hand turn 12 km from Lower

*Above: A buffalo and her calf drink at the north bank of the Sabie river at the low-level bridge just west of Lower Sabie camp. **Opposite page:** A lion seen on a hot day along the S29 Mlondozi road.*

Sabie takes you down towards the river on the 4 km-long S79 Nwatimhiri loop and across a low causeway of the same name. Should you venture along this loop, remember that to resume this described route, you must turn back along the main road towards Lower Sabie for 2 km before turning right onto the Nwatimhiri road (see map on p. 194 for clarification).

Now, turning off to travel south-west, away from the H4-1 on the gravel S21 Nwatimhiri road, after 11 km you will find the well-shaded Nwatimhiri windmills on your right. For photographers this is a "mornings only" place, because any later in the day you will find yourself staring straight into the sun when looking towards the water trough. Elephants frequently call here because

they use the adjacent dry river bed as a path to and from the Sabie river.

Five and a half kilometres further on, there is the new Nhloteni windmill on the left, perfectly positioned for photography right beside the road and the drinking place of many animals, especially white and black rhino. The waterhole is dedicated to the memory of Noel Bourhill (24) who was killed in a road accident in the Park while returning home to Nelspruit, after playing in a cricket match at Skukuza.

Another 2 km brings you to the tall conical Siyalo hill, said to be the home of klipspringers (although I have yet to see one there) where you can park on a small loop road that circles around a shady umbrella thorn tree. Almost 6 km further on you reach the S114 gravel road to the south from Skukuza which offers a number of alternative short drives to explore.

You may now wish to make a short detour to the north (those are the Renoster hills that tower near you, where a fine collection of Bushman paintings have been found) to the peaceful Shirimantanga outcrop where the ashes of the Park's first warden, Col. James Stevenson-Hamilton, and his wife were scattered. The memorial is situated a little to the north of the S112 link road leading to the tarred H3 main road to the west (see map on p. 194 for clarification).

Returning to the S114 and continuing south can be very interesting, especially if you turn off right to follow the river down the S23 Mbyamiti loop road. Immediately before reaching the S113 link road (to the main tarred H3 road) there is a short loop on the right which leads to a very pretty waterhole. Note the many buffalo weaver nests – even in the windmill. Continuing southwards on the S23, passing frequent mudwallows (after rain), there is an excellent waterhole in the river bed, the Mbyamiti weir, situated just before you rejoin the S114. It is ideal for morning photography.

Allow yourself a good three hours for the return drive to Lower Sabie (or Crocodile Bridge), taking either the easterly H5 Randspruit road which follows the tracks of the old Selati railway (traces of which can often be seen clearly on the south side of the road) or the S26 Bume road to the east as far as the S108 link road, where you turn left to join the H5 – obligatory for Lower Sabie residents otherwise they will find themselves having to drive all the way to Crocodile Bridge.

Whichever road you choose, it is worthwhile making a detour along the S102 to the Mpondo dam. Reedbuck can usually be seen in the long grass beneath the dam wall as well as waterbuck, cheetah and sometimes wild dog. Another delightful spot is the little shady loop that overlooks the Gaisenga waterhole near the eastern end of the S26.

The final leg of 4 km along the H5 brings you to the Gomondwane windmill and the junction with the H4-2 asphalt road between Lower Sabie and Crocodile Bridge. Travelling time from this point to either of the camps is a leisurely one hour.

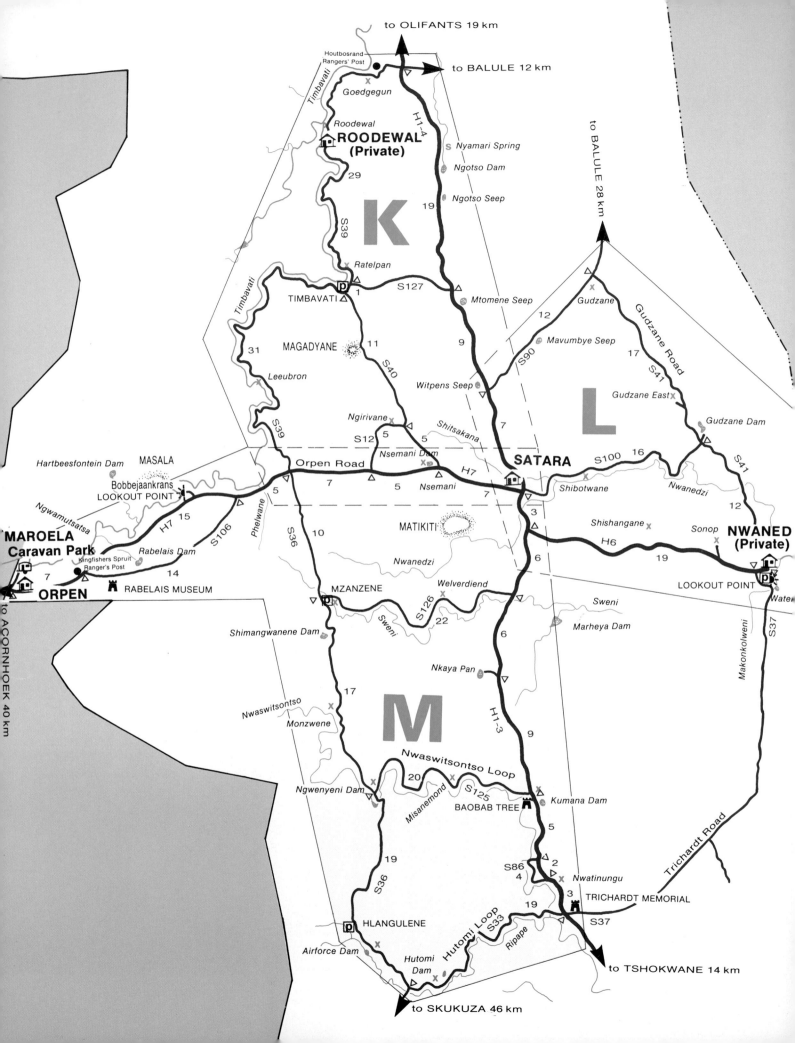

SATARA/ORPEN GATE AREA

Satara camp

Long ago, years before the Kruger National Park was born, an Indian surveyor whose name is not known was working his way through the dusty Eastern Transvaal bush, peering through his theodolite at a distant marker pole and scribbling down readings before moving on to measure another boundary in the wilderness.

He was one of the team of land surveyors working their way through the hot, almost trackless Lowveld dividing a great part of it into ranches and farms to be tamed and settled by the burghers of the young Transvaal Republic.

Completing the borders of one large parcel of virgin land, he scrawled its number in a notebook in his Hindi language – *Satra*, meaning "seventeen". Today it is not a farm but Satara, one of the focal points of a great national park.

Satara is not built on a hillside commanding spectacular views or on the banks of a river where herds of game come down to drink. This camp spreads itself in an area of flat, open grasslands enjoying its own special spaciousness among the knobthorn, leadwood, marula and a variety of *Acacia* trees that surround it.

Second only in size to the Park's "capital" camp, Skukuza, Satara is a firm favourite with many visitors with its six huge circles of huts, wide green lawns and tranquil atmosphere.

Much of its accommodation has been given a face-lift recently with many older huts being demolished and new ones built. All huts and cottages now have bathroom facilities and two are adapted for the handicapped. Guest capacity is 433 visitors in 148 huts, 10 family cottages and three guest houses – and this excludes campers and caravanners, who now have an extended area set aside for them with two new ablution blocks and coin-operated laundries. Day visitors are catered for in their own shaded area to the left of the main gates, complete with barbecue and picnic facilities.

A new, much bigger shop is being planned for Satara and the present shop will be converted into an information and educational centre to be staffed by a full-time officer.

Satara's spaciousness is much appreciated by young families as there is ample space for children to play their evening games – a welcome diversion after being confined to cars for much of the day – with a minimum of disturbance to others.

In the centre of the camp is an outdoor cinema, the best in the Park, with terraced seating and a well-designed projection room. It doubles as an information centre with a display of photographs and a static exhibition.

The entrance to the main building is flanked by two giant marula trees and as well as the usual shop, snack bar, restaurant and reception office, there is a comfortable lounge with fireplace and a private dining-room for VIPs which also serves as a small conference centre. From the long veranda shaded by flame creepers, rich lawns spread to the boundary where, just over the fence, there is a drinking trough favoured by warthogs as a mudwallow.

In the centre of the lawn is a rockery and fountain with a bronze statuette of a caracal sculpted in a hunting pose. A nearby tablet commemorates the many cities, towns and businesses that have contributed funds to the Park over the years.

The most popular tourist roads leading away from Satara seem to begin at the crossroads 2 km to the south of the camp. It is at this point that the early morning convoy of visitors gathers to make their daily decision: to travel south to explore the areas near Tshokwane,

west towards Orpen Gate, or east in the direction of Nwanedzi camp following the river of the same name.

The plains immediately around Satara are often full of animals and even before moving from the crossroads there is an excellent chance of seeing giraffe (often with young), wildebeest, jackal, buffalo, hyena, almost certainly warthog, and lion – all within a stone's throw of the camp.

ROUTE K

Having decided to explore the western sector of this part of the Park, you drive from the crossroads along the tarred H7 Orpen Gate road (note the number of trees near the roadside that have been doomed by ring-barking by elephants). After 7 km you reach the gravel S40 on the right, the direct road to the Timbavati picnic site. Opposite it, to the left,

is a short road to the Nsemani creek, a good spot for fish eagles. Continuing on the tar you pass the Nsemani dam, rich in waterfowl, from where you can take a short detour to the right to a windmill and water trough close to the dam. This is worth a stop as it is often used by elephants, large herds of impala and waterbuck, and photography is easy with a close, unimpeded view of the trough.

Five kilometres further along the main

Left: A yellowbilled hornbill at the Leeubron windmill offers a range of amusing expressions.
Above: A more demure painted snipe in the Gudzane river drift deals with a troublesome itch.

road is the Ngirivane windmill loop to the right (described more fully later) and then after another 7 km you reach a crossroads; a left turn will take you down the long S36 gravel road to the south of the Park via the Mzanzene and Hlangulene picnic sites (also to be described later). A right turn will take you on the winding S39 route that follows the often dry Timbavati river. Take this road north through well-wooded country but be wary of kudu that sometimes rush out onto the road when startled.

There are several short diversions on the S39 to the very edge of the river which are worth exploring. Thereafter you will reach the Leeubron windmill on the left, which I rank as one of the three best waterholes in the Park for photography.

As the name Leeubron ("lion spring") implies it is frequented by lions and at times there is an almost constant parade of impala, kudu, wildebeest and zebra herds. Herons, saddlebilled storks and hamerkops are often present and francolins and hornbills are easily photographed – very tame as they beg for titbits.

Senior ranger Ted Whitfield reports that one of the area's white lions, a lioness, also uses this waterhole.

You should try to be in position here early, not later than 9.30 a.m. in the winter months and much earlier in the summer, when many animals drink and the light is perfect for photography –

especially in the dry months before the spring rains.

Another hour of leisurely driving north and east brings you to a crossroads where you will find the Timbavati picnic site – an ideal stop for breakfast or a barbecue cooked on the ever-ready spread of hot coals. Toilet facilities are at present rather quaint and somewhat primitive but as always in the Park, spotlessly clean. Tables and chairs are set out beneath the spreading branches of a shady nyala tree, and a contest usually ensues above and around you between the flocks of glossy starlings, francolins and hornbills seeking scraps of food.

The birds here are extremely tame and even the ungainly hornbill will accept an offering from the hand – much to the delight of children, and the firm disapproval of Park rangers.

A thorn tree in the corner of the car park, the one furthest from the river, has a deep hole low down in the trunk about 6 cm in diameter. In the summer months, when there is a minimum of disturbance around the picnic site, the hole is used by a pair of nesting hornbills.

From the picnic site, you can return to Satara by travelling east on the S127 to the main H1-4 tarred road or south down the S40 to the tarred H7 Orpen road.

However, with plenty of time in hand, you should continue to follow the Timbavati river northwards on the S39. By taking this route, you soon reach Ratelpan ("honey badger pan") where, true to the waterhole's name, I have sometimes seen honey badgers.

Another 8 km northwards leads to a scattering of beautiful flat-topped *Acacia* trees – the umbrella thorn. Their seed pods are a delicacy to elephants and to get to them the mighty beasts simply push over whole trees. The crumbling remains of some of the victims can be seen scattered about. In order to protect them, Park authorities have circled a few of the trees with carpets of broken stones, which are painful to the soft feet of the marauding elephants, thus discouraging them.

Continuing northwards, the road passes an amazing rockface "wall" on the right which is worth studying through binoculars for plant and small reptile life.

Soon you will catch a glimpse of the luxurious, private Roodewal camp hidden in the trees on the left. One of the first sponsored camps, it consists of three beautifully designed thatched huts and a cottage, which together sleep a maximum of 19 visitors, who have to book it in its entirety. A highlight of this camp is a lookout platform resembling a treehouse, which is built into a huge nyala tree. It hangs over the bank of the Timbavati river above a waterhole – a perfect setting for the city-weary to while away contented hours in communion with unspoilt nature.

An impala ram reflected in the still water of Leeubron waterhole west of Satara.

On the right at this point is the Roodewal windmill, which attracts giraffe, buffalo and lion and is a favourite bathing place for elephant.

Just before this road joins the H1-4 main road to Satara and the north there is a turnoff on the left to the ranger's post of Houtbosrand, sited high on a hill overlooking the Timbavati river. On the right at this junction is the Goedgegun waterhole with a natural-looking artificial trough which is beautifully placed for photography in the afternoon.

Reaching the main north-south road you may go directly southwards back to Satara. But if the day is hot and noon is approaching, I would suggest you return back along the S39 Timbavati road, turn southwards at the picnic site crossroads and follow the direct S40 route to Satara. After 11 km, make a point of stopping at the Ngirivane waterhole which can be seen away from the road to the right. Here two windmills (donated by Mr and Mrs G. G. Kay) supply a drinking trough and small dam, a favourite place for elephants to enjoy a noontime bathe. Visitors may park on three sides of the dam, very close to the giants as they splash about in the heat of the day; a huge sycamore fig conveniently provides shade on the western end of the dam wall. Other than the elephants, the thirsty can include lion, wild dog, giraffe, buffalo, waterbuck, zebra and wildebeest (huge herds at times) and, of course, the usual streams of impala. I saw my first-ever black mamba here and at least two large water leguaans are resident. Waterbirds are usually in abundance. In my view, this spot ranks first in the entire Park for game viewing and photography.

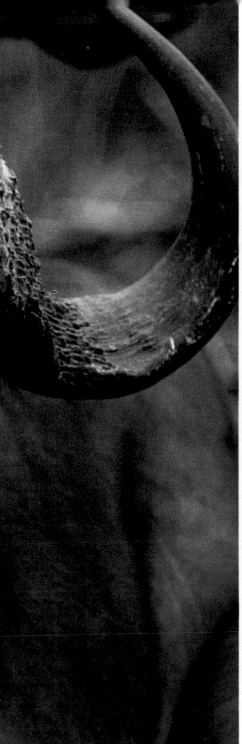

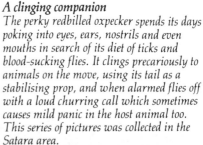

Continuing south, the S40 meets the tarred H7 Orpen Gate road and the journey east back to Satara along it can be excellent in the evening light. If the gate-closing time permits, giraffe will often be in position to be photographed against a spectacular sunset.

ROUTE L

Runner-up in popularity for a single road in the Park is almost certainly the gravel S100 Nwanedzi river route which leads to the area east of Satara. (The busiest in the Park is the Skukuza-Lower Sabie road.) The meaning of Nwanedzi (Shangaan) is "reflections of the moon".

Leaving the crossroads you will soon see the Nwanedzi river on the right. With the exception of the last 2 km, the road follows the river closely as it winds through open country with rich riverine vegetation along its banks.

The entire 16 km of this river road is a paradise for game. Leopard sightings are very common, although mainly in the early morning or late evening. I once watched as a cheetah, which had killed an impala right at the edge of the road, was frightened off her kill by the large gathering of eager tourist cars. Minutes later a big leopard dashed boldly through the jam of cars, grabbed the carcass, and then held it aloft for a moment for all to see before rushing back across the road to feed in the safety of the long river reeds.

On another occasion motorists were torn between waiting at a cheetah kill or moving on to a lion kill 6 km away – another example of the activity to be seen along this road.

The flesh eaters

Top: This impala being held aloft by a leopard was in fact killed by a cheetah which abandoned its meal when it was disturbed by chattering tourists. The leopard, having scented the kill, dashed through the traffic to snatch up the ready meal and make off with it. Above: Vultures hastily tear at a carcass before hyenas and jackals arrive.

Right and far right: Another cheetah and its kill, a young kudu, and a close-up of the cat's face as it feeds.

Above: A saddlebilled stork hunts silently along the edges of the Ngirivane waterhole to be rewarded with a fine catch – a tilapia (right) which it proceeded to swallow whole.

The open country on this route makes for easy photography and the game is often very "obliging". Waterbuck, zebra, wildebeest and impala are common, and are sometimes seen feeding together with giraffe in the background. On windy days, waterbuck lie down to escape the chill, unperturbed by traffic passing close by. Bushbuck browse beneath the giant fig trees on the far bank of the river, moving with characteristic quivering, alert caution as they feed quietly.

Secretary birds stride out across the veld, pausing to stomp a foot to flush a reluctant grasshopper, and the lilac-breasted rollers flash their amazing blues and purples – a beauty somehow spoilt by their harsh, screeching call.

The S100 passes the prominent Shibotwane windmill on the right (erected thanks to the generosity of Vic and Linde Venner), and then, towards the end of the road, there are two small river loops. The first of these offers shade where visitors can sit for a while to watch small, basking crocodiles, motionless grey herons and giant and pied kingfishers feeding.

Reaching the T-junction, you are confronted with the old dilemma: north on the S41 to the twin Gudzane windmills, or south along the S41 to Nwanedzi camp. As always on any road in the Park, it is a matter of luck for the big-game seeker. But for the birdlover, the decision is easier – they should travel south. However, they too should divert for a moment to the north for a short distance to the concrete causeway that crosses the Gudzane river where sandpipers, crakes and a pair of unusually bold painted snipes grub among the

reeds without a care for the camera.

The Gudzane dam is also rewarding: fish eagles are common and herons and hamerkops stand in line along the concrete dam wall seizing small fish that dart in the shallow overflow. Colourful crested barbets and Burchell's glossy starlings are very tame and big tawny plated lizards hunt for insects among the rocks or lie basking in the sun.

Travelling south, a careful lookout should be kept towards the distant Nwanedzi river on the right: cheetah are quite common and large herds of waterbuck are resident in the area.

Before reaching Nwanedzi camp and picnic site, the road crosses two river causeways. The first is the narrow and heavily reeded Gudzane river which flows from the dam. The second crosses the Nwanedzi river and offers a small

loop to the right which overlooks a long pool. Bird life there is excellent. Open-billed storks dig for freshwater mussels, and herons, crakes and snipes are plentiful. Water leguaans can often be seen snoozing, beautifully camouflaged, along low boughs of the river trees.

Nwanedzi picnic site next to the camp, a mere 3 km from the Mozambique border, is delightful with its open-sided thatched shelter overlooking the Sweni river flowing through a gorge below. A path climbs to a tiered lookout hut with lovely views upstream across the Park and downstream towards Mozambique. On the back wall of this little building is a mural showing 20 varieties of waterbirds painted by Chris Ebersohn. The lookout is dedicated to the memory of Johan Schmidt (20) and John O'Donnell (29), who died in a motor accident; the

Above: An impala ram stands guard. **Top and right:** *Younger members of the herd rest in the December heat. These pictures were taken north of Satara on the S39 Timbavati road.*

Top left and right: *A steenbok ewe sticks out its tongue while its mate looks about warily. Lambs are seldom seen because the mother steenbok leaves its offspring hidden by day, returning at night to suckle it. The pair were photographed on the S36 north of the Mzanzene picnic site.*
Left: *An elephant has a scratch at the Ngirivane waterhole while two others* **(above)** *wrestle in the muddied water.*

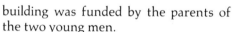

Two of the more common residents around Satara are giraffe (left) and impala (above). Here two rams practise the ritual of territorial dominance.

building was funded by the parents of the two young men.

In the spring after the first rains wild flowers of many colours bloom on this rocky hill, giving photographers the chance to get down among them – and to find picture-worthy insects too.

Leaving the picnic site you will notice to your right the entrance gate to the private 15-bed Nwanedzi camp, a favourite because of its delightful riverside setting. Before returning to Satara, I would suggest a final stop at the Sweni river causeway, to be found a short distance along the gravel S37 Trichardt road which turns off south towards Skukuza from the main tarred Nwanedzi road just over 1 km from the picnic site.

For about 100 m above the causeway, the river spreads out to form a large pool, home to an abundant variety of bird life. Whitefaced whistling ducks call to each other as they fly in formation, swinging up and down above the water before settling to feed. Their presence is sometimes resented by nesting moorhens, which paddle across to bully the gentle ducks into moving further up-

river. Woollynecked and black storks are often seen here, as are African jacanas, which strut across lily pads searching for food. Fairly large crocodiles bask in the sun on the islands. There is a small loop road just upstream from the causeway, which is an ideal lookout spot.

The return to Satara can be either directly west on the 20 km H6 tarred road (a short diversion down to the Sonop windmill and a pause at the Shisangane waterhole may be worthwhile) along which sightings of zebra herds, wildebeest and, quite often, lion, which are common here, are frequent. The alternative, if time will allow, is to return the way you came – back north along the S41 – but to now continue beyond the Gudzane dam towards Ngotsamond camp (Balule) for a variety of game, including two massive eland bulls which occasionally visit the Gudzane windmill intersection. You can then return to camp via the S90 which joins the main tarred road north of Satara. This detour is 48 km long, so to do it justice three hours should be allowed to reach camp. A shorter trip which is usually profitable in the late afternoon is to follow the Nwanedzi river along the S100 again, for another look along this exceptional road.

ROUTE M

Being the second largest camp in the Park and often full to capacity, Satara can be frustrating for those who wish to get away from the crowd and find solitude.

One's best bet in this case is to explore the areas in the far west towards Orpen Gate and south-west on the gravel S36 in the region of the Mzanzene and Hlangulene picnic sites, areas which are not very popular with visitors.

It is advisable to leave Satara early, carrying a picnic lunch, and to travel along the tarred road towards Orpen Gate for 20 km until you reach the Timbavati crossroads where the description of this route begins.

Continuing west from the crossroads, the Timbavati river may be glimpsed on the right after 2 km and for the next 13 km the road follows it closely. You soon cross the Phelwane river bridge and elephants can sometimes be seen feeding on the palms and reeds in the dry bed below.

A kilometre further on is a lookout point known as Bobbejaankrans (Baboon cliff) with a sheer drop to the river below. A pair of klipspringers have made their precarious home along the rocky ledges below, but it requires a careful search over the discarded beer tins, cigarette boxes, plastic bags and sweet wrappings to see them.

Moving on past the S106 Rabelais

dam turnoff to the left, you reach another river view a short distance off the road, where you can park in the shade of a fine jackal-berry tree. Careful scrutiny of the far bank of the Timbavati river, preferably through binoculars, can reveal a variety of game: steenbok, waterbuck, impala and especially kudu, including a number of fine bulls.

After passing another two river lookout points, the road travels through beautiful parklike country and then crosses the Ngwamutsatsa river before reaching the western turnoff to Rabelais dam. The road then traverses rather flat, featureless country for the last 7 km to the delightful little camp of Orpen Gate.

Just before the camp is a gravel road on the right which takes you 3 km to the Maroela caravan park. Shaded by a range of Lowveld trees, a giant jackal-berry, Transvaal saffron and a broad-pod false-thorn among others, this little camp is situated on the bank of the Timbavati right up against the western fence of the Park and enjoys a delightful bushveld atmosphere.

Picnickers should note that this little park is strictly reserved for the resident campers and caravanners, so a morning tea break should be taken at Orpen Gate camp close by, before driving back the way you came.

Travelling back towards Satara, you should now take the 15 km gravel detour south of the main road past the Rabelais dam. One kilometre down this road on the right is an intriguing little white rondavel. It is all that remains of Rabelais camp, which was closed many years ago when the Park boundary was moved to its present position further to the west. It will soon be converted into a mini-museum in memory of Mr J. H. Orpen, a surveyor and member of the National Parks Board, and his wife Eileen who donated large tracts of land to the Park and also, at their own expense, cleared much of the Park's western boundary line of bush.

The dam itself is a large "rambling" stretch of water with a network of short detours winding along the edges of the shallows, where hundreds of waterbirds pick through the water plants and mud wallows for food. Buffalo and warthog

Right: A waterbuck cow pulls a face while chewing the cud.

219

Two of the more colourful bird species to be seen around Satara are the lilac-breasted roller (above) and the crested barbet (left). Barbets often become very tame. This particular one frequents the Gudzane dam on the S41.

are often there, lolling in the mire during the heat of the day, and crocodiles sun themselves on the banks of the long earth wall. Saddlebilled storks, hamerkops and herons hunt small fish while fish eagles eye the water for larger ones. There is a constant stream of game arriving and departing, especially before mid-morning.

Rejoining the main H7 tarred road, you continue towards Satara until you reach the Timbavati crossroads again. At this point, turn south on the S36 gravel road towards Skukuza for the second part of the day's journey, allowing at least five hours to complete the full 102 km drive back to Satara at a leisurely pace.

The first leg to the Mzanzene picnic site can be rather dull and uneventful, except for a windmill 3 km from the start on the left which attracts game, including a few fine sable antelope. I once spent three fascinating hours there following a pair of courting steenbok for more than 2 km along this stretch of road.

After 10 km a link road turns off to the east. It is the newly opened S126 which follows the Sweni creek for 22 km to the H1-3 main road – an alternative route back to Satara. It is an upgraded firebreak road, and is excellent for lions. The Welverdiend windmill, 10 km from the S36, is a fine viewpoint for zebra, wildebeest, giraffe, large buffalo herds and lion.

But if you choose to continue south on the S36, after 1,5 km you reach the Mzanzene picnic site just off the road on the left, a good place to stop for lunch. The site is well wooded and shady and, in the heat of the day, a much better picnic stop than the Hlangulene site 21 km further on, which has facilities such as fires, hot water and toilets but is very exposed to the hot sun and might thus be very uncomfortable.

Continuing southwards, you soon reach the Shimangwanene dam where streams of impala and kudu use the earth wall as a crossing place, but the scene is rather too distant for good photography.

Another 7 km brings you to the two windmills at Monzwene where a short loop road gives excellent views of the drinking trough. A criss-cross of game paths brings a variety of animals, including unusually large herds of giraffe, to drink.

You soon meet the Nwaswitsontso river which runs alongside the road on the right for some distance, including a short stretch where huge riverine trees cast refreshing shade over the road.

Just before the Ngwenyeni dam on the left is yet another link road to the H1-3 main tarred road, which it joins just north of the well-known Baobab Tree landmark. Numbered the S125, this splendid drive winds along the north bank of the Nwaswitsontso stream and a future highlight for this road is a planned lookout point at the Misane-mond twin windmills about 8 km from the S36 turnoff. Of special note along this road are the three small river loops, each circling a magnificent shade tree. The first, travelling eastwards, is a nyala tree and the other two are sausage trees. A keen eye will notice signs of branches having been cut from the last of these and also an old spike, probably a lamp hook, driven into the trunk. This was the first site selected for the Park's by-products depot (for culling) which was abandoned because of uncertain water supplies.

However, should you stay on the S36 and continue southwards, you will come to the Hlangulene picnic site and then, after 2 km, the Lugmag (Air Force) dam where antelope graze on the sweet green grass along the water's edge. At times, a remarkable line-up of terrapins sun themselves along the base of the dam's earth wall.

Five kilometres further on you turn left onto the S33, which follows the winding Ripape river for 19 km to the tarred H1-3 road to the north and back to Satara.

While travelling on the H1-3, if you have time, it may be worthwhile detouring along the short S86 Nwaswitsontso loop which is a favourite haunt of white rhino. Also interesting is the Baobab Tree loop opposite the beautiful parklike setting around the Kumana dam where very large herds of game sometimes gather.

Orpen Gate camp

This pleasant little camp is situated about one third of the way up the Park's western border and is the quickest entry gate for visitors wishing to go direct to Satara from the north-western Transvaal.

It is a pretty camp with rectangular thatched huts, two with three beds and ten with two beds, and a common ablution block and kitchen but no restaurant, although there is a small shop which is also the office. The camp grounds are decorated with flower-filled rockeries and shaded by marula, *Acacia* and other indigenous trees.

There is no campsite at Orpen but caravanners can book in there to stay at the Maroela caravan park nearby.

The countryside surrounding Orpen is typical flat, open bushveld favoured by plains animals such as zebra, wildebeest and impala and also by elephant.

Below, left: A boomslang leaves its tree hole near the Satara camp gates. *Below, right:* A tawny plated lizard at Gudzane dam. *Right:* A brownhooded kingfisher at the Nwanedzi river drift on the S41 east of the camp.

Two elephants bulls in semi-serious combat near the Ngirivane windmill west of Satara.

OLIFANTS/LETABA AREA

Olifants camp

Olifants is the camp with perhaps the most spectacular views. It is sited high above the Olifants river with splendid scenery southwards and eastwards across the winding waterway, thickly bushed plains and mountains dimmed by distance.

It is said that for the visitor who seeks a complete rest, there is no need to leave the camp. Armed with only a good pair of binoculars, a great variety of game can be observed by simply scanning the banks of the river below.

To blend with the environment, most of the camp's buildings are painted dark green, with the base of the huts either of natural stone or painted a similar grey.

Of interest are the two pairs of elephant tusks projecting from the stone walls on either side of the entrance gates, which periodically visitors rush to report missing or stolen. Although extremely realistic, the tusks are artificial and in fact it is passing elephants who are the culprits: they either rip them from the walls and carry them away never to be seen again, or crush them unceremoniously underfoot.

For people interested in botany, the camp is blessed with a rich assortment of indigenous trees, flowering shrubs and aloes. A selection of Lowveld plants is also on sale at the camp shop.

The camp was opened on 3 June 1960, by the then Administrator of the Transvaal, Mr F. H. Odendaal, but 12 years later on 24 February 1972, the entire main building, which includes the reception area, restaurant and shop, was burnt to the ground. The camp was closed for 16 months for rebuilding and the original thatched roof was replaced with red shingle tiles.

As with most camps, Olifants is excellent for birdwatching. A fairly regular visitor is the redwinged starling which arrives in winter to sup nectar from the aloe blossoms. In the spring sunbirds gather to feed on the mass of crimson flame creeper flowers and a colony of nesting weavers offers comical close-ups for photographers near the entrance to the main building. Visitors who relish a barbecue are advised to keep a watchful eye on their meat because many a length of prime sausage has been whisked away by those summer migrants, the yellow-billed kites, that swoop in with incredible swiftness when backs are turned for a moment.

Often overlooked by visitors to this camp is the small but very useful information centre opposite the reception area. It is dominated by a huge mural of a prehistoric mammoth painted in 1977 by National Parks Board artist Chris Ebersohn. Photographs and graphics illustrate the story of the elephant, from the make-up of the herd to problems related to poaching. This building is to be demolished in the future to make way for a larger camp shop, and a new, upgraded information centre will be created in the present shop in the main building.

Olifants could be described as the lazy man's camp. Not only can you absorb the full flavour of the wild by armchair viewing of the great variety of animals and birds that frequent the river below, but you can wander the camp itself, exploring the plants and rockeries for the other fascinating world of beetles and bugs, rodents and reptiles.

Whether you decide to remain in camp or to take to the road, if the weather is clear the reward for the early riser is a magnificent display of sunrise colours: indigo to bright red in the sky, the distant hills deep purple, a ground mist of deep blue and as the sun climbs higher, the clouds lighting up in tones of pink, gold and yellow – and, added to this, a morning concert of birdsong.

ROUTE N

The entire road network around Olifants can be fully covered in a four to five hour morning, even driving slowly and stopping often, and one can return to camp in time for lunch. The afternoon can then be spent re-exploring specific areas that may have taken your fancy. The following is a suggested day's outing.

Enjoy an early breakfast in camp and load up with refreshments for the morning's trip (there are no picnic sites with facilities *en route*), then head out on the main H8 tarred road, keeping a sharp lookout for the herds of kudu that often cross over just beyond the camp gates.

I would suggest you ignore (for the moment) the first S93 turning right but after 4 km turn south off the tar down the rough gravel S92 road towards the Olifants river. Game paths are numerous in this area, bringing large herds of buffalo and impala, kudu and giraffe down to drink. It is interesting that only

in very recent years have giraffe moved north of the Olifants river in any numbers – previously they were a rarity.

This seems to be a favourite hunting ground of bateleur eagles and I have often seen them along this road perched in trees at the roadside quite contemptuous of passing traffic and thus easy to photograph.

Before reaching the low-level bridge which crosses the river on the left, there are several short loop roads overlooking the river and from these many animals can be observed feeding on the riverside vegetation.

At the intersection on the left where the S90 leads on to the bridge, you should look carefully along the wide, sandy river bank; it is often frequented by big herds of impala feeding on the new grass shoots and old buffalo bulls that love to rest on the warm sand, quietly chewing the cud.

Possibly because of its combination of soft sand along the flat river banks and then hard, very stony ground rising

steeply into the hills – terrain that will slow a fleeing animal – this place has known several lion kills recently. I have photographed lions feeding on a giraffe and vultures picking at the remains of a buffalo and have heard recent reports of at least two other kills here.

The crossing over the old bridge should be made very slowly with binoculars and cameras at the ready. If there is no oncoming traffic on the narrow causeway, stop for a while to scan the sandbanks, pools and rocks for the variety of river life. In the summer months European and redbreasted swallows line the concrete edging, giant and pied kingfishers hover and splash, goliath herons stand sentinel in the shallows of the fast-flowing water, and wattled plovers call noisily to one another. There are snipes, crakes, jacanas and crocodiles too. I once watched fascinated as a large leguaan shepherded small fish into the corner of a shallow pool and, by using its tail and body to form a dam, trapped the minnows by carefully and methodically

Left: Although the bateleur is a widespread resident, this colourful eagle seems to be more accessible to photographers along the Olifants river. *Above:* A pair of klipspringers photographed on the S91 just beyond the Olifants low-level bridge turn-off. *Right:* The female in close-up.

closing the "net" and then picked off each morsel at leisure.

Travelling south on the S90, you change into a low gear to climb the steep bank out of the Olifants river and then immediately on the left is a short side road that swings past the tiny, delightfully informal Balule camp.

Opposite the turnoff to Balule, on the right just upstream from the bridge, is one of the largest and most exciting Iron Age archaeological sites yet to be found in the Park. When the staff and students of the University of Pretoria's archaeological department have finished their investigation there, the diggings will become a feature of the new Ngotsamond caravan camp. Park engineers report that when Ngotsamond camp is opened, Balule will be closed and probably become another Park museum.

Four kilometres beyond the river is the S89 turnoff on the right, but for the moment I would suggest that you continue southwards past the Hlahleni river where the road forms a dam for a beautiful stretch of water, full of lilypads and shaded by huge trees – most picturesque with the reflections in the still water.

Finally, after 8 km, you reach the barren Bangu windmill – an excellent spot to find herds of giraffe, zebra, and kudu before retracing your tracks back to the S89 intersection where you should turn west. Drive on for 5 km before crossing the Ngotso river immediately below a small dam, a favourite roost for squadrons of marabou storks, and then another 3 km to the tarred H1-4 road to the north.

Turning north onto the tarred H1-4,

you will soon cross the high-level bridge over the Olifants river (please do not be tempted to get out of the car, as so many people do) and a short distance further you will find on your left the short, tarred loop to the Nwamanzi lookout point high on a hill overlooking the river.

Signposts there announce that you can leave the vehicle "at your own risk" to stretch your legs with a stroll along the edge of the hilltop. Binoculars will magnify the distant mopane forests to the west and the wide river below which is inhabited by very big crocodiles, snoozing hippos, waterbuck, elephant and other wildlife.

This is an ideal spot to enjoy a picnic tea, but extreme care should be taken if

there are baboons about. The resident troops are now bold enough to snatch food at random and any move to repulse these marauders can result in serious injury. If baboons do arrive, it is wiser to get back into your vehicle and to conceal all food until they have moved off.

Continuing your trip northwards, you soon pass the main tarred H8 access road to Olifants camp turning away to the right and then suddenly, spread out before you, is probably the most beautiful stretch of riverine "parkland" in the entire Park. Impala herds feed quietly on the short green grasses against a backdrop of giant wild fig trees that line the swiftly flowing river. In single file waterbuck splash through the shallows, and elephants gather with their trunks

Left: A buff-crested korhaan shrieks its piercing territorial call. Above: A black-bellied korhaan in typical posture while issuing its mating call.

and scythe with their tusks the succulent river reeds. Two gravel loop roads take you right down to the water's edge where you can sit for a while in the shade, savouring one of the most typical, lovely scenes of natural Africa.

Further upstream the river swings away to the west and you should continue north for another 10 km, past the cliffs of the Shamiriri hill on the right – a stony outcrop rich in aloes and klipspringers – before turning east on the gravel S46 towards the Letaba river. At this point a sharp watch may reveal a fine herd of sable antelope which ranges this area.

On reaching the intersection with the S93, you should turn right on the S93 to follow the Letaba river downstream. This road winds and dips through rugged countryside back to your starting point, Olifants camp. There are several loop roads turning down to the river and a number of roadside lookout points which should be explored as this area offers a good chance of spotting leopard and also nyala. Care should be taken not to get too close to elephants, especially the breeding herds that roam the thick mopane between the Olifants and Letaba rivers.

One kilometre before arriving at the S93-S44 fork 9 km after the Letaba turnoff, there is a magnificent baobab tree on the right (note the damage to the bark caused by elephants).

At the fork you can either take the more direct 7 km route back to camp by continuing on the S93 or you can turn left onto the S44 for 15 km, which follows the Letaba river before swinging west to follow the Olifants back to camp.

After lunch, you may wish to laze in the shade for an hour or two to while away the midday heat before setting off again by car.

Unless there is a specific reason to backtrack – such as the remains of an impala left in the fork of a tree where you could wait for the return of a leopard – I would suggest that you again travel down the S92 along the banks of the Olifants river to the low-level bridge below Ngotsamond and Balule.

However, after scouting the river loops and the area around the bridge (I have always found some attraction there), I would suggest you make a point of now continuing straight on past the low-level bridge turn-off to Ngotsamond and explore the S91 up to the tarred H1-5 main road, a stretch which you would have bypassed on the morning trip. Two kilometres after the bridge turnoff is a big rocky hill on the right which is home to several klipspringers that sometimes feed in the thickets at the roadside, but it takes a sharp eye to spot them.

On reaching the tar, you should call again briefly at the Nwamanzi lookout point before moving on to the "parkland" just beyond the main turnoff to Olifants camp. The afternoon colours are superb, as is the range of game. Zebra and wildebeest clop across the road to frolic and roll in the dust, warthogs trot swiftly towards the river with tails erect like radio antennae and elephants splash in the water.

A good way to end the day is to head back to camp on the H8 main road until you reach the S93. Turn north on it, and then immediately swing eastwards (to the right) and follow the Olifants river downstream.

After 2 km the road, the S44, passes a pretty grove of fever trees; at 5 km it has a short loop to the right which overlooks the river and from which hippo and waterbirds can be seen; and at 7 km there is a high viewpoint where you may leave your car. The view is spectacular. Straight ahead to the south the gently undulating bushveld is rich in late afternoon shades of colour and to the west the Olifants camp can be seen nestling on its bluff above the river among trees. To the east, cliffs rise vertically from the river and beyond

A pair of wild dog puppies at play. Packs of these big-eared carnivores seem to favour the sandy banks of the Olifants and Letaba rivers to enjoy their games.

them in the distance the low Lebombo mountains lie blue in the evening light. Right below the viewpoint hippo, waterbirds and sometimes klipspringers can be seen feeding along the bank.

If time permits you may continue on the 20 km loop (allowing yourself at least 45 minutes to return to camp) which takes you north-west beside the reed-fringed, broad bed of the Letaba river.

Two kilometres before you reach the S93 intersection a riverside loop swings down close to the river's edge. Elephants feed among the reeds, nyala can sometimes be seen and saddlebilled storks are numerous – I once counted 10 flying majestically upstream.

Balule camp

A short distance due south of Olifants camp, just across the low-level bridge over the Olifants river, is Balule camp.

Originally known as Olifants River Pont and described on the official maps as a "caravan camp", this five-hut camp (there are six but the extra one is for staff) is close to the south bank of the river, but does not actually overlook it. There is space for no more than ten

caravans. It is much favoured by traditionalists who enjoy simple bush life, a communal campfire and a chance to sit peacefully listening to the surrounding sounds of the night, almost as part of the veld.

The quaint huts are typical of the Park's original rondavels and have no windows. Ventilation is through a 20 cm gap between the top of the wall and thatched roof; if the night is warm, occupants simply leave the door open. In one corner of the camp is a big baobab tree – a landmark for some distance.

Night life includes regular visits along the fence from a pair of civets and the inevitable patrolling hyenas.

There is no electricity at Balule (visitors are supplied with candles or paraffin lanterns) and therefore no refrigerators, air-conditioners or electric fans. However, there is a paraffin-powered deep-freeze for communal use. Ablution facilities are simple, clean and quite adequate.

The future of Balule is presently in doubt. The Parks Board is considering closing it when the nearby Ngotsamond caravan camp is opened. Its name is already excluded from official tourist maps.

229

Letaba camp

Letaba is the delightful "halfway house" that welcomes weary motorists travelling the full length of the Kruger National Park and a favourite weekend retreat for residents of the busy mining town of Phalaborwa that borders the Park 50 km to the west.

The fourth largest in the Park, the camp is beautifully sited high on a bank above a great sweeping bend in the Letaba river, with views across 300 m of sand and water, mopane forest and the distant blue Lebombo foothills to the east.

Guests enjoying a meal or drink on the patio of the restaurant, or sitting in the shade along the perimeter fence, are often treated to close-up views of elephant, buffalo, waterbuck and the shy bushbuck feeding on the riverine undergrowth directly below. Baboons and vervet monkeys are frequent visitors, sometimes becoming a menace when people feed them. As a deterrent an electrified fence has been added to the boundary, with a kick strong enough to startle any raiding primate without hurting it. The fence has been strategically placed out of the reach of people.

The camp is renowned for its many fine trees. Other than the usual mopane, which are predominant in the area, there are outstanding examples of the lala palm (especially at the far southern end of the camp), the magnificent deep green Natal mahogany, weeping boer-bean, fever-berry, and a big apple-leaf tree with its purple flowers forming a colourful carpet on the lawns. There are several enormous wild figs with great buttress roots twisting above ground liberally carved with the hearts and initials of uncaring visitors.

At the entrance to the restaurant stands probably the finest Natal mahogany in the Park, its far-reaching branches giving shade to several cars parked beneath it. In the spring the approach to the restaurant is ablaze with the brilliant red of flame creepers, which almost completely envelop their long-dead, gnarled tree host.

Letaba has the most pleasant restaurant-lounge complex in the Park. Great glass doors, high thatched ceilings with slowly revolving fans, a rock pool with a trickling fountain built along the inside of one wall and the deep veranda are coolly inviting, particularly in the height of summer.

The present shop and reception area are inadequate to cope with the increasing number of visitors to Letaba, so a new administration building is being planned.

Soon after sundown, when the gates have been closed for the night, hyenas begin their nightly patrol around the camp's fence in search of leftover barbecue bones. Local comedians are the flocks of helmeted guineafowl lining up like aeroplanes waiting for take-off, before noisily flapping up into a sycamore fig to roost for the night.

Bird life in the camp is prolific and varied. The rare mourning dove is always present, as are the conspicuous and rowdy blackheaded orioles. Brown-headed parrots can be seen in the wild fig trees along the river and can be approached to within a few metres.

The river banks near the camp also attract large flocks of Egyptian geese, especially in the early spring. In one sighting, 183 were counted nearby and 165 were seen in another flock just below the boundary fence.

In the local dialect Letaba means "river of sand", understandable when one sees the great expanse of river sand, in which the water is a mere trickle for most of the year.

Enthusiasts who rise with the birds and are away through the gates as they open will agree that an equally fitting name for this camp is eTimhisini – "place of the hyena", in the language of the Shangaan people.

ROUTE O

Leave camp early and take with you a packed breakfast to enjoy while you explore the many highways and byways north of Letaba. Head out to the main Phalaborwa road intersection just 1 km to the west of the camp gates. It is at this junction (and also a few hundred metres down the Phalaborwa road) that you should meet the clans of hyenas. With their strange lolloping gait, whole families emerge from dens in the culverts beneath the tar to stare with bulging thyroid eyes at the passing traffic before slinking away into the deep bush.

Letaba-based rangers report that a small herd of eland also sometimes frequents these crossroads, but the slightest build-up of traffic makes them trot away to safety in the mopane.

Turn north up the main H1-6 road towards Shingwedzi and note that for the first 5 km there is a 40 km/h speed restriction. The Park helicopter is often parked on the right waiting to fly off on its daily business.

To the left of the road giraffe feed off the tips of the thick mopane scrub and on the right big trees conceal the steep banks down to the wide sands of the Letaba river. Several convenient riverside loops offer good viewing of elephant and waterbuck that frequently pad their way across the wide sands to drink from the main channel.

After 3 km the main road swings away from the river before crossing the high-level bridge, but you can continue along the river bank by following the 4 km-long S95 loop that turns off the main road to the right.

Immediately before the bridge you should take the S47 Mingerhout dam road on the left, to follow the Letaba river westwards for 13 km. This is one of the most exquisite river roads in the Park for game-viewing – and photography. But a word of warning: be extremely cautious after a rainstorm; approximately 3 km from the turnoff the road roller-coasters steeply down into the dry Nwanedzi riverbed (yes, there are two Nwanedzi rivers – the other is near Satara) and up the other side. I have known motorists to be stranded in the middle, unable to drive up either bank because of the slippery conditions and having to be rescued by a four-wheel-drive vehicle – but not before

A young hyena stares boldly at the camera near the gate of Letaba camp, while two others enjoy an early morning romp. Hyenas can usually be seen just beyond the camp fence at night.

causing heavy damage to the road in their futile attempts to get out.

The roadside forest is dense for much of the way, so constant attention is needed to spot game along this stretch. At times even the river close by is hidden from sight despite the 17 river loops and view-roads within the 13 km leading to the Mingerhout dam. However, glimpses of it may reveal herds of buffalo trekking over the sands. Elephants swing silently across the road,

appearing suddenly from thickets, and there are usually kudu, waterbuck, giraffe and bushbuck to be seen. Scan the forks of the very big mopane and apple-leaf trees, predominant in the area, for signs of a leopard – such as the remains of a kill hanging up there. Sure-footed klipspringers live in the granite outcrops situated on both sides of the road 2 km before the Mingerhout dam.

There is an excellent viewpoint overlooking the dam wall; visitors should not be tempted to get out of their cars. Use binoculars to scrutinise the sandbanks edging the long pool below the concrete wall for crocodiles: a senior ranger in the Park, Ben Lamprecht, reported 57 in a single count. The old causeway, which can still be seen below the dam, once linked the S47 with the S48 across the river further to the north, which gave tourists an alternative road northwards to the Tsende river and beyond (see map on p. 224). It is now unusable because of flood damage. This is why the S47 is still known as the Tsende loop, although there is now no connection with the river of the same name further north.

Motorists wishing to return directly to Letaba should swing southwards from the dam and travel along the S47 via the Nwanedzi river drift and the S131 back to camp.

Here I would suggest that you ignore the recently opened extension which continues to follow the Letaba river westwards. This new road is extremely rough and stony in places, and except for two attractive riverside areas towards the halfway point, it has very little to offer.

It is generally more fruitful to return along the S47, following the river in an easterly direction back to the H1-6 main road, because of the ever-changing vista of wildlife to be seen along this route.

On reaching the main H1-6 tarred road you should now turn left to go north, cross over the Letaba river bridge and then within 1 km turn right off the tar onto the S62 gravel road, which eventually takes you to the high lookout point above the Engelhard dam.

From the S62 there are three turnoffs down to the Letaba river: the first, after 1 km, leads to a point where you can look upstream to the bridge, the second 3 km further on takes you to the edge of the Engelhard dam where hippos, crocodiles, and nesting grey herons can be

The view from the Olifants river lookout point, 25 km south of Letaba, near the H8 turn-off to Olifants camp.

seen and, after another 2 km, the third turns off to run beside the Makhadzi river which forms a deep inlet into the dam. This last loop is splendid. Buffaloes splash in the shallows of the river as they feed on the water grasses, hippos honk and snort, kingfishers hover and dive, weavers chatter, herons stalk, and elephants rumble – an excellent spot to sit quietly in the shade for a while. At the end of this road is the grave of Mrs Ledeboer, wife of the first ranger at Letaba.

Back on the S62, the road swings up to a lookout point giving a panoramic view over the dam wall towards Letaba camp. Of interest is the fish "ladder" at the concrete spillway which enables fish, heading upstream to spawn, to "climb" up through the tumbling water from the river into the dam above.

Returning to the main road and heading north again on the H1-6, you may find ostrich and also a well-known bull elephant that displays a single mighty curved tusk (the other broke off long ago) who ranges in the area just north of the Letaba river bridge.

The main road now runs through rather monotonous, flat mopane veld, past the Malopanyana and Middelvlei windmills (excellent for tsessebe) and the Shawane waterhole (for buffalo, zebra, elephants) to the Mooiplaats picnic site. On the way you can divert along the Tsende loop road (S97 into S48), though it tends to be disappointing because it is difficult to see into the river from the high, grassy banks. Another diversion to try is the S49 further north, which travels past the Mooiplaas windmill to the right of the main road.

The Mooiplaas picnic site is a delightful spot. Blackened kettles with water for tea or coffee simmer over an open

fire, there are clean restrooms and two huge open-sided rondavels offer cool, breezy shade for picnickers. One shelter overlooks a deep pool on the Tsende river where animals gather to drink.

Before you return to Letaba for lunch, I would suggest a call at the Pioneer dam a little to the north of Mooiplaas, which is the site of the new medium-sized Pioneer camp to be completed in the future.

The dam is only 2 km off the main road and offers superb bird life, including occasional flamingoes. The big untidy nests in the fine baobab tree at the roadside just before the dam are those of the redbilled buffalo weavers, which live in colonies.

ROUTE P

An excellent full morning's drive of between four and five hours is to follow the main tarred H-9 which travels due west of Letaba to the Sable dam and the

Above: Mourning doves at Letaba can be very tame and easily photographed. Right: A magnificent kudu bull at rest along the S47 river road north of the camp.

Masorini museum near the Phalaborwa Gate.

A word of warning however, gained from bitter experience, is that this road should be avoided for game viewing on peak season travelling days such as public holidays, long weekends and the first and last days of school holidays, when much of the game sensibly moves away from the roadside to escape motorists who thunder along completely ignoring the 50 km/h speed limit.

Drive west from Letaba, crossing the main north-south road before continuing for less than 3 km to the first dirt road intersection. To the right is the start of the S131 old main road to Phalaborwa and on the left (the S69) is a sharp turn down the hill to the Nhlanganini stream. When it flows, the drift is an ideal spot for photographing waterbirds (malachite kingfishers, sandpipers and sandplovers) and elephants that seem to enjoy using the shallow, sandy waterway as an easy-on-the-feet thoroughfare.

This loop continues through good elephant and buffalo country for a further 4 km to a muddy waterhole and windmill on the left. This is a haven for elephants to cool off in the heat of the afternoon. The murky pond is also home

The extraordinary sight of a hippo that became a temporary island of terrapins at a waterhole on the S69 south-west of Letaba.

to scores of terrapins and I once saw a benevolent hippo, unable to submerge completely, covered from snout to tail in the crusty reptiles, all taking advantage of their temporary "island" to sun themselves.

Just before the loop rejoins the H-9 Phalaborwa road, you again cross the Nhlanganini stream and then also the Maswedzudzu river. Drive carefully through these drifts because they conceal rough, loose rocks under the water. Both causeways form shallow weirs, alive with tiny fish which often attract saddlebilled storks and a variety of herons. Close scrutiny of the rocky hill on the left between the rivers should usually reveal klipspringers.

Swinging left as you rejoin the main road, you cross the Maswedzudzu bridge and travel 2 km to the next loop turning away to the left. It is a short diversion that takes you close to a length of almost sheer cliff face where, I am told by rangers, the rocks are dotted with caves much favoured by hyenas as breeding dens.

Your next stop is the Winkelhaak dam (also known as the Nhlanganini dam), which can be viewed from two lookout points. It is excellent for bird life and game, but binoculars are a must. Move on past the imposing Shilawuri hill where the S96 link road skirts off right to meet the S131 further north. The S131 runs parallel to the H-9 main

Letaba/Phalaborwa road.

Elephant, buffalo, kudu and the rare roan antelope may be seen at the Rhidonda pan and Swartklip windmill, which are the next attractions, one on each side of the road. The Rhidonda windmill was another generous gift to the Park made by Mr and Mrs G. G. Kay.

Passing the halfway point, a new link road, the S132, swings north to also join the old S131.

Continuing west, the flat landscape ahead is broken by a number of rocky, conical koppies, two of which are on the left of the road opposite the Erfplaas windmill – known for its roan and sable antelope. Then 3 km further on are the giant Vudogwa outcrop and the smaller

*Left: Buffaloes stand sentinel in the early morning mist. **Right:** A tree squirrel surveys the chilly winter scene. Both pictures were taken on the S47 Mingerhout dam road.*

Masorini hill, both on the left, and away to the right towers the Shikumbu hill.

The Masorini archaeological museum should not be missed. The complex includes a car park, picnic and barbecue areas and restrooms. There is an information centre in a hut containing displays of photographs, clay models and artefacts collected on the site.

Masorini hill was once a favoured tribal dwelling site with new villages being built atop the old from the time of the Stone Age through the Iron Age, until it was finally abandoned during the last century by ancestors of the present baPhalaborwa tribe.

These people made a living forging iron implements such as spears, arrowheads and simple agricultural tools in primitive smelting furnaces. They left behind enough evidence of their lifestyle to enable scientists to accurately rebuild the village into the "living" museum that is Masorini today.

Periodically during the day guides escort visitors through the rebuilt village, past the excavated remains of a smelting furnace (a working replica is demonstrated) to dwelling huts and a granary.

The guide takes you up the hill to where Stone Age people once lived, along a winding path which leads from one terrace to another past various living areas and finally to the upper terrace where the chief presumably lived. The views from the hilltop are breathtaking.

Continuing towards Phalaborwa Gate, you turn off to the south within 1 km on the Sable dam loop. Ostrich may be seen in this area, sometimes in groups or with chicks, and at the dam itself – built just downstream of the source of the Tshutshi stream – there are kudu, sable antelope and sometimes tsessebe.

There is little more to see on the remaining 7 km to the border.

On your return to Letaba you should explore the network of new roads that branch north and south of the re-opened S131.

The S131 turns off a short distance from Phalaborwa Gate and travels north through dense bushveld before swinging east past the Ngwenyeni windmill

(usually of little interest) for 15 km to where it crosses the new tarred H14 north road.

Continuing eastwards for another 4,5 km, you cannot miss a small, attractive rocky outcrop to the right of the road. The resident family of klipspringers is usually visible although shy. On the crest of the hill is a fine example of a common tree *Euphorbia*.

After another 7 km you pass the S132 to the north (of little interest) and then after 1 km the southern sector of the S132 turns off to the right. This is

Brilliant flame creeper blossom in Letaba camp provides a meal for a resident masked weaver.

The rest of the journey back to camp is often uneventful, although after rain the mud wallows to the right, 2 km past the S96 Shilawuri loop intersection, can be fun – especially for bachelor buffaloes.

To avoid confusion, the description of the new tarred H14 road from Phalaborwa to the north is now presented separately.

This road, which is a credit to the Park's engineers and planning staff, is designed to give a more direct access from the Eastern Transvaal to Punda Maria, Shingwedzi, Boulders and the proposed Pioneer camp. The road traverses very attractive countryside and the loops to windmills and lookout points are varied and carefully selected to include splendid scenery.

The H14 begins its journey north exactly 7,5 km from the Phalaborwa Gate on the H-9 to Letaba. The towering Shikumbu mountain rises above the rolling bushveld and after 5 km the road passes close to its western slopes. Chiefs of the baPhalaborwa tribe once built their kraals on the summit. Note the sheer cliffs on the northern side.

After 7 km the road crosses the Ngwenyeni river and just before the bridge is a turnoff to the right leading to an excellent waterhole close to the confluence of the Ngwenyeni and Shicindzwaneni rivers. For the next 10 km, before reaching the second bridge over the Ngwenyeni, the road closely follows the river and offers two more river loops to explore, plus a roadside viewpoint which overlooks a bend in the river from a cliff-top. Still pools attract a variety of game including large herds of waterbuck and an outstanding sycamore fig offers shade to impalas resting in the heat of the day.

Having travelled 28 km, you cross the Letaba river, using the temporary low-level bridge until a high-level bridge is completed slightly to the west. Across the bridge, the road follows the banks of the Shipikani river for 5 km (with another glimpse after 8 km) through predominantly red bushwillow veld going along due north before it swings eastwards past the turnoff to the private Boulders camp to join the H1-6 north road.

A number of loops are planned for this section of the road including one to a viewpoint overlooking a splendid waterhole 12,5 km north of the Letaba bridge.

Before the road descends down to the Tsende river, it crosses a ridge which offers lovely views across the Park towards the north and east.

ROUTE Q

My idea of absolute bliss in the Kruger National Park is to sit quietly at the water's edge – be it a stream, dam or windmill, from sunrise until the heat reaches fidget-point, watching the array of wildlife coming and going to drink or feed.

Such a place, especially in the early spring when the water-level is at its lowest, is the south bank of the Engelhard dam on the Letaba river almost immediately downstream from the camp.

To get there, follow the gravel S94 that turns south just beyond the camp gates (take either road at the fork as they link up again after a short distance) and then, after 3 km, turn left along the S46 to follow the river to the Engelhard dam loop, also on the left. Special care should be taken while travelling this river road because elephants are inclined to loom suddenly from the unusually thick mopane veld at the roadside.

The dam, one of the largest in the Park, was built in 1970 with funds donated by American industrialist and chairman of Rand Mines, Charles William Engelhard, and is dedicated simply "to his concern for the preservation of wildlife". He died suddenly on 2 March of the following year of a heart attack, at the age of 54.

At the dam large herds of impala graze along the grassy flats at the water's edge, a sight that is especially charming in early December when the lambs frisk with new life. Animal traffic includes buffalo, kudu, waterbuck and almost always elephant, splashing and feeding along the water's edge. Crocodiles cruise the shallows and towards the distant north bank, hippos grunt and blow.

For birdlovers there are goliath, squacco, purple, grey and greenbacked herons, wattled, crowned and blacksmith plovers, storks, snipes, crakes and jacanas. Whitefronted bee-eaters swoop, dive and settle for a moment, sometimes posing perfectly to be photographed. According to ornithologist Kenneth Newman, six pairs of yellowbilled storks nested at this dam in the spring of 1971 – a rare occurrence for the Kruger Park.

known as the Marumbeni loop and it may be worthwhile driving for 4 km down it to the twin windmills of the same name. I have not seen any game using this waterhole.

It is probably more interesting to continue travelling east for 5 km on the S131 towards Letaba and then swing off to the right to the Nwanedzi windmill situated on the banks of the usually dry Nwanedzi stream. There is excellent shade beneath two spreading apple-leaf trees where you may sit in comfort during the heat of the day.

Seen around Letaba
A pot-pourri of wildlife scenes around Letaba:
Top: *A buffalo herd crossing the Letaba river sands.* ***Top right:*** *A buffalo grazing.* ***Above:*** *A bushbuck female resting.* ***Right:*** *Lions "smiling" for the camera and feeding.* ***Far right:*** *A leopard, one of the many that lurk in the heavy undergrowth along the river banks.*

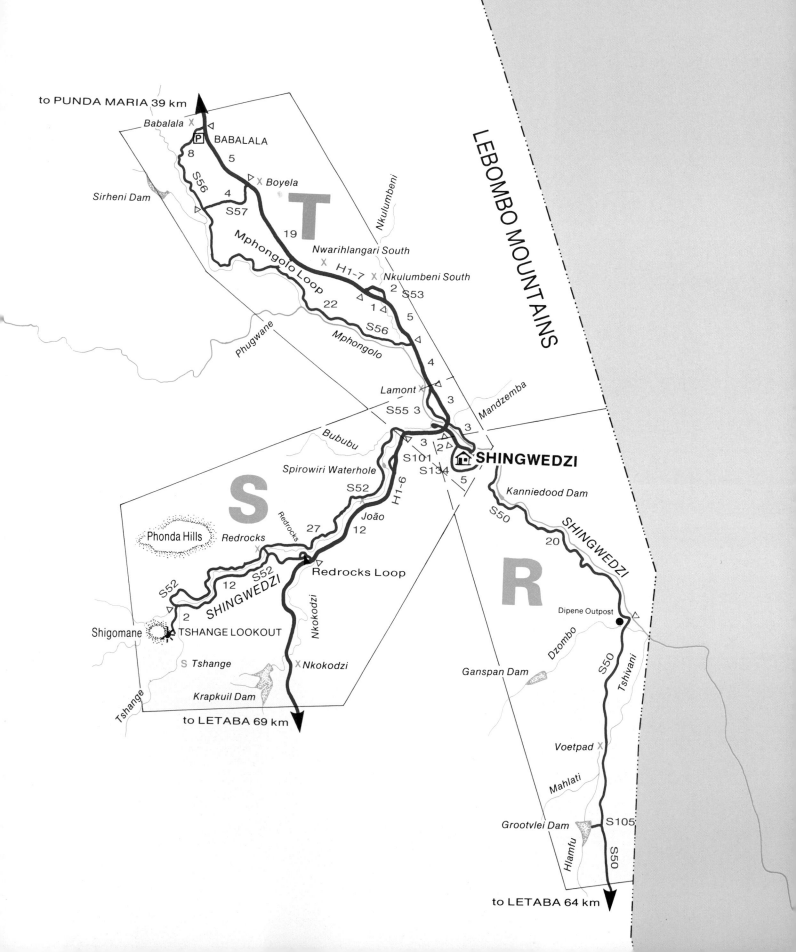

to PUNDA MARIA 39 km

Babalala X

P BABALALA

8

5

S56

4 Boyela X

Sirheni Dam

S57

19

Mphongolo Loop

Nkulumbeni

T

Nwarihlangari South X

H1-7 X Nkulumbeni South X

2 S53

22

1 5

S56

Phugwane

Mphongolo

4

Lamont X

3

Mandzemba

S55 3

Bububu

3

3

S101

2

S134

SHINGWEDZI

Spirowiri Waterhole

H1-6

5

Kanniedood Dam

S52

João

S50

20

SHINGWEDZI

Redrocks

27

12

S

R

Phonda Hills

Redrocks X

S52

S52

12

Redrocks Loop

SHINGWEDZI

Nkokodzi

Dipene Outpost

2

Shigomane

TSHANGE LOOKOUT

Ganspan Dam

Dzombo

S50

Tshivani

S Tshange

Nkokodzi X

Krapkuil Dam

Tshange

to LETABA 69 km

Voetpad X

Mahlati

Grootvlei Dam

S105

Hlamfu

S50

to LETABA 64 km

LEBOMBO MOUNTAINS

SHINGWEDZI AREA

Shingwedzi camp

The medium-sized camp of Shingwedzi is situated in the northern sector of the Park surrounded by flat mopane countryside, on the south bank of the beautiful river of the same name.

Shingwedzi, which means "the place of ironstone", is best known for the huge herds of elephant that range through the area and for the individual giant tuskers that have helped make the Park internationally famous.

Like its sister camp Punda Maria, further to the north, Shingwedzi is blessed with a delightful atmosphere characteristic of the old frontier Africa of yesteryear.

In spite of extensive hut reconstruction and the addition of new staff housing, guest cottages, a swimming pool and a magnificent new main building, most of the camp's original charm has been retained. A remarkable feat by Park engineers was the remodelling of the old mud and whitewash huts that were built in 1933 when the camp was first opened. While carefully supporting the original thatched roofs and reshaping them, builders have added new central brick retaining walls, verandas and two-bedded lofts above the rafters – a delight for children – without disturbing the old facade. The roofs of all the more modern huts, which are presently tiled, are to be thatched.

At the end of 1983 a new main building overlooking the Shingwedzi river was opened which boasts one of the biggest single spans of thatch in the southern hemisphere. Under this spectacular roof are reception area and offices, restaurant, information centre and museum, self-service cafeteria, a veranda as big as a tennis court, a supermarket, storerooms, coldrooms, restrooms and modern kitchen. (Catering staff have given their solemn assurance that the quality of that amazing delicacy, the Shingwedzi buffaloburger, will be maintained.)

Upstream of the complex is a delightful picnic area for day visitors with barbecue sites next to shady lawns linked by rustic wooden bridges. Downstream from the new main building, also overlooking the river, is a new, privately sponsored luxury guest house with three bedrooms, one with an *en suite* bathroom and the other two with showers, and a huge lounge-diningroom. A second similar guest house is to be built later.

Campers and caravanners are delighted with the newly extended area reserved for them close to the swimming pool, which was opened in the summer of 1983. This was the first pool to be built (except for Pretoriuskop) after the Board's decision to eventually add pools to all camps.

Shingwedzi camp is a photographer's paradise. Guests may stroll about focusing on woodpeckers, weavers (including the redheaded variety) and Scops owls as well as the usual glossy starlings, hornbills and francolins that poke and scratch around the camp in search of titbits. There are also scores of squirrels, some so tame that they take food from the hand, and plant lovers will enjoy the wild flowers which include aloes and the famous display of Shingwedzi impala lilies.

ROUTE R

I am often asked, as a photographer, to make the difficult choice of selecting a single place in Southern Africa where, in one week, I can be assured of a good selection of wildlife pictures including scenic views, flowers and plants, animals, insects and birds. My answer is always: the area around Kanniedood dam on the Shingwedzi river in the Kruger National Park.

The concrete causeway that crosses the Shingwedzi river a stone's throw downstream from the camp marks the headwaters of the Kanniedood dam. From this causeway, which also forms a weir for the long pool in front of the camp's veranda, you can watch impala, bushbuck, waterbuck, elephant and nyala

Among the residents of Shingwedzi camp are redheaded weavers (above) and the beautifully camouflaged Scops owls (below).

arriving to drink. Egyptian geese honk and hiss, guzzling in the shallows. Herons and hamerkops hunt fish and openbilled storks poke about the submerged stones for freshwater molluscs. Pied kingfishers hover with wings beating, bodies upright, heads motionless, watching – to suddenly dive with a splash followed by swift flight skyward, feathers streaming water, to favourite perches where they quickly eat their tiny catches, returning to flutter again at station with a *kwik-kwik* call of triumph above the water.

For me a superb sight from this causeway is the sunrise, especially in early summer when the warm reds of the sun touch the great trunks of the riverside sycamore figs. The water mist lies low, wispy and blue and for a moment the slanting rays highlight the flying insects, creating the illusion of millions of fireflies above the dark water. On luckier days the scene is enriched by early rising waterbuck standing in silhouette with the light highlighting their dewy fur fluffed up against the morning chill.

A sight for the fortunate few can be seen at the end of the short turnoff to the left along the river at the northern end of the causeway. The vertical earth bank is riddled with the deep, round holes of giant kingfisher nests. In spring

the raucous calling of the males to attract prospective, but reluctant, mates to view their labours of love can be heard for miles around. So eager is the male to woo a female to his diggings that at times he darts headlong into a shallow, unfinished tunnel, almost knocking himself out.

Return across the causeway and climb the steep south river bank, turning left, away from the camp, down the S50 which follows the river downstream to the Kanniedood dam wall and beyond.

The first point of interest is less than 200 m from the camp. Having negotiated a deep, dry stream bed you emerge into open mopane veld with a series of huge fig trees towering along the high river bank. Beneath the first of these trees a scattering of bleached elephant bones can be seen – all that remains of that

*Two late afternoon silhouettes, (**below**) a playful elephant and (**bottom**) a pair of marabou storks, in the river pool below Shingwedzi camp.*

mighty elephant named Shingwedzi, famous for his massive ivory, who died there on 16 January 1981 (see p. 123-4). This immediate area is now often frequented by another giant, João, who also had magnificent tusks until they were recently mysteriously broken off. Incidentally, careful examination of the fig trees may reveal a family of giant eagle owls; a favourite nesting place is in the fork of the lower boughs of one of them.

Continuing, you should explore every one of the 10 turnoffs and river loops, as each offers a different view of the Kanniedood dam. Proud nyala bulls with their slow, dainty tread keep guard over their females and calves, whose colour is a very different rich chestnut striped with white. Bushbuck go about their secretive business, showing themselves

Feathered fisherfolk of Shingwedzi
Below: *A surprised greenbacked heron caught these three fish simultaneously and then swallowed them all.*

Above: *With lightning speed, a great white egret stabs at a fish.*
Right: *A grey heron cocks its head at the water and a hamerkop hunts in the background.*

Right: A great white egret stands motionless, waiting for prey to come within range.

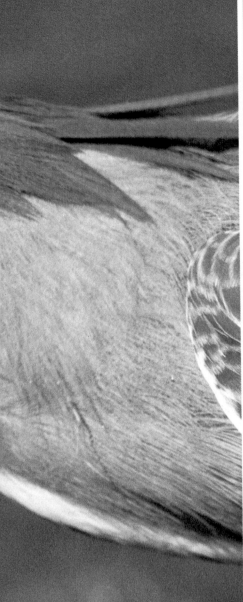

Above: An early morning silhouette of a goliath heron positioning itself on a log rising out of the Kanniedood dam.

ove: Waiting patiently above a *ingwedzi* river pool is a pied kingfisher.

momentarily between dense riverine thickets and impala graze in their hundreds on the wet green grasses. Waterbuck swing their woolly heads to stare across white ringed rumps, their upper lips curling quaintly in curiosity. Squirrels stationed in pairs outside their tree-hole dens groom and chatter with their mates. There are crocodiles, and hippos and elephants sport and splash in the water on hot days. Leopards are common in this area but these stealthy felines are seldom seen.

Bird life is spectacular. Numerous dead trees rise starkly from the water, their leafless branches forming scores of natural perches for waterbirds and others. The fisherfolk include goliath, greenbacked, squacco and purple herons, yellowbilled and great white egrets, a whole range of storks from white to woollynecked, darters and even the occasional pelican. There are whitefronted and little bee-eaters and sometimes a lone Burchell's coucal – well known to some photographers – which poses while singing its strange "bottle-pouring" song.

A waterside viewing platform just 3,5 km from the camp's filling station gate overlooks the dam. Only four vehicles may park under the hide at a time. From there can be seen spectacular bird life, elephant coming to drink and bathe on the far bank, and hippo.

So content are the hippos here that their number has increased from an original six, counted soon after the dam was built in the late 1960s, to more than 20 in 1986.

After travelling only 9 km from camp you pass the long grassed earth dam wall and if you are lucky you will find impala or waterbuck grazing on top of the wall, presenting stark silhouettes against the sky – although a fairly long telephoto lens is required for the best results.

For a further 14 km the road follows the Shingwedzi river and there are seven more river loops to investigate after the dam (which can reveal fine buffalo herds), before reaching the 23 km (to Shingwedzi) signpost where the road swings away from the river, to run

Shingwedzi is a haven for many of the Park's big tuskers, such as this giant photographed emerging from the Kanniedood dam.

southwards parallel to the Mozambique border, which lies less than 2 km to the east.

Two small drifts are crossed – the Dzombo and Tshivani streams – and a police outpost is passed on the left before you reach what I suggest should be the end of this journey, the Grootvlei dam along the short S105 to the west. If time allows, it is worthwhile waiting here for a while for tsessebe, zebra, giraffe and very big kudu bulls that may arrive to drink – as well as the usual array of resident waterbirds.

ROUTE S

There is the temptation to while away the hot early afternoon hours at the camp swimming pool, but I suggest that you endure the heat, gather up cameras and binoculars, and head south-east from camp down the main tarred H1-6 highway.

Above: A bushbuck ram glances up unconcernedly at the camera from its browsing. **Right:** *Hyena cubs often emerge to play in the evening from dens in stormwater pipes beneath the H1-6 main road south-west of Shingwedzi.*

Driving through the main gates, you may be lucky enough to see one of the only three recorded nesting sites of the yellowbilled kite in the Park, a few metres beyond the camp's main gate and water purifying plant.

For 3 km the tarred road runs close to the river and the magnificent trees along the south bank. A number of shady roadside resting places offer views of elephant, waterbuck and buffalo plodding through the heavy sand to drink from the scattered pools.

On reaching the main crossroads (the high-level bridge over the Shingwedzi river is to the right), a search around the rotting, fallen timber there can reveal a colony of dwarf mongooses. Although

these gregarious little fellows are great wanderers, I have often seen them at this spot, busily hunting for food until disturbed, when they rush to hide. Their curiosity, however, soon overcomes fear and one by one their heads appear, little black eyes staring boldly at the camera, before they again scurry about their chores.

Interrupt your journey briefly to cross over the main road onto the short gravel byway that leads to the viewpoint overlooking the confluence of the Shingwedzi and Mphongolo rivers – a place where waterbirds are prolific and which is specially favoured by fish eagles. My only sighting of a python in the wild was here.

Back on the main road, travelling towards Letaba, you pass through open country with the Shingwedzi river, much smaller and less obvious, away to the right. Animal life includes large herds of impala plus waterbuck, warthog and those inseparable compatriots, zebra and wildebeest.

Three kilometres west of the bridge you pass the S52 loop to the Tshange viewpoint (ignore it for the moment) and then further on you quickly detour around the little S101 loop with its fine view overlooking the Shingwedzi river. Pause to check for activity at the João windmill on the right before passing the southern sector of the S52 loop where the Nkokodzi river "flows" beneath the road to join the Shingwedzi.

Carry on due south, with the Nkokodzi river – intermittently seen – away to the left. You will notice that the mopane scrub is much thicker and that, as with much of the territory around Shingwedzi, the water table is unusually high, forming a number of surface pools along the roadside. It is only during very dry periods that these disappear.

· This terrain is ideal for elephant (the famous tusker called Shawu ranged in this area). There is plenty of food, clean surface water and many mud wallows. Here your decision to forego a swim back at camp may well be rewarded with excellent close-up views of elephant, sometimes in very big herds, happily splashing, squirting, rolling and wrestling in the mud like boisterous children.

Before turning back, you should go a little further along the H1-6 road, exploring as far as the Nkokodzi windmill, not only for elephant but for other

A roan antelope cow chews the cud at the roadside near Nkokodzi windmill on the main H1-6 southwest of the camp. Roan are fairly common throughout the Park's north.

game including herds of roan antelope which are often in this area, sometimes resting unperturbed at the roadside contentedly chewing.

Heading back towards camp, you should now turn off to the west to explore the long gravel S52 loop that skirts both the south and north banks of the Shingwedzi river. At least two hours should be allowed to do justice to this 44 km detour and to allow enough time to get back to camp before the gates close.

The road follows the south bank of the river through excellent elephant territory. Exactly 1,3 km from the turn-off is a loop road to the right which should not be missed. It is no longer than 2 km and leads to a viewpoint overlooking the strange, brick-red rock formations in the river bed known as Redrocks.

Then 3,5 km from the main road is a very pretty resting place overlooking a natural rock weir in the river that seems to hold water in even the driest times. Buffalo love to wallow in the long pool and during the summer months of 1985 a lone hippo took up residence there – offering very easy photography.

Immediately after this is a link road to the north bank through the usually dry river bed which offers a short-cut back to camp if necessary. Just to the west of this link road is a windmill and water trough, situated between the road and the river, which is very good for game, especially in the drier months.

As the S52 continues, leisurely winding along the river, nyala, kudu and waterbuck are certain to be seen. There are also several low-lying clearings in the mopane offering natural mud wallows which attract both elephant and warthog.

Before reaching the point where the loop turns back, you cross the dry Tshange river causeway and then on the left is the 2 km turnoff to the Tshange viewpoint. Here you may leave your vehicle and clamber among the two rocky outcrops to enjoy spectacular views across the endless mopane veld towards the remote ranger's post of Shangoni in the west. It was in this wilderness area that the most famous of all Park elephants, Mafunyane, reigned before his death in October 1983. During this great elephant's twilight years, when all the big tuskers were threatened by ivory poachers from Mozambique, the

A grey heron spreads its wings to dry in the warm sunshine below the S52 river drift across the Shingwedzi river near the Tshange lookout point.

Sunrise over the Shingwedzi river photographed from the causeway below the camp.

whereabouts of Mafunyane were kept secret.

Back on the S52, you should now cross to the north bank of the Shingwedzi, fording a concrete causeway which forms a weir, an excellent place to photograph herons, darters and hamerkops feeding, preening and drying their wings. Old buffalo bulls often enjoy this spot, sometimes blocking the road by lying in the shallow water overflowing the causeway.

You now begin the longer leg of the S52, travelling east for 27 km through well-wooded countryside, until you again cross the Shingwedzi at its confluence with the Bububu river just before you rejoin the tarred road back to camp.

If you reach the S52-H1-6 main road junction in the late afternoon, I suggest that you carefully, and as quietly as possible, search the edges of the tarred road for the stormwater culverts that pass under it. There is one just to the south of the junction and at least two

more within the next kilometre or two towards Letaba. They are used by spotted hyena as breeding dens and towards sundown the cubs emerge to play in the dust, offering perfect opportunities for photographers – the finishing touch to a wonderful afternoon.

ROUTE T

The area north of Shingwedzi camp on the road to Babalala picnic site is the best elephant country in the Park and at times the most exciting. Often while driving there, especially along the Mphongolo river road, the problem is not finding elephants, but avoiding them, as breeding herds of 50 and more are common.

To make the most of this drive, I suggest a mid-morning departure from camp (use the earlier hours to explore the Kanniedood dam) making the best of the hotter times of day when the elephants are most likely to be near the road on their way to drink in the river.

Carry a picnic lunch so you can remain in the area throughout the afternoon.

Depart across the concrete Shingwedzi river causeway opposite the camp and follow the gravel north bank road for 3 km to the main H1-7 road to the north. While negotiating the dry bed of the Mandzemba river, note the yellow-green bark of the huge fever tree there, which makes an excellent background for photographing the many resident bush squirrels.

On the left as you reach the crossroads is the high-level Shingwedzi bridge and straight ahead is the S55 river loop, which is a worthwhile short detour to see impala, waterbuck, steenbok and lone elephant bulls. The latter can be highly amusing at the Lamont windmill as they stretch up to "steal" a drink from the tall concrete reservoir there.

Returning to the main road from the loop and heading north, you quickly reach the gravel S56 Mphongolo loop which turns off to the left and then runs

Left: A frumpish Burchell's coucal at the Kannie-dood dam. An emerald spotted dove (above) and an African hoopoe (right) photographed along the S56 Mphongolo loop.

for 27 km to Babalala picnic site.

For 3 km this road winds through the mopane, out of sight of the river. Soon after negotiating the steep banks of the Nkulumbeni stream, you suddenly find the Mphongolo river on the left. At this point you will notice on your right a lovely area reminiscent of a peaceful English deer park, especially when herds of waterbuck or impala are grazing there.

The road meanders beside the river offering a choice of several shady river-view resting places on the left. The forests on the route are dominated by huge, very old jackal-berry and nyala trees and some surprisingly big mopane trees. You will almost certainly see kudu and waterbuck, which may panic if they are "trapped" in the narrow area between your vehicle and the river. In this situation, stop your car and let the animals cross the road to the "safe" side where they will soon resume feeding quietly.

This road is also very good for observing the antics of bush squirrels and for photographing birds. Using a medium telephoto lens, I was able to photograph an African hoopoe and the often elusive emerald spotted dove within a few metres of each other and at amazingly close quarters.

And then come the elephants. They appear suddenly and silently from out of the deep mopane or up the steep river

bank, their bodies still wet and black from the river or brown and muddied from a noon-day bathe. Alive to your presence, many will hurry across the road with bodies swaying, trunks swinging, until they are safely out of sight. Some elephants stroll past nonchalantly, heads held high, eyes straight ahead, seeming not to have noticed you but for the tell-tale tip of the trunk that points in your direction alert for the slightest whiff of danger. Others, bold "teenager" bulls or protective cows, resent your intrusion. With a great kick of dust and a clap of ears they will swing their trunks to face you, standing still and tall on straight front legs and tiptoes, heads held high, ears spread like great wings, trunks curled, defiantly posing in an image of great stature – and bluff, or so one hopes.

Then, as suddenly as they came, they are gone again, with only the distant sounds of snapping branches and rumbling bellies as they feed.

On the rest of the journey to the Babalala picnic site, you will pass the confluence of the Phugwane and Mphongolo rivers (where river sand is dug for road building) and then on the right you will find the first of two short cuts back to the main road, should you wish to return to camp directly.

But if you continue along the S56 northwards you can explore a network

of new loop roads that offer pretty scenery as well as nyala, waterbuck, sable and roan antelope and even more elephants.

At Babalala (meaning "old man sleeps") you can stretch your legs and enjoy a picnic lunch and, if you are quiet enough, you will probably see a family of reed-buck grazing just beyond the picnic site fence. Note the ram's alarm call – a sharp piercing whistle, a sound made by blowing through the nose.

And then it's back to Shingwedzi camp, along the main road or again through the fabulous elephant country of Mphongolo – if you are brave enough.

Should the elephants be seen in the river **(above)**, it is best to park and wait quietly for them to climb the bank and cross the road ahead.

Above: After bathing in the river, elephants often like to take a roadside dust bath.

...ger elephants sometimes shake their heads and spread ...ears in warning **(left and above)**. Others simply stop and ...curiously **(top)**.

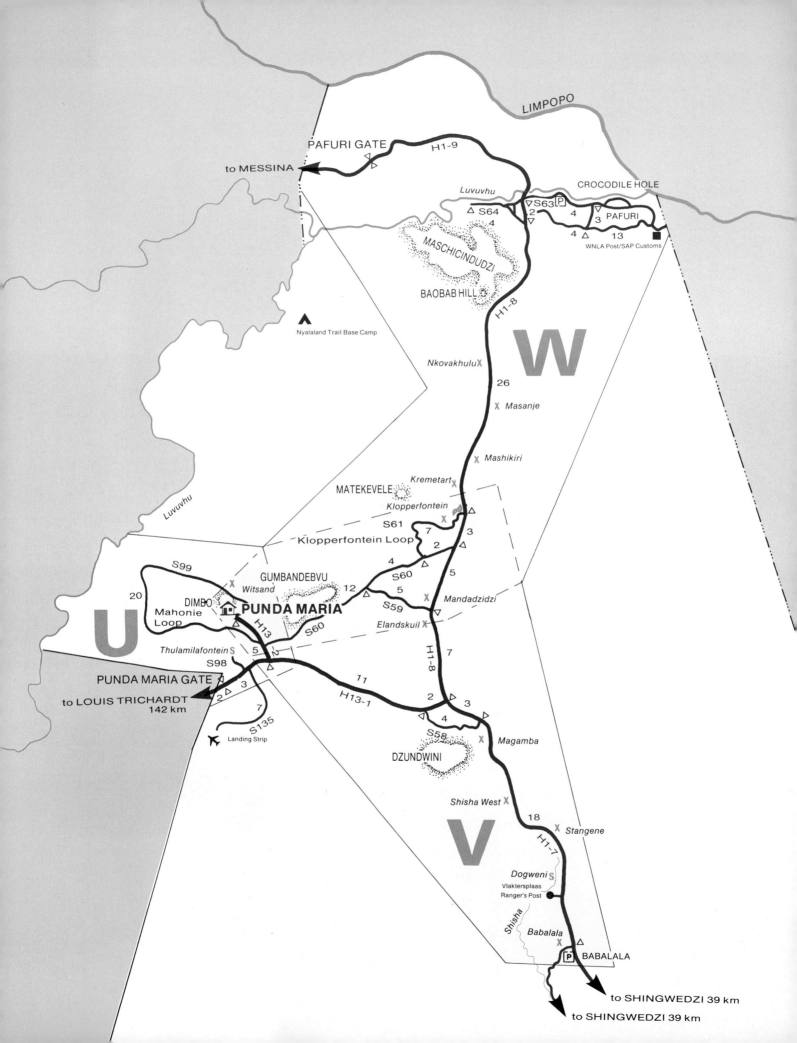

LIMPOPO

PAFURI GATE   H1-9

to MESSINA

CROCODILE HOLE

Luvuvhu   ▽ S63 P
△ S64   ▽   ▽   PAFURI
         4   2   4   3
                              13
              ▽ △
                    WNLA Post/SAP Customs

MASCHICINDUDZI

BAOBAB HILL
                    H1-8

                              W

Nkovakhulu X
              26

              X *Masanje*

              X *Mashikiri*

MATEKEVELE   *Kremetart* X
              Klopperfontein   △
         S61   7   3
Klopperfontein Loop   2   △
                    4   5
              S60
GUMBANDEBVU   12   5
Witsand   X   *Mandadzidzi*
DIMBO   S59   ▽
Mahonie   **PUNDA MARIA** *Elandskuil* X
Loop   H13   S60
                              H1-8   7
Thulamilafontein S 5
S98   2
PUNDA MARIA GATE   △   11
                    3   2   ▽   3
to LOUIS TRICHARDT 2 △   △   4   ▽
142 km   7   H13-1   *Magamba*
              S135   S58   X
         ✈ Landing Strip
                    DZUNDWINI

                              V
              Shisha West X
                         18   X *Stangene*
                              H1-7
                    Dogweni S
                    Vlaktersplaas
                    Ranger's Post ●
              Shisha *Babalala*
                         X   △
                         P   BABALALA

                         to SHINGWEDZI 39 km
                         to SHINGWEDZI 39 km

▲ Nyalaland Trail Base Camp

Luvuvhu

U
S99
20

PUNDA MARIA/PAFURI AREA

Punda Maria camp

There is no other camp quite like Punda Maria in the Kruger National Park. Neat rows of square, whitewashed traditional huts, mature shade trees rich in blossom and bird life and the soft pervasive aroma of woodsmoke – an atmosphere evocatively reminiscent of a long-gone era of ox-drawn wagons and hunters on horseback; a time when survival meant relying on the land for food, fuel and home comforts.

Punda Maria first became a ranger post in 1919 when it was placed under the charge of Captain J. J. Coetser. He named it Punda Maria partly after his wife – who found it difficult to cope with the harsh life-style in the bush and thus spent most of her time away from her husband – and partly after the Swahili word for zebra, *punda milia*.

Many years later, after consultations with the renowned historian, Dr W. H. J. Punt who has made a specialised study of the Lowveld pioneers, it was decided to change the name to Punda Milia because zebra were said to be the first animals seen in the area by Captain Coetser. In 1979, however, after representations to the National Parks Board by ranger Coetser's family, it was decided to restore the original name.

The camp was built in 1933/34 and most of the original huts are still in use. The roofs are thatched and the mud walls are compacted onto frameworks of thin poles. These and the poles that support the narrow surrounding verandas are cut from Lebombo ironwood trees and after 50 years they are still untouched by termites or the African weather.

Recently the interiors of the huts were renovated. Toilets and showers were added to all of them, but the outer walls were preserved in their original state. The huts are built in tandem and are remarkably cool, even in the hottest weather. This is said to be due to an old tradition of building two walls side by side as partitioning between neighbouring huts, leaving a narrow gap which acts as an insulator. This interesting old system of air-conditioning also offers better privacy. Along newly cut terraces on the hillside behind the old camp, 30 new huts are to be built, all identical to the original ones in style and materials.

A new double-storeyed main complex includes a reception office, restaurant and information centre, and a cafeteria on the veranda is planned for the future.

Accommodation will be increased to 47 two-bed huts, four three-bed huts with kitchenettes and two four-bed apartments with full facilities.

There is a special picnic site for day trippers near the main gate and the caravan and campsite has been extended with two new ablution blocks and two camp kitchens. Below the campsite a swimming pool is to be built and paths have been opened up along the koppie behind the camp for walks. These paths are aptly named the Paradise Flycatcher Trail, because strollers are almost certain to see this beautiful bird. The trail follows the camp water pipeline.

Two benches have been provided under shade along the way for those wishing to sit quietly and enjoy the solitude.

A natural-looking waterhole with a lookout platform has been completed on the camp boundary for day and night game viewing.

The camp's diesel power plant has been moved to behind a ridge to the west of the camp to dampen the noise. The camp's water comes from two boreholes which each deliver more than 4 000 litres (900 gallons) an hour – even during times of drought.

Left: A lion stands guard over a kudu kill on the S58 Dzundwini loop. The rest of the pride rests out of sight.

A charming tradition of Punda Maria, one that is usually found only in the smaller private camps, is the barbecue ritual in the evenings. At sunset red coals are spread by camp staff and guests gather around the fires to cook their meals, exchange experiences of the day and sip cold drinks.

The area is steeped in fascinating tribal history. A short distance to the east is a hill called Gumbandebvu, named after a chief who lived there long ago. He had a daughter, Khama, who was said to be blessed with the miracle of rainmaking. When the weather became fickle and

This much-photographed suni orphan was reared in captivity by an honorary ranger at Pafuri.

newly planted crops remained dry in the spring, tribesmen would gather from far and wide at Khama's kraal, bearing gifts of trinkets, livestock and snuff. A goat would be sacrificed, killed in such a way that its prolonged death cries would soon summon the rain spirits. Khama would then climb the hill, taking with her selected bones and other pieces of the sacrificed animal plus selected magic potions, where in a secret cave she would weave her spells until the rain clouds rolled in.

Today Gumbandebvu hill is still regarded by tribespeople as sacred and is said to be haunted.

Further to the north, rising above the Luvuvhu river, is another sacred and ghostly hill where the Lembethu people once lived high up in a fortified stone settlement. The tribe was ruthlessly ruled by a mad chief called Makahanja, whose favourite pastime was to order people who had committed even the most minor crimes, or had merely irritated him, to be flung from a cliff top to their deaths as food for crocodiles in the river far below.

Word of these grisly deeds soon reached the paramount chief of the area, who ordered his son (incidentally Makahanja's half-brother) to go out and kill the madman. With the drums throbbing in welcome, the visitor suddenly turned and speared Makahanja to death. Ruins of the settlement can still be seen today by visitors taking the Nyalaland Trail.

Those on trail are sometimes also taken to a site near Xantangelani, southeast of the trail camp, where there is the only known Bushman painting of an elephant in the Park.

Punda Maria may be reached either by motoring through the length of the Park or from the west through the Republic of Venda from the town of Louis Trichardt in the far northern Transvaal. Either way the journey is long, and weary travellers are usually content to see out what remains of the day in the camp, simply enjoying the welcome shade and the fragrance of the bush while watching the great variety of birds busying themselves.

ROUTE U

If the spirit is willing, and time allows (two hours at least) an excellent short drive can be enjoyed along the Mahonie loop, the 26 km S99 route that swings in a great circle behind the camp. Not only is it excellent for game and scenically rewarding, but it travels through the heart of a floral wonderland, the Punda Maria sandveld region, renowned for its rich variety of trees and shrubs – some of which are unique. The forests along this route include such varieties as the fragrant lavender tree (from whose leaves a tea can be brewed which is said to be a deworming agent), the wild custard apple whose fruit is enjoyed by primates, the broom cluster fig whose fruit grows only from the tree's stem, the decorative red-heart tree, and one found only in this area of the Park – the Rhodesian gardenia with beautiful scented white flowers speckled with red. Yet others are the spreading and shady myrtle bushwillow, the knobbly fig, the mobola plum, the pod mahogany with its red and black seeds, and the tall, longstemmed wild syringa.

To travel the Mahonie loop (named after the pod mahogany tree – *peulmahonie* in Afrikaans) in an anti-clockwise direction, you take the first turn left a stone's throw from the camp gates.

The road winds through thick mopane veld and swings west to skirt Dimbo koppie behind Punda Maria where, after 2 km, you will find the small Maritubi dam on the left. This picturesque little watering place often appears to be devoid of wildlife but, given a little time and patience it can be very rewarding.

Throughout the day an amazing var-

A paradise flycatcher feeding its chick was photographed in front of the main reception building at Punda Maria camp.

iety of birds gather there to fish, drink and hunt for water insects. Because you can park your vehicle quite close to the water, taking photographs is fairly easy. There are two four-footed visitors worth mentioning which frequent this dam: an unusually large water leguaan, over 2 m long and so sturdy that it is often thought to be a crocodile, and an old nyala bull, a strange sight because both its horns are broken off at the base. During times of drought, this waterhole dries up.

Continuing for another 4 km, you will reach the Witsand windmill down a short side road to the right which is excellent for kudu and nyala. I have been lucky enough to see a pack of wild dogs pause for refreshment here, a rare sight in the far north of the Park.

Back on the main road you soon swing south and it may be of interest that you are now travelling on the most westerly stretch of public road in the Park. It

twists and climbs through beautiful rugged countryside and from the high points there are magnificent views to the west across the Luvuvhu river into the neighbouring Venda Republic.

The rocky and well-wooded terrain offers sanctuary to two very rare antelope species for the Kruger National Park, the diminutive Sharpe's grysbok and the even smaller suni – but it takes keen sight and a lot of luck to spot them because of their shy and secretive habits.

Approaching the halfway mark of the Mahonie loop, you ford the crystal clear Tsila-vila stream, which the road follows for several kilometres. One kilometre past the ford, pull off to the left and pause for a while at the Matukwala dam. Although excellent for general viewing, this spot is frustrating for photographers at present, unless their equipment includes longer than usual telephoto lenses. However, one of the many planned hides to be spread through

the Park has been earmarked for this dam. The afternoon light and the background should then be superb for photography.

Continuing, you follow the stream meandering beside the road which attracts a variety of wildlife, especially to the vivid green of the water grasses where the flow slows to become marshland. Here sandpipers and shy black crakes poke about the shallows and impala, kudu and nyala may be seen where the road swings east back towards the camp.

Shortly after this the road splits briefly to skirt around a giant pod mahogany tree. Here kudu often wait patiently nearby for the brown seed pods to drop, which they love to eat. Away to the left a water trough attracts herds of impala

261

and beyond it can be seen the thatched rooftops of Punda Maria camp.

Two kilometres further on and 1 km before reaching the main tarred road back to Punda Maria is a magnificent marula tree surrounded by a "garden" of stones, placed deliberately to discourage destructive elephants. It is said to be the biggest marula in the Park.

ROUTE V

The roads that lead east into the Park interior from both Punda Maria Gate and the camp pass through thick mopane woodland dense enough, especially after summer rains, to make game viewing difficult. It can also be dangerous. While travelling on either the S60 gravel short cut past the Gumbandebvu koppie to meet the main H1-8 north road to Pafuri, or down the H13-1 tar to meet the same road *en route* to Shingwedzi further south, drivers should be especially alert for animals that may suddenly dash out into the road, confused by the sound of an approaching vehicle.

Although game viewing may be difficult, the mopane is so rich a food source that these forests abound with game. Travelling very slowly and quietly, stopping often with engine stilled and using ears as well as eyes will reveal the rustlings and rumblings of elephant, kudu, bushbuck, nyala and buffalo feeding close by. Seemingly oblivious to your intrusion, they may step out into the open to stare, breathtakingly close, before moving away to browse again out of sight.

Although this area boasts many attractive rocky outcrops and koppies, there is one landmark that cannot be missed, especially when travelling the main road from the south. It is not as obvious when motoring from Punda Maria because the dense mopane fringing the road often hides it from view. It is the Dzundwini hill rising from the flat veld like a squashed volcano. Beneath the northern slopes of this koppie runs the S58 loop that links the two tarred roads of the north, the main H1-7 and

the H13-1. This short 4 km stretch of dusty road can offer a wealth of interest.

Approaching it from the west, past the twin power lines that bring power from the giant hydro-electric scheme of Cahora Bassa in Mozambique, and turning right onto the loop, you are immediately aware of lovely park-like scenery. Here the mopane thins to scrub, revealing the bigger trees more distinctly. Soon the hill looms above the roadside to the right. I have occasionally seen leopard here in the early morning. They always seem to cross the road towards the hill, probably returning to their lairs after a night's hunting. To the left, a green marsh conceals a rising spring, a favourite wallow for buffalo and warthog and a drinking place for a variety of game, lions among them.

The visitor who wants to stop and watch here for any length of time will find at the roadside a very obvious and perfectly placed sausage tree offering excellent shade. In the spring when it is in blossom, families of nyala gather beneath it, not to keep cool but to relish the big magenta flowers as they fall.

For the tree enthusiast, a careful scrutiny along the slopes of Dzundwini hill will reveal a few specimens of the propeller tree, very rare in the Park and found at only one other locality – in the hills around Pafuri. The tree is so named because the fruit sprout a shuttlecock of thin dragonfly-like wings, about 7 cm long, which cause the seeds to spin like propellers as they "parachute" to the ground.

A little over 1 km from where the S58

Dzundwini road leaves the H1-7, a new road turns off southwards which climbs in bends up the flanks of the hill. It leads to a Park network radio aerial on top of the hill but visitors may go only about two-thirds of the way up, where there is a fine view of undulating plains and distant low mountain ranges to the east and south-east. Nyala may be seen along this road.

On reaching the H1-7 north-south "highway", a run down to the Babalala picnic site can be worthwhile, if time allows. The veld in this area is flat and generally open and sightings of rarer antelope like roan, sable and tsessebe are almost guaranteed – as well as of zebra, wildebeest, buffalo and elephant. Some very large elephant breeding herds may be seen along here and should the veld be freshly burned, the sight of one of these groups moving fast is a magnificent spectacle with great clouds of black soot and dust rising behind them. While resting at Babalala, you should remember to look out over the vlei bordering the picnic site for the resident family of southern reedbuck that regularly show themselves.

Heading north again back to the Dzundwini loop, but now continuing towards Pafuri, you pass the tarred H13-1 turnoff to Punda Maria on the left and then, after 6 km, the Elandskuil waterpoint on the left. Soon after this, you pass the gravel S59 link road turnoff and then the Mandadzidzi windmill, also on the left. Between these two waterholes may be seen the king of all antelope, the mighty eland. Although a herd

of about 120 regularly visits this area, this fawn to grey, ox-like beast is seldom photographed because, in spite of its formidable size and long wicked twisting horns, it is extremely timid and trots away quickly at the sight or sound of even a distant vehicle.

Shockingly obvious here are the two long lines of pylons that march across the veld carrying power from Cahora Bassa into the grid in South Africa – an ugly and probably unnecessary intrusion within this National Park.

To return to camp, you can take your pick of three gravel roads that head west, the S59, the S60 or S61 further north. I would suggest the latter, known as the Klopperfontein loop.

Within 1,5 km after turning down this loop there is a small dam on the right where you can park virtually at the water's edge. This is a spot where you

The brilliant red of the flame creeper is at its best in early spring.

should make yourself comfortable, settle down and stay for a while. It is very attractive with an expanse of lilypads attracting insect-hunting waterbirds like the African jacana while the bigger hamerkops, herons and saddlebilled storks arrive to hunt frogs and fish in the shallows. Game can be plentiful and elephant sometimes come to the far bank to drink and bathe, offering a perfect opportunity for photographers. A surprisingly large crocodile suns himself here usually at the dam wall, and just beyond the wall a high rocky outcrop is home to a family of klipspringers. The Klopperfontein windmill itself is disappointing, but for those who enjoy photographing baobab trees, there is a fine example just to the east of the waterhole at the roadside.

ROUTE W

Those who know Pafuri well, those who have worked there and the many visitors who make the long trek there from Punda Maria and Shingwedzi camps, usually agree that it is the most beautiful part of the Kruger National Park.

Although Pafuri is hot and dry (it has the lowest mean average rainfall in the entire Park) its beauty lies in the rich green riverine forests that flank the Luvuvhu river, watered mainly by the chocolate-brown floodwaters that in the summer rush down, full of topsoil, from the cultivated lands in the neighbouring Venda Republic and Transvaal province.

Its beauty also lies in its amazing variety of wildlife and vegetation – birds, animals, trees and shrubs – many of

The majesty of Pafuri
These pages offer only a taste of it.
Right: *The forest of fever trees reflected in the summer floodwaters.* ***Below:*** *A nyala bull staring at the camera.*

Above: *The imperious mien of a black eagle that soars above Maschicindudzi mountain.*
Right: *A rather bewildered kudu "teenager".*

264

Top: *An impala ram stands tense on the Luvuvhu river bank.* **Left:** *The rotund bulk of a hippopotamus.* **Above:** *The comical crested guineafowl with its "hippy hairstyle".*

which are rare or unique to Pafuri and the Nwambiya sandveld region close by.

For me, the beauty begins at Baobab hill.

Your day there will be a long and fascinating one, so I suggest you rise with the sparrows, having packed food and drink the night before.

The road to Pafuri travels northwards along the tarred H1-8. To reach it from Punda Maria it is best to follow the gravel S60 short cut that travels past Gumbandebvu hill, but should you so wish, you can enjoy the comfort of a tarred surface all the way by driving down the main H13-1 from Punda Maria Gate to meet the H1-8.

Having reached the main north road, you travel through flat mopane scrub past the Kremetart waterhole on the left (note the very pretty koppie behind it), the Mashikiri and Masanje windmills on the right and then finally the Nkovakhulu windmill on the left, where cheetah, sable and roan antelope and even

occasionally eland or giraffe (very rare this far north) might be seen. A selection of splendid baobab trees stands out starkly against the early morning sky like gaunt prehistoric relics, perfect for photography.

Exactly 8,8 km before the Pafuri bridges, the road winds up to a hill on the left, on the crest of which stands a lone, prominent baobab tree. This is Baobab hill, for centuries the historic landmark of tribesmen and, much later, for those colourful characters – the poachers, smugglers and other fugitives who once roamed this inhospitable area and found refuge at Crooks' Corner from the long arm of the law. It is said that below this hill to the north is a crystal clear spring which was once a favourite campsite of these early travellers and is now an exclusive drinking place for game.

Beneath the big baobab a roadside plaque records that this was the first outspan for Africans recruited at Pafuri between 1919 and 1927 to work on the

The view of the south bank of the Luvuvhu river from the main high-level bridge looking east. The antelope is a nyala bull.

Witwatersrand gold mines. They travelled by foot and on oxwagons and donkey carts to board trains far to the west.

Opposite, a second plaque points out another baobab on the other side of the road. In 1891 surveyor-writer G. R. von Wieligh, who surveyed the Transvaal-Mozambique border, paused here with surveyors Vos and Machado. Machado carved his name on the baobab's trunk and the plaque requests – "Please do not follow his example."

From Baobab hill the road drops quickly towards Pafuri with the mopane becoming a bold forest of colours, especially in the early spring. Game becomes more prolific: kudu run off with their spring-loaded gait, buffalo stand boldly staring and lions sometimes block the roadway, stretched out to enjoy the early morning sun.

The famous tree of Baobab hill at sunrise in early summer.

As you approach the Luvuvhu river, glimpses of the wide green of the forest can be seen ahead. You then pass two side turnings, one to the west, which is a feeder road to the western section of the S64 Nyala road along the Luvuvhu river and the other to the east, which is a direct route to the old Mozambique border post, now a police station.

Arriving at Pafuri, you come to two high-level spans of bridges over the Luvuvhu river. Across the river is the newly opened road which travels north and west through rolling hill country and a parkland of big mopane trees and baobabs, the main feature of this northernmost stretch of the Park. After travelling 15 km, the recently opened Pafuri Gate is reached, which actually lies several kilometres inside the Park boundary, and is the entry point for visitors from Messina and Zimbabwe.

This previously restricted area only became part of the Park in 1968, when it was exchanged through the then Department of Native Affairs for an area of similar size on the western boundary, south of Punda Maria. Although game was sadly depleted at the time, populations are recovering and include the biggest herd of eland in the Park.

Returning to Pafuri and the Luvuvhu river, the bridge itself gives views upstream and downstream of magnificent riverine forest. The banks in the shade of giant trees are a favourite place for a variety of animals, which includes bushbuck, impala, nyala, waterbuck and large troops of baboons and vervet monkeys. If you are fortunate, you may see one of the samango monkeys, which were

recently re-introduced to the area. Hippos soak themselves in pools upstream and honking Egyptian geese and silent, smooth-flying herons patrol above the water. With the bridge railings helping to hide your car, very good pictures can be taken of nyala and bushbuck browsing among the shrubs. On the north bank is a mass of flame creeper, a splash of brilliant red in the early spring, a favourite place for vervet monkeys to sun the morning chill from their bodies.

Your choice of roads at the crossroads is easy if you have thoughts of breakfast; you can turn east to follow the gravel S63 riverside road downstream for 3,1 km to the new Pafuri picnic site. The attendant there is always welcoming: he stokes the fire to ensure piping hot water and cheerfully humps his great black kettles across to top up your mugs. At his special barbecue place he will spread glowing red coals for visitors to grill their steaks and bacon, fry eggs, make toast and brew coffee.

Above the gentle sounds of the river can be heard a chorus of bird calls, from the ringing duets of tropical boubous and the shriek of big Cape parrots to the melancholy cry of the trumpeter hornbill. You may see birds such as the beautiful Narina trogon, spectacularly clad in luminous blue, green and crimson; the shy and secretive but equally spectacular gorgeous bush shrike; the Natal robin with its brilliant orange; and the lightweight of the barbet family, the yellowfronted tinker barbet, among dozens of others. Fairly common visitors to the mudbanks opposite the picnic site

are whitecrowned plovers. Entire days can be spent there doing nothing other than scanning the tree canopies for the hundreds of bird species that have made Pafuri famous – as well as nursing a stiff neck.

Continuing east on the S63 river road downstream towards Mozambique, you can experience the full impact of Pafuri's wildlife. Although the forest is heavy, you may see bushbuck, kudu, impala, leguaan (some of them real giants), squirrel, nyala by the score, buffalo, bushpig and lion. The sheer size of the forest gives it a magical atmosphere: trees seem bigger and greener than any others in the Park – leadwood, sausage, fig, fever, nyala, ana and jackal-berry trees line the road. Baobabs stand majestically and the undergrowth is splashed with the colours of big yellow and maroon wild hibiscus flowers and wild cucumbers, bright orange when ripe. The strange creepers that grow over and sometimes completely cover quite large trees and shrubs, and look very much like army camouflage nets, are a species of wild cucumber.

Pass the link road that loops to the right (it leads back to the main road) and then there are several short loops to the river on the left. Again keep a sharp lookout for samango monkeys, which were recently translocated from cultivated forests further to the west in the

Although there was no other bull to challenge him, this nyala went through a typical aggressive display of horning the ground after drinking in the Luvuvhu river.

Top right: *Crocodiles tear at the carcass of a nyala in the Luvuvhu river.* **Top:** *The sinister reptiles spread themselves along a sandbank.* **Above:** *A tree squirrel eats its meal overlooking the river.*

Above: *Hippos snooze and yawn away the hours of daylig*
Right: *An impala ram leaps gracefully away after drinki*

Northern Transvaal. I have yet to see one.

The last of these loops turns down to the river and follows it for a short distance to an old picnic site that was abandoned during hostilities between neighbouring countries in the 1960s. All that now remains are the ruins of the concrete water tank.

Here you should stay for two or three hours. Almost side by side hippo and crocodile snooze in the sun on the sand banks. From the deep forests on the far bank, bushbuck and impala tread slowly down to drink with eyes watchful and ears pricked for the slightest sign of danger. Sadly, the serious drought of 1982-83 took a heavy toll of the hippo population along the Luvuvhu river.

On the mudflats, scores of crocodiles laze with mouths agape, showing the yellow within. Some are stunning in their size – one giant is more than 5 m in length and is believed to be the biggest in the Park. Nyala and kudu browse quietly nearby and vervet monkeys and chacma baboons forage with their babies clinging to them.

Whitefronted bee-eaters, which nest in the mudbanks, swoop and dive like fighter planes, snatching seemingly invisible insects out of the air. Goliath herons nonchalantly stroll among the crocodiles. Like ghosts, fishing owls glide silently between the trees and settle in a flurry of feathers to stare with big unblinking, fathomless black eyes. This is also the domain of the crowned eagle, the most powerful of all raptors, which preys on monkeys and even small bushbuck and is known for its spectacular stunt-flying when courting. Common too, are flocks of the comical crested guineafowl with their quaint, untidy hairstyles.

You may complete the S63 loop by

driving to within a few hundred metres of the Mozambique border, with Crooks' Corner hidden across the river to the left. Turning south you will drive through a magnificent stand of fever trees stretching away in a belt on both sides of the road. A "no entry" road to the left leads to the Pafuri police post and to an old labour recruiting camp no longer in use.

Returning to the Pafuri bridge, you can now explore the S64 Nyala road west of the crossroads. It travels past the old picnic site, through a sandy dry riverbed flanked by fine baobab and very big nyala trees (good for game spotting because the clearing allows an open view through the usually thick bush), past the entrance to the tsetse fly research camp on the right and then past the steep, craggy hillside of Maschicindudzi koppie on the left. On the slopes and the crest of this hill are big stands of baobab trees interspersed with their "cousins", the bulbous multi-stemmed impala lilies with their beautiful, delicate pink and white flowers. Black eagles nest high on the cliff ledges and if you are lucky you may see these giant raptors soaring overhead.

After rains, there are often pools at the roadside which bushpigs and warthogs reduce to mudwallows. Clinging to twigs and foliage above the ponds are what look like balls of frothy white soapsuds, the nests of grey tree frogs. A number of females deposit a liquid; by using their back legs, they churn it to foam in which eggs are laid. On hatching, the tadpoles drop from these nests into the puddles below.

After only 4 km the road ends at a very big jackal-berry tree. Keep a sharp lookout for bushpig in this area. Although usually nocturnal, they are sometimes seen here by day.

Whitefronted bee-eaters busy themselves above their nests in the mudbanks of the Luvuvhu river near the old Pafuri picnic site.

WILDERNESS TRAILS

Car travel is essential in the Kruger National Park because of its size. The car brings the great range of different environments and camps within easy reach and is a fine "platform" from which to view wildlife – but it is still a barrier between man and nature.

People sitting in a metal, glass and vinyl box are effectively denied the full flavour of the world outside. They can see it, but cannot hear it or smell it. There is only one way to fully savour the "taste" of the bush and feel the beat of its pulse, and that is to walk through it.

In the Park that can be done. Four wilderness trails have been opened on which groups of eight people at a time can spend three nights, two full days and most of a third day living in the bush as close to nature as it is possible to get.

The idea took hold in the Park after the success of many hiking trails in other parts of South Africa of unusual beauty and interest, such as the Otter Trail in the lush Tsitsikama National Park forest on the southern Cape coast, the Klipspringer Trail in the spectacularly harsh surroundings of the huge Augrabies Falls National Park and the Fanie Botha Trail along the rim of the Great Escarpment in the Eastern Transvaal.

There were some misgivings about letting people hike in the Park. This was, after all, wild Africa full of untamed and potentially dangerous animals. But nowhere else in South Africa could people experience the wild in the same way as their forefathers had done who had to travel it on horseback and on foot, when game was as much a part of the scenery as the trees and mountains and sky.

The first trail, the Wolhuter Trail, was started in mid-1978, followed by the Olifants, Nyalaland and Bushman Trails, and others are planned.

Each trail in the Park is in a distinctly different part of the Park. The Nyalaland Trail in the far north has a unique variety of plant and bird life and some animals which are rare or non-existent in other regions. The Olifants Trail has breathtaking views along two rivers and the best concentrations of game. The Bushman Trail in the far south reveals scores of Bushman paintings and places where Iron Age man lived, and the rugged countryside of the Wolhuter Trail, also in the south, is the main area for white rhino.

The early doubts about letting people hike in the Park have vanished. There have been some scares, but of many thousands of trail visitors none has ever been hurt by a wild animal. The trails have been carefully chosen and an experienced ranger is always in command.

The public, clearly, have never been in doubt; the trails have become possibly the Park's major attraction, so that bookings have to be made a year in advance, although latecomers can often be accommodated owing to cancellations. So popular are they now that some people book for two or even three trails in succession, sometimes the same one.

None of the trails is arduous. Unlike other South African trails where the object is to complete a set distance in a certain number of days – which can mean marching up to 20 km a day up and down mountains – the purpose of the Park trails is to learn to know the wilderness, not to hike to get fit.

Each trail has a base camp in the bush and from this the ranger and his party set out every day, along no specific path. The routes they take are arranged informally to try to meet the interests and needs of the people in the group and they might go in any direction from the camp – straight out through the grassland, bush or forest, pausing when they come across an elephant or a tortoise or a column of Matabele ants or an unusual plant or simply a beautiful scenic view.

The rules on all trails are that visitors must be between the ages of 12 and 60 and must strictly abide by the instructions of the rangers. The only other rules, the rangers say, are: "Thou shalt not argue about religion or politics; thou shalt not bring transistor radios, and thou shalt not have any more rules."

The trails are held at each trail camp

twice a week throughout the year, except for a short break over the Christmas period. They begin on Mondays and Fridays and trailists must meet at the specified main rest camp at 3 p.m. on those days to be briefed by their ranger. They then travel out to the bush camp in the trails vehicle, taking with them only their clothing and personal effects.

All food and drink – except liquor, which the trailist must provide – is provided by the Park and prepared by the staff at the trail camp. People with special dietary needs must bring their own supplies. If people have particular food preferences or problems they should discuss these with the ranger at the main rest camp and, if necessary, they can buy additional food at the shop there to take with them. Everything else needed on trail – rucksacks, cutlery and crockery, water bottles, towels, beds and bedding – is provided by the Park.

The rangers themselves selected the sites for the trail camps and also built them: little groups of huts or tents skilfully laid out to blend into the bush in such a way that for the whole duration of a trail visit the wilderness is right on the doorstep.

The first evening out in the bush is spent around the campfire with the night and the wild pressing in close while the visitors have supper and discuss with the ranger what they want to see and do in the following days. They do not *have* to go walking in the veld: some people choose to stay in the camp to enjoy the solitude, the bird life, the fragrance of the trees and grass, the sunshine, or just to read a book.

Life at a trail camp follows a simple routine. Visitors are woken early in the morning with tea or coffee and when they have packed their rucksacks with camera, binoculars, sun barrier cream and whatever else they want to take along, they either head out from the camp on foot or board a vehicle to go to a starting point further afield.

In summer, because of the heat, they set out before sunrise and have a snack out on the trail, returning at about 11 a.m. for a hearty brunch and to rest during the hottest hours of the day before going out again in mid-afternoon. In winter the routine is to breakfast first, then spend the whole day out in the veld with a packed lunch and to return to a hot shower before dinner –

*Above: A founder ranger of the Wilderness Trails in the Park, Mike English, with one of the unique Bushman camp huts behind him. **Opposite page:** Waterbuck are usually seen while walking on trail.*

usually a campfire *braaivleis* (barbecue).

Visitors have to be prepared to take a chance with the weather. If they are caught in summer showers they have to find whatever shelter they can under trees or beneath rocky overhangs, but if the rain is heavy the ranger's vehicle is used as a mobile shelter from which the visitors take short walks. On one Nyalaland Trail outing the visitors cheerfully heaved and shoved their vehicle waist-deep through a flooded stream to get back to camp.

Out in the veld the accent is on enjoyment and learning. The trail turns the Kruger National Park into a vast outdoor classroom, a kind of living museum where under the friendly tutelage of the ranger people can see, feel, smell and savour some of nature's limitless spectrum of life.

For most people the trail is much more than a walk in the wild. For the harassed city dweller, especially, it can be a total relief, a recharging of the soul.

To many visitors it is more a spiritual than a physical experience. James Clarke, one of South Africa's best-known writers on nature and conservation, described it:

"For days one sees nothing that is not primaeval, nothing that was not there, just as one sees it now, when thousands of years ago Stone Age people hunted in the hot river beds.

"Everything one sees, smells and hears is relevant to the experience – unlike in a city where one has to shut off one's senses or be driven mad by the artificial stimulation of the traffic roar, the neon signs, the nameless odours and the anonymous crowds.

"There is no greater medicine for the urban prisoner than a spell of living the way we used to live for millennia."

Another visitor from the city gave this graphic description of a Nyalaland Trail:

"We tasted the water of the Tshalungwa hot spring and saw two cream-backed bateleurs drinking. In the base camp there is a boabab tree about a thousand years old which has a natural cave in its huge trunk. This serves as a perfect miniature habitat for five different bat species and two different kinds of spinetails.

"We used the binoculars to take a close look at a black eagle pair feeding their chick on a *kranz* (cliff) not far from the base camp. On trails along riverbeds, game paths and elephant paths we saw buffalo, impala and we even had a close look at a glistening python, and scores of animals, trees and birds that we got to know well.

"At night we sat around the campfire and recounted the events of that day while listening to the sounds of the night – the Scops owl, the threatening rasping cough of a leopard followed by the frightened snorting of an impala."

The ranger will introduce his group to what is called "the newspaper of the veld" – the tell-tale tracks in the sand left by animals, birds, reptiles and insects, which reveal to the experienced eye what has been happening in the past day or even several days. An animal's spoor can tell whether it was trekking or grazing, moving fast or slowly and sometimes, depending on other spoor, whether it was pursuing or being pursued.

There is often that matchless experience of following the ranger who has spotted, or picked up the spoor of, an elephant, giraffe or white rhino, and moving quietly, cautiously to within a stone's throw of the animal.

The trails teach the lesson that the wilderness is not a single raw entity but an intricately complex symbiosis of a multitude of life forms and activities evolved over millions of years. Understanding its everyday patterns is a challenging and stimulating skill. For those attuned to it the particular tang of dust, grass and pollen in the air, the feel of a leaf, the brittleness of a pod, the flight patterns of birds and butterflies, the bustle of termites and a broken stem are all words in a wonderful wild vocabulary.

The Wolhuter Trail

This was the first trail to be established, in mid-1978, and was named after the famed father and son rangers Harry and Henry Wolhuter. Harry Wolhuter became a legend in his own lifetime by killing a lion which attacked him, using only his sheath knife (see p. 108).

The gathering place is Skukuza and the trail camp is sited midway between the Berg-en-dal (hill and dale) and Pretoriuskop main camps and is surrounded by about 400 km² of wilderness ranging from riverine forest to monolithic, dome-shaped granite outcrops.

Spreading trees and shrubs shading the tented campsite are alive with birds, some of which, like hornbills and francolins, are now so accustomed to people that they feed in and around the camp. There is a waterhole at the camp's edge.

At night while visitors are having their meal they may well be joined by two regular wild guests. A genet often arrives to beg a chicken drumstick and usually takes his meal up onto a low bough where he lies feeding, quite unconcerned by the flashing of cameras. Later, a much more wary civet slips under the camp fence to snoop around

A civet is a regular late-night visitor at the Wolhuter trails camp in the south of the Park.

the kitchen area in search of a snack, giving photographers a rare opportunity to capture this shy creature on film.

The surrounding rugged hills and plains are the favoured home of white rhino and reedbuck and most of the other mammal species in the Park can be found here: impala, lion, giraffe, wildebeest and sable. On one memorable trail, 37 white rhino sightings were recorded.

The Olifants Trail

The starting point of this trail is at Letaba camp in the middle of the Park and the trail camp lies south-east of Olifants camp, about 4 km from the confluence of the Letaba and Olifants rivers. It is regarded by many as scenically the most beautiful of all the trails. Perched against a ridge and nestling among great white syringa trees some 100 m above the south bank of the Olifants river, the camp gives superb views along the river and of the Lebombo mountain range away to the east.

Just upstream from the camp is a stretch of river known simply as the "potholes", where the water rushes through narrow ravines. Thousands of years of ceaseless erosion by the water have sculpted fascinating holes and hollows in the rocks. These "potholes" also form a natural barrier for fish heading upstream to spawn so thousands of them crowd the pools below waiting for rain and for the river to rise. The fish, in turn, attract scores of crocodiles which lie along the rocky banks of the river within full view of the camp – more than 90 have been counted lying along the sandbanks less than 300 metres downstream.

Fish eagles are very common and several nest within sight of the camp along the river to the east. White-crowned plovers are often seen and sightings of other rarities have been recorded such as violeteared waxbills, ospreys, a lone sooty falcon (extremely rare), a breeding pair of jackal buzzards and familiar chats (six were recorded at the confluence of the rivers), among many others.

The variety of walks from this trail camp is wide and includes the river banks and rhyolitic foothills of the Lebombo mountains (which form the Mozambique border), the plains and the

Trails ranger, Don English, right, and a visitor stroll through the Olifants trail camp during a rest period.

forests. A highlight of the trail, usually reserved for the last evening, is to drive and then walk to a cliff-top 70 metres above the river confluence where visitors are allowed to simply enjoy the view, the solitude and the splendour of an African sunset.

The nights are full of the sounds of the bush including the soft, measured chomping of hippos grazing around the camp. Leopards have been known to stroll between the huts at night and on one memorable evening two lions had to be chased from the camp when they tried to settle in as uninvited guests.

The Olifants Trail is regarded by many as the best for game viewing with elephant breeding herds, giraffe, kudu, waterbuck, wildebeest and zebra, sometimes in large numbers. During one winter walk in 1985, the trail group encountered no less than five different prides of lions – which prompted trail ranger Dave Chapman to write in his diary, "We had lion fever today . . ."

The Nyalaland Trail

Punda Maria is the starting point and the trail camp lies to the north-east near the Luvuvhu river, on the bank of the Madzaringwe stream in the shade of a giant jackal-berry tree, three nyala trees and with a fine leadwood tree sheltering the kitchen.

This wilderness area is spectacularly beautiful with breathtaking scenery. More importantly, it is one of the leading botanical areas in the world which, apart from the southern sector of South Africa's Cape Province, has the greatest known variety of plant species growing anywhere on earth – and this, together with its topography of river and stream, sandstone formations and rocky hills and ridges, accounts for its almost equally impressive display of bird life.

The nests of black eagles can be seen near the trail camp and photographers who dare risk the wrath of these raptors can sometimes get near enough to record them. During 1985, for some inexplicable reason, a pair of black eagles changed their nesting habits and moved from the sheer cliffs north of the trails camp to a nearby baobab tree in which to build their home. The nest can be seen clearly through binoculars from the camp.

Black storks also inhabit the area, as does the exquisite crimson and green Narina trogon, the African finfoot, up to five kinds of rollers, seven types of hornbills (including the rare silvery-cheeked variety), the crowned eagle and the mysterious Pel's fishing owl – an impressive red-brown giant whose bare legs, long claws and rough soles adapt it to catching its chief diet of fish. Mottled spinetails nest in the stem of the baobab

tree in a corner of the camp and pennant-winged nightjars and the secretive bat hawk are often seen dipping between the huts at dusk. On one summer trail, made up of local ornithological club members, 186 species of birds were recorded. A count of 120 bird species on a single trail is not unusual.

Among the mammals that range in the area are roan and sable antelope, eland, Sharpe's grysbok, grey duiker and the very rare suni, plus the usual predators like hyenas and leopards. A single white rhino has been seen several times; in its loneliness, it has attached itself to a herd of buffalo.

Visitors can walk through a forest of baobabs, explore the unspoilt riverine forests and enjoy superb views from the Xantangelani sandstone ridge and the high cliffs overlooking the Luvuvhu river. The best known of these viewpoints is the Makahanja ruins, once the cliff-top kraal of the legendary chief known as Makahanja the Cruel (see p. 260).

The Bushman Trail

Berg-en-dal camp in the far south of the Park is the starting point of this trail.

Visitors studying the tourist map of the Kruger National Park are often surprised at the close proximity of the Bushman and Wolhuter trail camps – just 7 km apart in the south-east of the Park.

The tented Wolhuter trail camp.

SSSSHH !!

The reason is that the newest camp has been specially sited in the heart of an area steeped in archaeological interest. Evidence of primitive man is everywhere – Stone Age tools, the remains of Iron Age smelting furnaces, ancient clay storage pots and most of the 108 known Bushman painting sites within the Park.

It is the taste of the ancient past which gives this camp its special flavour, although it too has a wide range of animal life.

The landscape is rugged and hilly with granite outcrops which make it particularly beautiful at sunrise and sunset. The diversity of game to be seen includes kudu, giraffe, plains reedbuck, mountain reedbuck and, just possibly, grey rhebok (vaalribbok), a number of which were recently released on a plateau 5 km from the camp. White rhino sightings are almost a certainty.

The camp itself is delightful. Opened on 15 August 1983, its accommodation is twin-bed rondavel-type huts constructed of timber and thatch, designed and built by rangers Mike English and Johan van Graan (the latter having been involved in the building of all four trail camps). The huts are built on stilts and the cleverly designed windows have huge wooden flaps opening inwards and upwards to allow for maximum fresh air to pass through. There are no curtains, but the thatched eaves of the roofs drop to the level of the lower sills to give complete privacy. Unlike those in other trail camps, the huts are sited well apart. The camp boasts its own waterhole.

What to take on trail

The correct choice of footwear is most important. Boots or good walking shoes – preferably with ankle support – are recommended. The shoes should have previously been "worn in" and the soles should be reasonably thick. All-wool socks should be worn for complete com-fort. Clothing should be loose enough to allow freedom of movement, and neutral in colour, such as khaki or green.

While in camp, a tracksuit is a good choice, as are tennis shoes, sandals or even a favourite pair of slippers to relieve weary feet. Other than the usual basic toiletries (and *don't* forget anti-malaria tablets), all that is needed is a jersey, raincoat, a hat with a brim wide enough to protect neck and ears, a torch, binoculars and, if desired, camera equipment.

Visitors not accustomed to harsh sunshine or who have sensitive skins are advised to take dark glasses, a good sunscreen lotion, shirts and blouses with long sleeves and long trousers. An absolute must is a good bird reference book such as *Newman's Birds of the Kruger National Park*, or *Roberts' Birds of Southern Africa*.

Portable radios are prohibited on trail but visitors are welcome to bring tape recorders to record bird calls and other sounds of the bush.

KNOW YOUR TREES BY NUMBERS

Visitors to the Kruger National Park will observe that many trees have numberplates attached to the stems. Those interested in botany may identify a particular species from the following, numbered according to the National List of Trees:

22	*Phoenix reclinata*	Wild date palm
23	*Hyphaene natalensis*	Lala palm
28	*Aloe bainesii*	Tree aloe
39	*Celtis africana*	White stinkwood
42	*Trema orientalis*	Pigeonwood
43	*Chaetachme aristata*	Thorny elm
47	*Ficus sansibarica*	Knobbly fig
48	*Ficus thonningii* (= *F. petersii*)	Common wild fig
50	*Ficus sur* (= *F. capensis*)	Broom cluster fig
55	*Ficus ingens*	Red-leaved rock fig
57	*Ficus natalensis*	Natal fig
60	*Ficus salicifolia* (= *F. pretoriae*)	Wonderboom fig
62	*Ficus tettensis* (= *F. smutsii*)	Small-leaved rock fig
63	*Ficus abutilifolia* (= *F. soldanella*)	Large-leaved rock fig
64	*Ficus glumosa* (= *F. sonderi*)	Mountain rock fig
65	*Ficus stuhlmannii*	Lowveld fig
66	*Ficus sycomorus* subsp. *sycomorus*	Sycamore fig
70	*Pouzolzia hypoleuca*	Soap nettle
71	*Urera tenax*	Mountain nettle
75	*Faurea saligna*	Transvaal beech
76	*Faurea speciosa*	Broad-leaved beech
101	*Olax dissitiflora*	Small sourplum
102	*Ximenia americana*	Blue sourplum
103	*Ximenia caffra*	Large sourplum
104	*Portulacaria afra*	Porkbush
105	*Annona senegalensis*	Wild custard-apple
106	*Hexalobus monopetalus*	Shakama plum
107	*Monodora junodii*	Green-apple
108.2	*Uvaria lucida* var. *virens*	Large cluster-pear
110	*Xylopia odoratissima*	Savanna bitterwood
120	*Gyrocarpus americanus*	Propeller tree
122	*Boscia albitrunca*	Shepherd's tree
122.1	*Boscia angustifolia* var. *corymbosa*	Rough-leaved shepherd's tree
126	*Boscia mossambicensis*	Broad-leaved shepherd's tree
127	*Boscia foetida* subsp. *rehmanniana*	Stink shepherd's tree
129.1	*Cadaba natalensis*	Natal worm-bush
132	*Maerua angolensis*	Bead-bean
133	*Maerua cafra*	Bush-cherry
135	*Maerua rosmarinoides*	Needle-leaved bush-cherry
146	*Parinari curatellifolia*	Mobola plum
148	*Albizia adianthifolia*	Flat-crown
149	*Albizia amara* subsp. *sericocephala*	Bitter false-thorn
150	*Albizia anthelmintica*	Worm-bark false-thorn
152	*Albizia brevifolia*	Mountain false-thorn
153	*Albizia petersiana* subsp. *evansii*	Many-stemmed false-thorn
154	*Albizia forbesii*	Broad-pod false-thorn
155	*Albizia harveyi*	Common false-thorn
157	*Albizia tanganyicensis* subsp. *tanganyicensis*	Paperbark false-thorn
158	*Albizia versicolor*	Large-leaved false-thorn
159	*Acacia albida*	Ana tree
160	*Acacia ataxacantha*	Flame thorn
160.1	*Acacia borleae*	Sticky thorn
161	*Acacia burkei*	Black monkey thorn
162	*Acacia caffra*	Common hook-thorn
163	*Acacia welwitschii* subsp. *delagoensis*	Delagoa thorn
163.1	*Acacia davyi*	Corky thorn
164	*Acacia erubescens*	Blue thorn
164.1	*Acacia exuvialis*	Flaky thorn
167	*Acacia gerrardii* var. *gerrardii*	Red thorn
168.1	*Acacia grandicornuta*	Horned thorn
172	*Acacia karroo*	Sweet thorn
174.1	*Acacia luederitzii* var. *retinens*	Belly thorn
178	*Acacia nigrescens*	Knob thorn
179	*Acacia nilotica* subsp. *kraussiana*	Scented thorn
180	*Acacia polyacantha* subsp. *campylacantha*	White thorn
183.1	*Acacia robusta* subsp. *clavigera*	Brack thorn
185	*Acacia senegal* var. *leiorhachis*	Slender three-hook thorn
185.1	*Acacia senegal* var. *rostrata*	Three-hook thorn
187	*Acacia sieberana* var. *woodii*	Paperbark thorn
187.2	*Acacia swazica*	Swazi thorn
188	*Acacia tortilis* subsp. *heteracantha*	Umbrella thorn
189	*Acacia xanthophloea*	Fever tree
190	*Dichrostachys cinerea* subsp. *africana*	Sickle bush
190.1	*Dichrostachys cinerea* subsp. *nyassana*	Large-leaved sickle bush
191	*Newtonia hildebrandtii* var. *hildebrandtii*	Lebombo wattle
192	*Xylia torreana*	Sand ash
193	*Elephantorrhiza burkei*	Sumach bean
197	*Burkea africana*	Wild seringa
198	*Colophospermum mopane*	Mopane
200	*Guibourtia conjugata*	Small copalwood
202	*Schotia brachypetala*	Weeping boer-bean
203	*Schotia capitata*	Dwarf boer-bean
207	*Afzelia cuanzensis*	Pod mahogany
209	*Piliostigma thonningii*	Camel's foot
212	*Cassia abbreviata* subsp. *beareana*	Sjambok pod
213	*Cassia petersiana*	Monkey pod
215	*Peltophorum africanum*	Weeping wattle
216	*Cordyla africana*	Wild mango
219	*Calpurnia aurea* subsp. *aurea*	Natal laburnum
222	*Bolusanthus speciosus*	Tree Wistaria
223	*Baphia massaiensis* subsp. *obovata*	Sand camwood
226	*Tephrosia sericea* (= *Mundulea sericea*)	Cork bush
230	*Ormocarpum trichocarpum*	Caterpillar bush
231	*Dalbergia armata*	Thorny rope
232	*Dalbergia melanoxylon*	Zebrawood
234	*Dalbergia nitidula*	Glossy flat-bean
236	*Pterocarpus angolensis*	Transvaal teak

236.1	*Pterocarpus lucens* subsp. *antunesii*	Thorny teak
237	*Pterocarpus rotundifolius* subsp. *rotundifolius*	Round-leaved teak
238	*Lonchocarpus capassa*	Apple-leaf
240	*Ostryoderris stuhlmannii*	Wing bean
241	*Xanthocercis zambesiaca*	Nyala tree
244	*Erythrina latissima*	Broad-leaved coral tree
245	*Erythrina lysistemon*	Common coral tree
248	*Erythroxylum delagoense*	Small-leaved coca tree
249	*Erythroxylum emarginatum*	Common coca tree
251	*Balanites maughamii*	Green thorn
252	*Balanites pedicellaris*	Small green thorn
253	*Zanthoxylum capense*	Small knobwood
256	*Calodendrum capense*	Cape chestnut
260	*Vepris reflexa*	Bushveld white ironwood
262	*Toddaliopsis bremekampii*	Wild mandarin
265	*Clausena anisata*	Horsewood
267	*Kirkia acuminata*	White seringa
269	*Kirkia wilmsii*	Mountain seringa
275	*Commiphora edulis*	Rough-leaved corkwood
277	*Commiphora harveyi*	Red-stem corkwood
278	*Commiphora marlothii*	Paperbark corkwood
279	*Commiphora merkeri*	Zebra-bark corkwood
280	*Commiphora mollis*	Velvet corkwood
283	*Commiphora neglecta*	Sweet-root corkwood
285	*Commiphora pyracanthoides* subsp. *pyrocanthoides*	Common corkwood
285.1	*Commiphora pyrocanthoides glandulosa*	Tall common corkwood
287	*Commiphora schimperi*	Glossy-leaved corkwood
289	*Commiphora tenuipetiolata*	White-stem corkwood
291.1	*Commiphora zanzibarica*	Lebombo corkwood
292	*Ptaeroxylon obliquum*	Sneezewood
293	*Entandrophragma caudatum*	Mountain mahogany
297	*Turraea nilotica*	Bushveld honeysuckle tree
298	*Ekebergia capensis*	Cape ash
301	*Trichilia emetica*	Natal mahogany
303	*Securidaca longipedunculata*	Violet tree
308	*Pseudolachnostylis maprouneifolia*	Kudu-berry
309	*Securinega virosa*	White-berry bush
311	*Phyllanthus reticulatus*	Potato bush
312	*Phyllanthus cedrefolius*	Forest potato bush
314	*Drypetes gerrardii*	Forest iron-plum
315	*Drypetes mossambicensis*	Sand iron-plum
317	*Hymenocardia ulmoides*	Red-heart tree
318	*Antidesma venosum*	Tassel berry
320	*Cleistanthus schlechteri*	False tamboti
322	*Bridelia cathartica*	Blue sweetberry
324	*Bridelia micrantha*	Mitzeeri
325	*Bridelia mollis*	Velvet sweetberry
327	*Androstachys johnsonii*	Lebombo ironwood
328	*Croton gratissimus* var. *gratissimus*	Lavender fever-berry
329	*Croton megalobotrys*	Large fever-berry
330	*Croton sylvaticus*	Forest fever-berry
331	*Croton zambesicus*	White lavender fever-berry
334	*Alchornea laxiflora*	Venda bead-string
341	*Spirostachys africana*	Tamboti
345	*Euphorbia confinalis*	Lebombo Euphorbia
346	*Euphorbia cooperi*	Transvaal candelabra tree
348	*Euphorbia evansii*	Lowveld Euphorbia
351	*Euphorbia ingens*	Common tree Euphorbia
352	*Euphorbia keithii*	Swazi Euphorbia
355	*Euphorbia tirucalli*	Rubber Euphorbia
360	*Sclerocarya birrea* subsp. *caffra*	Marula
362	*Lannea discolor*	Live-long
363	*Lannea stuhlmannii*	False marula
371	*Ozoroa engleri*	White resin tree
375	*Ozoroa paniculosa*	Common resin tree
376	*Ozoroa insignis* subsp. *reticulata*	False currant resin tree
384	*Rhus gueinzii*	Thorny karree
387	*Rhus leptodictya*	Mountain karree
391	*Rhus pentheri*	Common crow-berry
392	*Rhus pyroides*	Common wild currant
399	*Maytenus heterophylla*	Common spike-thorn
399.2	*Maytenus mossambicensis* var. *mossambicensis*	Black forest spike-thorn
402	*Maytenus senegalensis*	Red spike-thorn
403	*Maytenus undata*	Koko tree
404	*Catha edulis*	Bushman's tea
410	*Cassine aethiopica*	Kooboo-berry
416	*Cassine transvaalensis*	Transvaal saffron
422	*Apodytes dimidiata* subsp. *dimidiata*	White pear
423	*Allophylus decipiens*	False currant
427	*Atalaya alata*	Lebombo krantz ash
430	*Deinbollia oblongifolia*	Dune soap-berry
433	*Pappea capensis*	Jacket-plum
435	*Stadmannia oppositifolia* subsp. *rhodesica*	Silky plum
437	*Dodonaea viscosa* var. *viscosa*	Common sand olive
438	*Hippobromus pauciflorus*	False horsewood
447	*Ziziphus mucronata*	Buffalo-thorn
448	*Ziziphus rivularis*	False buffalo-thorn
449	*Berchemia discolor*	Brown ivory
450	*Berchemia zeyheri*	Red ivory
455	*Heteropyxis natalensis*	Lavender tree
458	*Grewia bicolor*	Witrosyntjie
459	*Grewia caffra*	Climbing raisin
459.1	*Grewia flava*	Velvet raisin
460	*Grewia hexamita*	Giant raisin
461.1	*Grewia microthyrsa*	Lebombo raisin
462	*Grewia monticola*	Silver raisin
463	*Grewia occidentalis*	Cross-berry
463.2	*Grewia villosa*	Mallow raisin
467	*Adansonia digitata*	Baobab
469	*Dombeya cymosa*	Natal wild pear
470	*Dombeya kirkii*	River wild pear
471	*Dombeya rotundifolia* var. *rotundifolia*	Common wild pear
475	*Sterculia murex*	Lowveld chestnut
477	*Sterculia rogersii*	Common star-chestnut
481	*Ochna natalitia*	Natal plane
483	*Ochna pulchra*	Peeling plane
486	*Garcinia livingstonei*	Lowveld mangosteen
488	*Warburgia salutaris*	Pepper-bark tree
492	*Oncoba spinosa*	Snuff-box tree
493	*Xylotheca kraussiana*	African dogrose
498	*Scolopia zeyheri*	Thorn pear
501	*Homalium dentatum*	Brown ironwood
506	*Flacourtia indica*	Governor's plum

507	*Dovyalis caffra*	Kei-apple
523	*Galpinia transvaalica*	Transvaal privet
532	*Combretum apiculatum* subsp. *apiculatum*	Red bushwillow
534	*Combretum celastroides*	Trailing bushwillow
535	*Combretum engleri*	Sand bushwillow
536	*Combretum erythrophyllum*	River bushwillow
537	*Combretum molle*	Velvet bushwillow
538	*Combretum hereroense*	Russet bushwillow
539	*Combretum imberbe*	Leadwood
540.1	*Combretum paniculatum* var. *microphyllum*	Flame creeper
541	*Combretum collinum* subsp. *gazense*	Rhodesian bushwillow
541.2	*Combretum collinum* subsp. *suluense*	Weeping bushwillow
545	*Combretum woodii*	False forest bushwillow
546	*Combretum zeyheri*	Large-fruited bushwillow
547	*Pteliopsis myrtifolia*	Myrtle bushwillow
549	*Terminalia phanerophlebia*	Lebombo cluster-leaf
550	*Terminalia prunioides*	Thorny cluster-leaf
551	*Terminalia sericea*	Silver cluster-leaf
553	*Eugenia zeyheri*	Wild myrtle
555	*Syzygium cordatum*	Water berry
557	*Syzygium guineense*	Water pear
562	*Cussonia natalensis*	Rock cabbage tree
564	*Cussonia spicata*	Common cabbage tree
569	*Steganotaenia araliacea*	Carrot tree
577	*Maesa lanceolata*	False assegai
579	*Sideroxylon inerme*	White milkwood
581	*Bequaertiodendron magalismontanum*	Transvaal milkplum
585	*Mimusops zeyheri*	Transvaal red milkwood
586	*Manilkara concolor*	Zulu milkberry
587	*Manilkara mochisia*	Lowveld milkberry
590	*Vitellariopsis marginata*	Bush milkwood
595	*Euclea divinorum*	Magic guarri
597	*Euclea natalensis*	Natal guarri
600	*Euclea schimperi* var. *schimperi*	Bush guarri
601	*Euclea undulata* var. *undulata*	Common guarri
605	*Diospyros lycioides* subsp. *lycioides*	Karroo bluebush
605.1	*Diospyros lycioides* subsp. *sericea*	Natal bluebush
606	*Diospyros mespiliformis*	Jackal-berry
607.1	*Diospyros natalensis* subsp. *nummularia*	Lebombo jackal-berry
610	*Diospyros villosa*	Hairy star-apple
611	*Diospyros whyteana*	Bladdernut
613	*Schrebera* var. *trichoclada*	Wild jasmine
614	*Chionanthus battiscombei* (= *Linociera battiscombei*)	Water pock ironwood
617	*Olea europaea* var. *africana*	Wild olive
621	*Salvadora angustifolia* var. *australis*	Transvaal mustard tree
622	*Salvadora persica*	Red mustard tree
624	*Strychnos decussata*	Cape teak
625	*Strychnos henningsii*	Coffee bitterberry
626	*Strychnos madagascariensis*	Spineless monkey orange
629	*Strychnos spinosa*	Green monkey orange
630	*Strychnos potatorum*	Black bitterberry
631	*Strychnos usambarensis*	Blue bitterberry
632	*Anthocleista grandiflora*	Forest fever tree
634	*Nuxia floribunda*	Forest elder
635	*Nuxia oppositifolia*	Water elder
639	*Acokanthera oppositifolia*	Common poison-bush
640	*Acokanthera rotundata*	Round-leaved poison-bush
640.2	*Carissa haematocarpa*	Karroo num-num
642	*Holarrhena pubescens*	Fever pod
643	*Diplorhynchus condylocarpon*	Horn-pod tree
644	*Tabernaemontana elegans*	Lowveld toad tree
647	*Rauvolfia caffra*	Quinine tree
650	*Wrightia natalensis*	Saddle pod
653	*Cordia grandicalyx*	Round-leaved saucer-berry
654	*Cordia ovalis*	Snot berry
656	*Ehretia amoena*	Sandpaper bush
657	*Ehretia rigida*	Puzzle bush
658	*Premna mooiensis*	Skunk bush
659	*Vitex amboniensis*	Plum finger-leaf
660	*Vitex harveyana*	Whorled finger-leaf
665	*Vitex wilmsii*	Hairy finger-leaf
667	*Clerodendrum glabrum*	Resin-leaf
668	*Holmskioldia tettensis*	Wild parasol flower
676	*Rhigozum zambesiacum*	Mopane pomegranate
677	*Markhamia acuminata*	Bell bean tree
678	*Kigelia africana*	Sausage tree
680	*Sesamothamnus lugardii*	Transvaal sesame bush
682	*Hymenodictyon parvifolium*	Wild firebush
683	*Crossopteryx febrifuga*	Sand crown-berry
684	*Breonadia salicina* (= *Adina microcephala*)	Matumi
687	*Tarenna littoralis*	Dune butterspoon bush
689	*Catunaregam spinosa* subsp. *spinosa* (= *Xeromphis obovata*)	Thorny bone-apple
689.1	*Coddia rudis* (= *Xeromphis rudis*)	Small bone-apple
690	*Gardenia amoena*	Thorny gardenia
691	*Gardenia volkensii* subsp. *volkensii* (= *G. spatulifolia*)	Transvaal gardenia
694	*Rothmannia fischeri*	Rhodesian gardenia
698	*Tricalysia capensis*	Forest coffee
702	*Vangueria infausta* subsp. *infausta*	Wild medlar
707	*Canthium huillense*	Bushveld rock alder
708	*Canthium inerme*	Common turkey-berry
710	*Canthium mundianum*	Rock alder
711	*Canthium obovatum*	Quar
715	*Plectroniella armata*	False turkey-berry
716	*Pavetta gardeniifolia* var. *gardeniifolia*	Common bride's bush
717	*Pavetta edentula*	Gland-leaf tree
721	*Pavetta schumanniana*	Poison bride's bush
722	*Pavetta zeyheri*	Small-leaved bride's bush
727	*Brachylaena huillensis*	Lowveld silver oak
731	*Brachylaena discolor* subsp. *transvaalensis*	Forest silver oak
733	*Tarchonanthus camphoratus*	Camphor bush
734	*Tarchonanthus trilobus* var. *galpinii*	Broad-leaved camphor tree

ACKNOWLEDGEMENTS

This book would not have been possible without the generous help and encouragement of a large number of people, many more than those listed below. I know that as time goes by others will be remembered, others who should be mentioned by name, and to them I apologise for not giving them the credit they so richly deserve.

I must begin by thanking my colleague and co-author Wilf Nussey who came to my aid with hardly a whimper, somehow finding the time as a busy newspaper editor to put much of this book together on his typewriter, and also Peter Dugmore, a life-long friend, who has contributed so much good business sense to this project – and future ones.

I am sincerely grateful to the National Parks Board of Trustees for allowing me the freedom to travel, photograph and gather information within the Kruger Park and also to Piet van Wyk, Head of Research and Information, for his initial guidance and encouragement.

Among those who work within the Park, I must particularly mention the Chief Warden, "Tol" Pienaar, and also Salomon Joubert and Johan Verhoef for their professional advice and kindness.

Many Park staffers became valued friends and among them I would like to make special mention of Mike English (and his wife André), Flip Nel, Ted Whitfield, Trevor Dearlove, Eric Reisinger, Johan van Jaarsveld, Louis Olivier and Pat Wolff.

I am indebted to many others, most of them Park staffers, but also people in other walks of life. Included among them in alphabetical order are: Dirk Ackerman, Roy Bengis, Mike Boshoff, Bruce Bryden, Hannes Eloff, Irene Grobler, Anthony Hall-Martin, Charles Hoffman, Hans Kolver, Ben Lamprecht, Koos Lombard, André Meyer, Thys Mostert, Michael Pepper, Ben Pretorius, Schoeman Rossouw, Johan van Graan, Hugo van Niekerk, Lyn van Rooyen, V. "Vossie" de Vos, Ludwig Wagner, Merle White and Tom Yssel.

Artists and illustrators made a valuable contribution to the visual impact of the book. They are: Ralph Rillman Miller, to whom I am specially grateful for his contribution of the superb portrait of President Paul Kruger; Bruce Paynter, for most of the other full-colour illustrations; Tony Grogan, Mynderd Vosloo and Mike Chase for sketches and graphics; Peter Slingsby for the maps; and finally cartoonist Gordon Linley, for his delightful lighthearted touch.

I sincerely thank Jennifer Paynter, Doreen Nussey and Neilah Miller for searching through the text for literal and other errors during the final production stages, designer Mel Miller for his contribution and again, my son, Bruce, for whistling his way through scores of colour trace-guides sketched while I completed the page layouts.

Others who gave valuable time to check and confirm facts and identify species were John Comrie-Grieg of the magazine *African Wildlife*, entomologist Dr Vincent Whitehead, arachnologist Dr Ansie Dippenaar and the team of botanists under John Rourke at Kirstenbosch Botanic Gardens.

A number of photographs had to be collected from "outside" sources. Historical pictures were supplied by the National Parks Board, Professor Hannes Eloff, Dudley Tomlinson and Gretel Reitz. The series of graphic sketches of Harry Wolhuter and the lion were drawn by the late C. T. Astley Maberly.

Other illustrations were supplied by the J.L.B. Smith Institute of Ichthyology (Liz Tarr), Hercules Els of the Department of Electron Microscopy at MEDUNSA, *The Argus*, Cape Town, the Transvaal Nature Conservation Division, Eric Reisinger, the late Hilda Stevenson-Hamilton and Melanie Baker of Ottawa, Canada. The colour pictures of the darting of the elephant, Mafunyane, are by Lorna Stanton and the "white" animals of Tshokwane by Pat Wolff, both on the staff of the National Parks Board. I am most grateful to these contributors.

There are scores of others, too numerous to list, to whom gratitude is owing – camp reception staff, Automobile Association technicians, Park secretarial staff and typists, researchers, librarians, and all those who helped in the physical preparation and production of this work.

Among this last group there are several I must mention. Most of the brilliant colour origination in this book was handled personally by Gunther Zimmermann, managing director of Unifoto (Pty) Ltd; the overseer of the typesetting was Paul Kruger of the same company. I salute them and their team.

I am also grateful to Bernard Mason at Printpak Cape (Pty) Ltd for his enthusiasm while watching over the printing of the book and to Ted Jenkins, works manager of *The Argus*, for allowing me the facilities of their ATEX editing system and for never failing to find me a vacant terminal when I needed one.

Finally, I thank my publishers, Macmillan South Africa (Pty) Ltd, and in particular Basil van Rooyen and Erroll Marx, not only for their true professionalism, but also for their patience and friendship.

David Paynter

BIBLIOGRAPHY

Research for this book spanned a number of published works listed in the following bibliography. Some of these are usually available at shops in the Park or from various National Parks Board offices and are marked with an asterisk.

Abshire, D. M. and M. A. Samuels (Eds.), *Portuguese Africa*, Pall Mall, 1969.

*Braack, L. E. O., *The Kruger National Park*, Struik, 1983.

Bulpin, T. V., *The Ivory Trail*, Books of Africa, 1981.

Bulpin, T. V., *Treasury of Travel Series*, No. 19, Mobil, 1974.

Cattrick, Alan, *Spoor of Blood*, Howard Timmins, 1965.

Custos, the official magazine of the National Parks Board.

*Fitzpatrick, Sir Percy, *Jock of the Bushveld*, Penguin Books, 1981.

Henriksen, Thomas H., *Mozambique, a History*, Rex Collings, 1978.

*Kloppers, Johan and G. van Son, *Butterflies of the Kruger National Park*, National Parks Board, 1978.

*McLachlan, G. R. and R. Liversidge, *Roberts' Birds of South Africa*, John Voelcker Bird Book Fund, 1980.

*Newman, K., *Birds of the Kruger National Park*, Macmillan, 1980.

*Onderstal, Jo, *The South African Wild Flower Guide 4, Transvaal Lowveld and Escarpment*, Botanical Society of South Africa, 1984.

*Pienaar, U. de V., *The Freshwater Fishes of the Kruger National Park*, National Parks Board, 1978.

Pienaar, U. de V., *The Kruger Park Saga (1898-1981)*, unpublished.

*Pienaar, U. de V., W. D. Haacke and N. H. G. Jacobsen, *The Reptiles of the Kruger National Park*, National Parks Board, 1978.

*Pienaar, U. de V., with N. I. Passmore and V. C. Caruthers, *The Frogs of the Kruger National Park*, National Parks Board, 1976.

Pivnic, H. L., *Dining Cars and Lounge Cars of the South African Railways*, S.A. Railways, 1985.

Potgieter, D. J., P. C. du Plessis and S. H. Skaife, *Animal Life in Southern Africa*, Nasou Limited, 1971.

Smithers, Reay M. N., *The Mammals of the Southern African Subregion*, University of Pretoria, 1983.

Stevenson-Hamilton, J., *South African Eden*, Cassell and Company, 1952.

*Van Wyk, Piet, *Field Guide to the Trees of the Kruger National Park*, Struik, 1984.

Wolhuter, H. C., *Memories of a Game Ranger*, The Wild Life Protection Society of South Africa, 1961.

INDEX

A

Accommodation 2, 61, 140
  see also Camps
Africalus aurens 14
Agama
  tree **13**
Albasini
  João 41, **42,** 43, 170
  ruins 171
Albino
  buffalo 131, **132, 133**
  giraffe **131**
  kudu 131, **132**
Aloes 30
Anglo-Boer War 45, 47-48
Animal products plant 74-75
Animals
  aggressive **118-119**
  battling **112-113**
  behaviour of **115-117**
  census: *see* Census
  dangerous 73, 101-111, 118
  see also various species
Ants
  army 22
  behaviour 22
  Matabele 22-23
Archaeology 37-39, 177, 227, 237, 276
Art: *see* Rock art; Bushman paintings
Arthropods 18-21
Askaris 122

B

Baboons
  chacma 7, **184**
  dangerous 101
Balule: *see* Camps
Baobab
  Hill **25,** 266
  trees 24, **25,** 221, 228, 229, 263
Barberton daisies 30, 171
Barbets **203,** 215
Bateleur: *see* Eagle
Bats 273
Bauhinia galpinii 30, 33
  see also De Kaap, Pride of
Bee-eater
  carmine 8
  European 8
  little **7**
  whitefronted **11,** 206, 269, **270-271**

Beetle
  dung 17, **18, 19**
  scarab 17-18
Berg-en-dal: *see* Camps
Binoculars **148,** 149, 225, 232, 236
  see also Photography
Birds 7-12, 195, **203,** 205, 206, **214, 220,**
    226, 230, 234, 236, 239, **144,** 248,
    261, 267, 275
  see also various species
Blue wildebeest: *see* Wildebeest
Boomkoggelmander 13
Botha, Dr Johann 96
Braack, Leo 18, 23
Bryden, Bruce **89,** 91
Brynard, Mr A. M. 67
Buffalo 5, **68, 132, 133, 190, 206, 212,**
    213, 231, **236, 240-241**
Bullfrog **14**
Bushbuck 5, **156, 205, 240, 249**
Bushman 37-39, 84, 177
  paintings **37, 38,** 196, 272, 276
  see also Rock art
Bushpig 267, 269
Bushveld
  Jock of the 171, 175
    camp: *see* Camps
Buthidae 18
Butterflies 20

C

Caldecott, Harry Stratford 56
Camps
  Balule 167, 218, 227, 229
  Berg-en-dal 2, **145,** 147, 165, **168,**
    169, 177-179
  Boulders **147,** 239
  Crocodile Bridge 147, 167, 195-207
  Doornplaat 147
  Jock of the Bushveld 147, 173
  Lanner Gorge 147
  Letaba **224,** 225-239
  Lonely Bull 147
  Lower Sabie 165, **194,** 195-207, **200**
  Malelane 147
  Maroela Caravan 219, 221
  Narina 147
  new 147
  Ngotsamond Caravan 147, 218, 227
  Nwanedzi 213-218
  Olifants 164, **224,** 225-229

Orpen Gate 221
Pioneer 147, 234, 239
Pretoriuskop 165, **168,** 169-177
Punda Maria 164, **258,** 259-269
Roodewal 147, 211
Satara 164, **208,** 209-221
Shingwedzi 239, **242,** 243-255
Skukuza 165, **180,** 181-193
Camping: *see* Caravanning
Caravanning 143-144, 182, 195, 219, 243,
  259
Cattrick, Alan 49
Census, animal **70,** 71, 97, **99,** 122
Centipede 19
Chapman, Abel 46
Cheetah 181, **191,** 196, 201, 205, 207,
  213, **214,** 266
Civet 205, **274**
Clarke, James 18, 273
Coetser, Capt. J. J. 259
Conservation 69-70
Coral tree **28,** 29
Cougal **255**
Crocodile
  attack by 102-103, **104-105,** 106
  biggest 269
  Nile 13, 232, 239, **268-269**
  rescue from 101-106
  River: *see* Rivers
Crooks' Corner **25,** 51-52, 266, 269
Culling 66, 71, **72-73,** 79, 97
  darts used in **74, 75**
  drugs used in 73-74
  helicopter used in 97
  methods 71, 73, 74
  other parks, in 72
  species involved in 72, 74-75
  utilisation of products 72
Custos 91, 96

D

Dams
  Engelhard 232, 239
  Gezantombi 196
  Gudzane 215, 218
  Kanniedood 164, 243, 245, 254
  Lugmag 221
  Manzimahle 187
  Maritubi 164, 260-261
  Mestel 171
  Mhlanganzwane 197

Mingerhout 164, 231, 232
Mlondozi 201
Mobeweni 177
Mpondo 207
Ngwenyeni 221
Nsemani 210
Pioneer 234
Rabelais 219
Sable 234, 237
Shiloweni 129, 189
Shimangwanene 220
Transport 115, 173
Winkelhaak 236
Dearlove, Trevor **118**, 119
Decoration, Wolraad Woltemade **101**, **106**
De Kaap, Pride of 30, **33**, 179
see also *Bauhinia galpinii*
Delagoa Bay: see Mozambique
Disease 50, 70, 79-80
anthrax 80, 98
foot-and-mouth 80
rinderpest 50
Dog, wild: see Wild dogs
Dorylinae 22
Dove
emerald spotted 255
mourning 234
Drugs
culling 73, 78-79, **80**
darts for **74**, **75**
Duiker
grey 5, 275
red 170

E

Eagle
bateleur 10, **226**
black v, 10, **264**, 269, 275
blackbreasted snake 10
brown snake 10
crowned 9, 269
fish 10, 206, 210, 274
martial v, 9, 10, **12**, 206
steppe 10
Wahlberg's 10
Ecosystem 69-70
fire, role of, in 85
Eel 34
Egret
black 8
great white **246**, **247**, 248
Egyptian geese 231, 244, **263**, 267
Eland 5, 231, 267
Elephants **4**, 5, **114**, 119-128, **144**, **150**, **172**, **176**, 183, **204**, **222-223**, 228, **233**, 243, **245**, 255, **256-257**
culling 71-74
counting 71-72

Hall of Fame 147
poaching **89**
see also "Magnificent Seven"
Eloff, Prof.
Fritz **67**, 126
Hannes 39
Engelhard
dam: see Dams
Charles William 239
English, Mike **273**, 276
Entrance gates: see Gates
Euphorbia 170
confinalis 201
cooperi 201
Explorers 40-42, 84, 196

F

Falcon, sooty 274
Fever trees 25, 28, 229, 254, **264**
Fig tree: see Sycamore
Fire 66, 70, 84, **85**, 146
role of, in ecosystem 85
Fish 34
barbel 34
bream 34
cat 34
eagle: see Eagle
killifish 34-35
ladder 234
lungfish 34, **35**
Fitzpatrick, Sir Percy 171, 173, 175
Flame
creeper 30, 209, **238**, **263**, 267
lily 30
Flamingoes 234
Flowers 30-34, **31**, **32**, **33**, 201
Fortune seekers 40-43
Francolin 7-8
Fraser, Maj. A. A. 50, **51**
Frogs 14-17
bullfrog **14**
dwarf puddle 14
golden reed 14
ornate 14
sandveld pyxie 14
treefrog 269

G

Game counting **70**, **71**, 97, **99**
Gates, entrance
Crocodile Bridge 142
Malelane 142, 177
Numbi 142, 170
Orpen 142, 209-221
Pafuri 142, 267
Paul Kruger 142
Phalaborwa 142, 234-239
Punda Maria 142

Gecko 13
Genet 274
Giraffe **7**, **138**, **185**, **188**, **218**, 231
Gloriosa superba 30
Gold 41
Gray, "Gaza" 47
Greenhill-Gardyne, Major 46-47
Green pigeons: see Pigeons
Grobler, Mr Piet 55, **56**, 187
Ground hornbill 8, **9**, **210**
Grysbok, Sharpe's 261, 275
Guinea fowl **265**, 269
Gymnogene 8

H

Hall-Martin, Dr Anthony 76, 90, **91**, 122, 124, 126
Hare, scrub **203**
Hartebeest, Lichtenstein's v, **76-77**
Healy, G. R. **53**
Helicopter **73**, 74, 97-98
counting, used in 97-98
culling, used in 70-71, 97, 122
elephant and 98
Heron 200, 248
goliath **247**
greenbacked **246**
grey 232, **251**
Hippopotamus 7, **234**, 248, **265**, **268**
culling 234
Honey badger 5, 211
Hoopoe **255**
Hornbill 7-8, **9**, **155**, **210**
see also Ground hornbill
Hyena 5, **186**, **187**, 192, 230, **231**, 236, **249**, 254
Hymenoptera 22

I

Impala 5, **6**, **84**, **115**, **164**, **170**, **178**, **197**, **211**, **213**, **216**, 239, **265**, **269**
lily 30, **31**, **262**, 269
Insects 17-20
Iron Age 37-38, **39**, 177, 227, 237, 272
Ivory 42, 52, 75, **89**, 90, 119-128

J

Jacana, African 8, 263
Jackal **203**
-berry 28, 269
Joubert, Dr Salomon 96

K

Kay, Mr and Mrs G. G. 236
Kaiser, Corrie 103-104
Killifish 34-35

Kingfisher 205, **221**, 244, **247**
Klipspringers 218, **227,** 228, 229, 238, 263
Knobel
    Rocco, 66, **67**
    Yvonne 169
Kolver, Hans 102-103, **104,** 105-106
Korhaan **228**
Kruger National Park
    accommodation and amenities 2, 139-141, 143-144
    see also Camps
    caravans: *see* Caravanning
    climate 139
    development 1-2, 61-67
    dimensions 1
    ecosystem 67, 69-70
    entrance gates 142
    fencing of 67, 70
    future 1-2, 67
    guide to 167-275
    history 37-67
    maps: *see* Maps
    motorists 57, 142-143
    overpopulation 70-82
    pollution 67, 82
    restaurants 140, **142**
    roads 57, 61, **63,** 66, **143, 144,** 145, 167, 239
    rules 145-146
    speed limits 143-146
    staff 1, 139-141
    temperature 139
    transport to 142
    travelling hours 143
    topography 139
    visitors 57, 62
    water: *see* Water
    wilderness trails 272-276
Kruger, Pres. Paul 37, **44,** 45-47
    gate: *see* Gates
Kudu 5, **132, 164, 190, 198-199, 202, 235,** 249, **264**

L

Lagden, Sir Godfrey 47
Lala palm 28
Lavender tree 29, 260
Leadwood 24, 28
Lebombo Mountains 40, 43, 201, 229, 274
Leeubron 211
Leeupan 187
Leguaan 13, **15, 36,** 261, 267
Leopard 5, 192, 213, **214, 232,** 241, 248, 262, 275
Library, Stevenson-Hamilton Memorial 182
Lichtenstein's hartebeest: *see* Hartebeest

Lilacbreasted roller 8, **220**
Lion 5, **61, 100, 116-117, 133, 193, 207,** 213, 226, **240-241, 260,** 262, 267
    mating **116-117**
    white 128-131, **130, 133,** 189, **195**
        lioness **129,** 211
        cubs 130, **133**
Lloyd, Mrs 62
Longclaw **203**
Lookout points 192, 201, 207, 212, 215, 227, 239, 251, 259
Lourenco Marques: *see* Mozambique
Lourie
    grey 7, 178
    purplecrested 7, 195, 206
Lowveld 30, 37
Loxodonta africana 122
Lungfish 34, **35**
Luvuvhu River: *see* Rivers

M

"Magnificent Seven" **89,** 90-91, 95, **114, 119-123, 125**
    Dzombo 90, 124
    João 91, 122, 124, 245, **248**
    Kambaku 124
    Mafunyane 95, 124, **125,** 126-128, 251
    Ndlulamithi 124
    Shawu **89,** 90, 124, 250
    Shingwedzi **120, 123,** 124, 245
Mahogany 29
Malaria 28, 35, 63, 141
    see also Mosquito
Mammals 5-7
Maps
    Lower Sabie 194
    master 166
    Olifants 224
    Pretoriuskop/Berg-en-dal 168
    Punda Maria/Pafuri 258
    Satara 208
    Shingwedzi 242
    Skukuza/Tshokwane 180
Maputo: *see* Mozambique
Marabou: *see* Storks
Marula 24, 262
Mashele, Judas **139**
Masorini **39,** 234, 237
McBride, Chris 128-130
Mdluli, Nombolo 110, **111**
Megaponera foetens 22
Memorial tablets 187, 207
Meyer
    Andre 38-39
    Marius 106
Millipede 19
Milner, Lord Alfred 46
Mongoose **5, 176,** 249

Monkey
    samango 7, 267
    vervet 7, **134-135**
Mopane 24, 28-29, 244, 262
    worms 29
Morning glory **30**
Mosquito
    Anopheles 17
Mountain reedbuck 5, **173,** 207, 255, 263
Mourning dove: *see* Dove
Mozambique 38, 40, 42, 43, 50, 67, 80, 90, 91, 92, 130, 175, 196, 249, 262, 266, 269
Museum
    Masorini 234, 237
    Selati Railway 181

N

Nahpe 173, 192
Narina trogon 8, 267, 275
National Parks Act 55-57
National Parks Board v, **57,** 182
Nephila 19
Ngirivani waterhole 212
Nicknames 111
Nkumbe 201
Numbi 169, 170
    see also Gates
Nussey, Wilf v
Nyala 5, 251, **254,** 261, **264, 266, 267**

O

Olivier, Louis 101-106
Orpen
    Eileen, Mrs **63,** 65, 187, 219
    gate area 209-221
    J. H., Mr **63,** 65, 219
Ostrich 7, 201, 234, 237
Otto, Piet 94-95, **97**
Owl
    giant eagle 245
    Pel's fishing 10
    Scop's 243, **244,** 274
Oxpecker **212, 213**

P

Pafuri v, **26-27,** 149, **258,** 259-269
    gate 267
Parabathus 18
Paradise
    flycatcher 205, 259
    whydah 8
Parasites 80, 81
Parrot
    Cape 267
    dove: *see* Pigeon, green

Paynter, David v
Parks Board: *see* National Parks Board
Photography 151-165
    awareness 152, 154
    colour 155
    driver during 157
    dust, effect of 157
    equipment 151, 152, 157, 159
    favourite places 155, 162, 164-165
    fixed-point, use of **85**
    flowers 155
    heat, effect of, on 158
    hints, general 159, 162
    light, effect of, on 154, 157, 162
    moisture 158
    nocturnal animals, of 155
    patience in 151, 152
    subject knowledge 152, 154
    techniques 152
    timing 152
    vibration 158
    wildlife 151, **152-165**, 196, 207, 211,
       212, 220, 225, 231, 243, 248, 254,
       255, 261, 263
Picnic sites 165, 185, 187, 205, 210, 211,
    215, 218, 220, 221, 226, 227, 234,
    254, 255, 267
Pigeon, green **8**, 206
Pienaar, Dr U. de V. 2, 67, **69**, 126
Plover **189**, **201**
Poaching 88-94, 266
    food, for 88, 90
    ivory 72, **89**, 90, 124
    profit, for 88, 90
    snares 94
Ponerinae 22
Pongola Reserve 46
Populations
    census 71
    count, aerial **70**, **71**
    numbers 70, 72, **99**
Portuguese 40
    see also Mozambique
Protopterus annectens brieni 35
Puffadder 14
Punda Maria: *see* Camps
Punt, Dr W. H. J. 259

R

Rabelais: *see* Dams
Rainfall 82-84
    cycles 83-84
    lowest 263
Rangers 47, 85-87, 88, **93**, 102-106, 118,
    189
Raptors 9-10, 201, 169
Ratel: *see* Honey badger
Redrocks 251
Reedbuck: *see* Mountain reedbuck

Reintroduction 75-79
    Lichtenstein's hartebeest **76-77**
    rhino
       black 78-79
       white 75, 78
Reitz, Col. Denys 54, **56**
Reptiles 13-17
Research: *see* Scientific research
Restaurants 140, **142**, 181, 209, 225,
    243, 259
Rhebuck, grey 79, 177, 176
Rhino
    black 5, 78-79, 206
    white 5, 75, 58, 169-170, **171**, 272
    shot 118
    trail 177
Rivers 81-83
    Crocodile 55, 81, 178
    Letaba 81, 228, 230-232, 234, 274
    Limpopo 52, 81
    Luvuvhu 52, 81, 263, **266**, 267, 269,
       275
    Olifants 81, 164, 226, 228-229, **233**,
       274
    Makhadzi 234
    Nwanedzi 165, 231
    Nwaswitsontso 221
    Sabie 81, 165, 184-185
    Shingwedzi 164, 243, 248-265
    Timbavati 164, 211-212, 219
Roan antelope 5, 236, **250**, 263, 266
Rochat, Mike 94-95, **96**
Rock art **3**, **37**, **38**, 39
    see also Bushman paintings
Rooihaarbossie **28**

S

Sabie
    Bridge **50**, 51, **54**, 55, 57
    Game Reserve 46-56
Sable antelope 5, **112-113**, **174**, **191**, 205,
    220, 236, 263, 266
Sagas 101-111
Sandenbergh, Col. J. A. B. 66, **67**
Satara 61, 62
    see also Camps
Sausage tree 28, 205, 221
Scientific research 69-75
    computer, by **79**
    fixed-point photography, used in **86**
    flying for 94
    staff 69
    station 69
Scorpionidae 18
Scorpions **18**
Secretary birds 215
Selati Railway 53, 55
Shark 34
Shirimantanga Hill 66, 192, 207

Skukuza: *see* Camps
Sickle bush **28**
Silver Leaves people 38
Snakes 14, 212, **221**
Sneezewood 30
South African Railways 53, 55
Spiders 19, **20**, **21**
    baboon 19
    jumping 19
    sun 19-20
Squirrels **152**, **153**, 164, 248, 267, **268**
Starlings 7-8, **154**, 215
Steenbok 5, **217**
Steinaecker, Col. Ludwig 46, 48, 49
    Horse 48
Stevenson-Hamilton
    children **64**
    grave 192
    Hilda **64**, **65**
    James, Col. 1, **47-57**, **60**, **64**, **65**, 66
    Memorial Library 70, 109, 182
Steyn, Mr L. B. 66, **67**
Stone Age 37, 39, 177, 237, 276
Storks
    black 218, 275
    marabou 8, 192, 205, 227, **245**
    saddlebilled 8, **215**
    woolly-necked 218
    yellowbilled 239
Suni v, 5, **260**, 261, 275
Sycamore fig 24, 25, **29**, 212

T

Tamboti 30
Terrapin 13, 236
Timbavati 128-129
    River: *see* Rivers
Tomlinson, Bert 111
Tomopterna krugerensis 14
Tortoise 13, 221
Tourism
    future of 147
    maximum 148
Tracking 91-92
Trails: *see* Wilderness trails
Translocation 66, 75-79
    see also Reintroduction
Trees 24-30, 179, 226, 230, 260
    jackal-berry 269
    knobthorn 197, 201
    lavender 260
    propeller 262
    see also various species
Trichardt, Louis 40, 189
Trollope, Harold **55**, 110-111
Tsessebe 237, 249, 234, 263
Tshokwane 129, 131, 165, **180**, 181,
    187, 189, 201

Tusks 119-128, **120-122, 125, 127**
   biggest 119

U

Umbrella thorn 25, 211

V

Van Niekerk, Hugo 94-95, **96**
Van Wyk, Piet v
Veld
   condition of **86**
   fire, effect on 84-85
Viewpoints: *see* Lookout points
Visitors
   information for 139-149
   restrictions 148
Voortrekkers 40, 45, 84, 170, 171
Vultures 10-11
   Cape **10,** 11
   hooded 11
   lappet-faced 11
   whitebacked 10-11
   whiteheaded 11

W

Wagon road **41,** 175
Warthogs **202-203**
Water 65, 66, 69-70, 81
   hyacinth 82
   pollution 82-83
   rivers providing 81
   see also Rainfall
Waterbirds 8, **117,** 201, 205, 210, **211,**
   218, 234, 250, 263
Waterbuck **191, 219,** 244, **253, 272**
Waterlilies 30, 227
Weavers 195, 207, **238,** 243
Wild dogs 5, **202,** 207, **229,** 261
Wildebeest, blue 5, **179**
Wilderness trails 272-276
   Bushman 39, 119, 272, **275,** 276
   Olifants 272, 274, 275
   Nyalaland 119, 269, 272, 273, 275
   Wolhuter 118, 272, 274, **275**
Wildlife Society of Southern Africa v, 52
Wolff, Pat 128-129

Wolhuter
   Harry Christopher 47, **107, 108, 109,**
      189
      knife of **108-109**
   Henry 67
   wilderness trail: *see* Wilderness trails
World War
   I 51-53
   II 65, 66
Woodpeckers 243

Y

Yellowbilled
   kite 249
   stork 239
Yssel, Tom 88, 102, **103,** 104-106

Z

Zambezi shark 34
Zebra 5, **160-161, 176, 185, 190, 191**